Poromechanics

Poromechanics

Olivier Coussy
Laboratoire Central des Ponts et Chaussées

Institut Navier, Marne-la-Vallée, France

John Wiley & Sons, Ltd

Telephone (+44) 1243 779777

Email (for orders and customer service enquiries): cs-books@wiley.co.uk
Visit our Home Page on www.wileyeurope.com or www.wiley.com

Other Wiley Editorial Offices

John Wiley & Sons Inc., 111 River Street, Hoboken, NJ 07030, USA

Jossey-Bass, 989 Market Street, San Francisco, CA 94103-1741, USA

Wiley-VCH Verlag GmbH, Boschstr. 12, D-69469 Weinheim, Germany

John Wiley & Sons Australia Ltd, 33 Park Road, Milton, Queensland 4064, Australia

John Wiley & Sons (Asia) Pte Ltd, 2 Clementi Loop #02-01, Jin Xing Distripark, Singapore 129809

John Wiley & Sons Canada Ltd, 22 Worcester Road, Etobicoke, Ontario, Canada M9W 1L1

Wiley also publishes its books in a variety of electronic formats. Some content that appears in print may not be available in electronic books.

British Library Cataloguing in Publication Data

A catalogue record for this book is available from the British Library

ISBN 978-0-470-84920-0

Contents

Preface

I am inclined to believe that engineers and engineering schools will play an important part in restoring unity and central viewpoint in the natural sciences. This is because modern engineering, by its very nature, must be synthetic. Specialization, carried to extreme, is a form of death and decay. M.A. Biot[1]

We define *Poromechanics* as the study of porous materials whose mechanical behaviour is significantly influenced by the pore fluid. By this definition, poromechanics involves a broad range of materials, from the rocks and soils that were the subject of the theory of poroelasticity developed by Maurice Biot more than half a century ago, to gels and biological tissues. Poromechanics is then relevant to disciplines as varied as geophysics, geotechnics, biomechanics, physical chemistry, agricultural engineering or materials science. If the porous materials and the fields concerned are many, their unity lies in the fact that they are all subject to the same coupled processes: hydro-diffusion and subsidence, hydration and swelling, drying and shrinkage, heating and build-up of pore pressure, freezing and spalling, capillarity and cracking. Accordingly, the main purpose of this book is to provide a unified and systematic continuum approach to poromechanics for engineers and applied physicists. The key concepts to reconcile continuum mechanics with the microscopic discontinuities inherent in porous media constituted by solid and fluid phases are twofold. The first concept is to consider the porous medium as the superposition of several continua that move with distinct kinematics, while mechanically interacting and exchanging energy and matter. Since the formulation of the constitutive equations of any solid requires its deformation to be referred to an initial configuration, the second concept is to transport the equations governing the physics of the superposed fluid and solid continua from their common current configuration to an initial reference configuration related to the solid skeleton. Based upon the pioneering work of M.A. Biot, these two concepts allow us to extend all significant achievements of the continuum mechanics of solids to poromechanics. Within the energy context of thermodynamics of open continua these two concepts also open up the way to explore new frontiers related to various thermo/hydro/chemo/mechanical couplings.

The book is intended to be self-contained. It only presupposes some basic mathematical knowledge of tensorial analysis and differential geometry. Starting from the basics it progressively extends the fundamental concepts of continuum mechanics to continuum poromechanics, by focusing attention on the coupling of the deformation of porous continua with various other physical processes. The book can possibly be divided in to

[1]Biot M.A. (1962), 'Science and the Engineer', acceptance talk for the Timoshenko Award Medal.

two main parts. The first part comprises Chapter 1 to 5, starting from the description of deformation and stress, going on to the thermodynamics of porous continua considered as open systems, and ending in poroelasticity, including the solution of evolution problems. The second part comprises Chapter 6 to 9 and explores various independent extensions of the first part. Chapter 6 extends the analysis to unsaturated poroelastic media by addressing the effects of the capillary pressure and that of the surface energy between the components. Chapter 7 offers a unified approach to the analysis of various fronts that can penetrate a porous medium. Finally, Chapters 8 and 9 extend the picture to inelastic behaviour, dealing first with poroplasticity and ending with hereditary behaviour.

The book can serve as a support for both graduate and advanced courses on continuum poromechanics, revisiting and completing an initial treatment of continuum mechanics. Indeed, there are essentially two volumes in one. Volume 1 is formed by the succession of all chapters in the first part and is intended for graduate students. Volume 2 is formed by gathering all the 'Advanced Analysis' sections of each chapter. It is intended for more advanced courses in order to provide sources of information and inspiration to future scientists in the field of poromechanics, whatever the particular discipline or application in mind. These developments include microporomechanics, deal with surfaces of discontinuity and wave propagation, provide some solutions to non-linear problems of poroelasticity, analyse the drying of weakly permeable materials, consider the localization of deformation, explore the frontiers between poromechanics and chemomechanics, poromechanics and mechanics of colloidal mixtures, etc.

Acknowledgements

When my editor asked me about a second edition of my book *Mechanics of Porous Continua*, I was immediately grateful to all the readers of the first edition for allowing such an exciting new adventure. It is now over 10 years since the first edition in French, *Mécanique des Milieux Poreux* (Technip, 1991) went to press. Writing the second edition was like renovating an old house: I started by drawing up new plans (the table of contents), I refurbished the rooms (the chapters), putting down partitions, making needless developments, and adding new accommodation (new sections), to welcome new friends—hopefully new readers. Finally, the renovated old house had turned into a new book, so I decided to change the front door—the title.

Poromechanics is the word we coined, Younane Abousleiman, Alex Cheng, Emmanuel Detournay, Jean-François Thimus and I, when we launched the first Biot Conference on the Mechanics of Porous Media in 1998. Since then a second Biot Conference was held in 2002 and the third one is already scheduled. I hope that this book will contribute to the grounding of this emerging field which combines the mechanics of solids and the mechanics of fluids and offers a bridge towards other large scientific fields, as well as being a transversal link between various domains where porous materials are concerned.

There is only one author on the title page, but this book is the result of many friendly collaborations. I am particularly indebted to four colleagues and friends for their often decisive contributions: namely, Patrick Dangla, with whom I have explored the unsaturated side of poromechanics; Luc Dormieux, whose achievements in microporomechanics are now well known; Robert Eymard, who introduced me to the numerical world of transport in porous media; Franz-Joseph Ulm, whose enthusiasm led us to explore the frontiers between poromechanics and chemomechanics. I also deeply acknowledge the contributions of my former PhD students, Elisabeth Bemer, Bertrand Fauchet, Thierry Lassabatère, Catherine Larive, Kefei Li, Marc Mainguy and Jean-Marc Picard. The existence of this book would not have been possible without the support over the years of the Laboratoire Central des Ponts et Chaussées. Finally, I dedicate this book to my wife, Sandra, and to my daughter, Flavia, who now may think that they will see more of me on Sundays.

Chapter 1

Deformation and Kinematics. Mass Balance

The aim of this chapter is to describe the deformation and the kinematics of a porous medium formed from a deformable skeleton and a fluid saturating the porous space. The underlying idea consists in approaching the porous medium as the superimposition of two continua, the skeleton continuum and the fluid continuum. The description of the deformation and of the kinematics of each continuum considered separately differs in no way from that of a monophasic continuous medium. Nevertheless, the skeleton deformation is eventually the one that can actually be observed so it is the one discussed in the following.

The laws of physics governing the evolution of a porous continuum involve the time rate of the physical quantities attached either to the skeleton or to the fluid whatever their further distinct movement. Accordingly the particle derivative is therefore introduced, allowing us to follow separately the motions of the skeleton and the fluid. A first illustration of its use is given at the end of this chapter by expressing the mass balance for the two superimposed continua.

1.1 The Porous Medium and the Continuum Approach

1.1.1 Connected and Occluded Porosity. The Matrix

A saturated porous medium is composed of a matrix and a porous space, the latter being filled by a fluid. The connected porous space is the space through which the fluid actually flows and whose two points can be joined by a path lying entirely within it so that the fluid phase remains continuous there. The matrix is composed of both a solid part and a possible occluded porosity, whether saturated or not, but through which no filtration occurs. The connected porosity is the ratio of the volume of the connected porous space to the total volume. In what follows the term 'porosity', used without further specification, refers to the entire connected porosity.

Poromechanics O. Coussy
 ISBN 0-470-84920-7

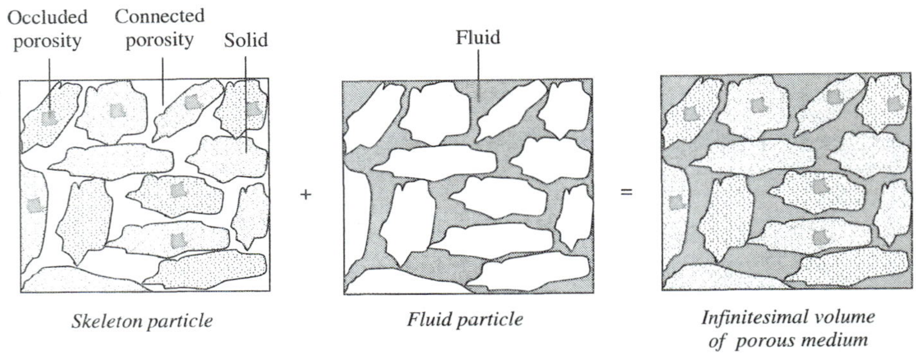

Figure 1.1: The porous medium as the superimposition of two continuous media: a skeleton particle and a fluid particle coincide with the same geometrical infinitesimal volume.

1.1.2 Skeleton and Fluid Particles. Continuity Hypothesis

A porous medium can be treated as the superimposition of two continua, the skeleton continuum and the fluid continuum. Accordingly, as illustrated in Fig. 1.1, any infinitesimal volume can be treated as the superimposition of two material particles. The first is the skeleton particle formed from the matrix and the connected porous space emptied of fluid. The second is the fluid particle formed from the fluid saturating the connected porous space and from the remaining space without the matrix.

A continuous description of a medium, which is heterogeneous at the microscopic scale, requires the choice of a macroscopic scale at which the inner constitution of matter is ignored in the analysis of the macroscopic physical phenomena. For instance, the porosity is associated with an elementary volume including sufficient material to be representative of the filtration process. More generally the hypothesis of continuity assumes the existence of a representative elementary volume which is relevant at the macroscopic scale for all the physical phenomena involved in the intended application. The physics is supposed to vary continuously from one to another of those juxtaposed infinitesimal volumes whose junction constitutes the porous medium. In addition, continuous deformation of the skeleton assumes that two skeleton particles, juxtaposed at a given time, were always so and will remain so.

1.2 The Skeleton Deformation

When subjected to external forces and to variations in pressure of the saturating fluid, the skeleton deforms. The description of this deformation differs in no way from that of a standard solid continuum and is succinctly developed below.

1.2.1 Deformation Gradient and Transport Formulae

At time $t = 0$ consider an initial configuration for the skeleton. In this configuration a skeleton particle is located by its position vector $\mathbf{X}$ of components X_i, in a Cartesian coordinate frame of orthonormal basis $(\mathbf{e}_1, \mathbf{e}_2, \mathbf{e}_3)$. At time t the skeleton has deformed and lies in the current configuration. In this configuration the particle whose initial position vector was $\mathbf{X}$ is now located by its current position vector $\mathbf{x}$ of components $x_i(X_j, t)$. We write:

$$\mathbf{X} = X_i\mathbf{e}_i; \quad \mathbf{x} = x_i(X_j, t)\mathbf{e}_i \tag{1.1}$$

with a summation on the repeated subscript i. In what follows this convention is adopted and, provided that no further indication is given, the index notation refers to a Cartesian coordinate system.

Deformation gradient and transport of a vector. In the initial configuration consider an infinitesimal material vector $d\mathbf{X}$ joining the skeleton particle located at $\mathbf{X}$ to the juxtaposed particle located at $\mathbf{X} + d\mathbf{X}$. After deformation $d\mathbf{X}$ becomes $d\mathbf{x}$ joining the same skeleton particles in their new positions, $\mathbf{x}$ and $\mathbf{x} + d\mathbf{x}$ (see Fig. 1.2). The vector $d\mathbf{x}$ can be obtained from $d\mathbf{X}$ by differentiating (1.1):

$$d\mathbf{x} = \frac{\partial x_i}{\partial X_j} dX_j\mathbf{e}_i \tag{1.2}$$

or, equivalently:

$$d\mathbf{x} = \mathbf{F} \cdot d\mathbf{X} \tag{1.3}$$

where:

$$\mathbf{F} = \nabla_X\mathbf{x}; \quad F_{ij} = \frac{\partial x_i}{\partial X_j} \tag{1.4}$$

In (1.4) ∇_X stands for the nabla operator relative to the initial configuration. $\mathbf{F}$ is called the *deformation gradient*. It transports any material vector $d\mathbf{X}$ onto its deformed $d\mathbf{x}$. Its inverse $\mathbf{F}^{-1}$ and its transpose ${}^t\mathbf{F}$ are respectively defined by:

$$d\mathbf{X} = \mathbf{F}^{-1} \cdot d\mathbf{x}; \quad d\mathbf{x} = d\mathbf{X} \cdot {}^t\mathbf{F} \tag{1.5}$$

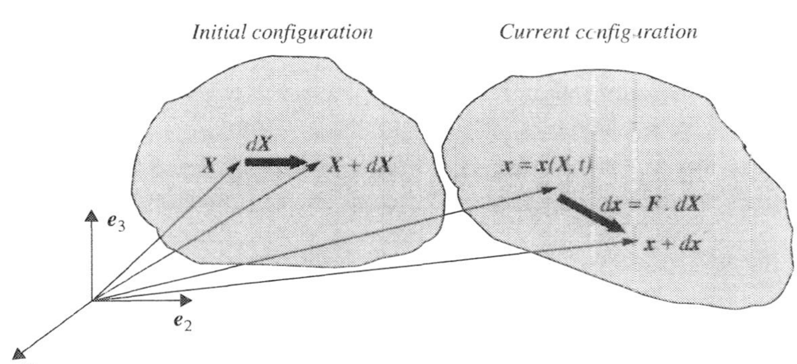

Figure 1.2: Deformation gradient $\mathbf{F}$ and transport of a material vector $d\mathbf{X}$.

and satisfy:

$$({}^t\mathbf{F})_{ij} = F_{ji}; \quad (\mathbf{F}^{-1})_{ij} = \frac{\partial X_i}{\partial x_j} \tag{1.6}$$

Deformation gradient and displacement. Let $\boldsymbol{\xi}$ $(\mathbf{X}, t)$ be the displacement vector of the particle whose initial and current positions are $\mathbf{X}$ and $\mathbf{x}$. We write:

$$\mathbf{x} = \mathbf{X} + \boldsymbol{\xi} \tag{1.7}$$

From definitions (1.4) and (1.7) deformation gradient $\mathbf{F}$ can be expressed as a function of displacement vector $\boldsymbol{\xi}$ according to:

$$\mathbf{F} = \mathbf{1} + \nabla_X \boldsymbol{\xi}; \quad F_{ij} = \delta_{ij} + \frac{\partial \xi_i}{\partial X_j} \tag{1.8}$$

where δ_{ij} is the Kronecker delta, that is $\delta_{ij} = 1$ if $i = j$ and $\delta_{ij} = 0$ if $i \neq j$.

Volume transport. The current infinitesimal volume $d\Omega_t = dx_1\, dx_2\, dx_3$ is equal to the composed product:

$$d\Omega_t = (d\mathbf{x}_1,\ d\mathbf{x}_2,\ d\mathbf{x}_3) = d\mathbf{x}_1 \cdot (d\mathbf{x}_2 \times d\mathbf{x}_3) \tag{1.9}$$

where $d\mathbf{x}_i = dx_i\, \mathbf{e}_i$ (with no summation). The linearity of the composed product with respect to the vectors it combines allows us to write:

$$d\Omega_t = (\mathbf{F} \cdot d\mathbf{X}_1, \mathbf{F} \cdot d\mathbf{X}_2, \mathbf{F} \cdot d\mathbf{X}_3) = \det \mathbf{F}\,(d\mathbf{X}_1,\ d\mathbf{X}_2,\ d\mathbf{X}_3) \tag{1.10}$$

As a consequence any initial material volume $d\Omega_0$ transforms into the material volume $d\Omega_t$ through the relation:

$$d\Omega_t = J d\Omega_0 \tag{1.11}$$

where $J = \det \mathbf{F}$ is the Jacobian of the deformation.

Surface transport. Consider a material surface dA, oriented by the unit normal $\mathbf{N}$. Throughout the deformation dA transforms into the material surface da, oriented by the unit normal $\mathbf{n}$. Since vectors $\mathbf{N}$ and $\mathbf{n}$ are not material vectors, they do not match in the deformation. Let $\mathbf{U}$ be any material vector in the initial configuration. The material cylinder of initial volume $\mathbf{N} \cdot \mathbf{U} dA$ transforms into the material cylinder of volume $\mathbf{n} \cdot \mathbf{F} \cdot \mathbf{U} da$ (see Fig. 1.3). According to (1.11) we write:

$$\mathbf{n} \cdot \mathbf{F} \cdot U\, da = J\, \mathbf{N} \cdot \mathbf{U} dA \tag{1.12}$$

Since (1.12) holds whatever the vector $\mathbf{U}$, we derive:

$$\mathbf{n}\, da = J\, {}^t\mathbf{F}^{-1} \cdot \mathbf{N} dA; \qquad n_i\, da = J \frac{\partial X_j}{\partial x_i} N_j\, dA \tag{1.13}$$

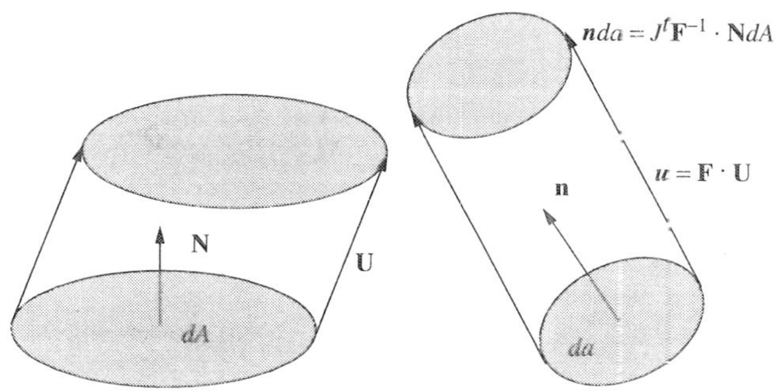

Figure 1.3: Transport of the oriented material surface $\mathbf{N}\,dA$ onto its deformed $\mathbf{n}\,da$.

Let $\mathbf{v}$ be any vector attached to the current configuration and let $\mathbf{V}$ be the associated vector attached to the initial configuration and defined in such a way that the flow of $\mathbf{v}$ through da coincides with the flow of $\mathbf{V}$ through dA. We write:

$$\mathbf{v} \cdot \mathbf{n}\,da = \mathbf{V} \cdot \mathbf{N}\,dA; \qquad v_i n_i\,da = J\,V_i \frac{\partial X_j}{\partial x_i} N_j\,dA \tag{1.14}$$

From (1.13) and (1.14) we derive:

$$\mathbf{V} = J\mathbf{F}^{-1} \cdot \mathbf{v}; \qquad V_i = J \frac{\partial X_i}{\partial x_j} v_j \tag{1.15}$$

Integration of (1.14) over the volumes Ω_0 and Ω_t matching in the deformation, followed by use of the divergence theorem and relation (1.11), gives the useful identity:

$$\nabla_x \cdot \mathbf{v}\,d\Omega_t = \nabla_X \cdot \mathbf{V}\,d\Omega_0; \qquad J \frac{\partial v_i}{\partial x_i} = \frac{\partial V_i}{\partial X_i} \tag{1.16}$$

1.2.2 Eulerian and Lagrangian Porosities. Void Ratio

Let n be the Eulerian porosity, so that the fluid occupies the volume $n\,d\Omega_t$ in the current configuration. Since the skeleton material volume $d\Omega_t$ changes throughout the deformation, porosity n does not properly quantify the volume change undergone by the porous space attached to the initial material volume $d\Omega_0$. In contrast to the Eulerian porosity n, which refers to the current volume $d\Omega_t$, the change in the porous space is eventually better captured by the Lagrangian porosity ϕ, which refers the current porous volume to the initial volume $d\Omega_0$ according to:

$$\phi\,d\Omega_0 = n\,d\Omega_t; \qquad \phi = Jn \tag{1.17}$$

For its part the current degree of compactness of a porous material is well captured by the void ratio e defined as the ratio of the current porous volume to the current volume of the matrix. Owing to its definition the void ratio e is a Eulerian variable, with no Lagrangian counterpart, and is expressed as a function of n in the form:

$$e = \frac{n}{1-n} \tag{1.18}$$

1.2.3 Strain Tensor

Deformation induces changes in both the lengths of the material vectors and the angles between them. The Green–Lagrange strain tensor $\mathbf{\Delta}$ measures these changes by quantifying the variation of the scalar product of two material vectors $d\mathbf{X}$ and $d\mathbf{Y}$ transforming the deformation throughout into $d\mathbf{x}$ and $d\mathbf{y}$. We write:

$$d\mathbf{x} = \mathbf{F} \cdot d\mathbf{X}; \quad d\mathbf{y} = \mathbf{F} \cdot d\mathbf{Y}: \quad d\mathbf{x} \cdot d\mathbf{y} - d\mathbf{X} \cdot d\mathbf{Y} = 2d\mathbf{X} \cdot \mathbf{\Delta} \cdot d\mathbf{Y} \tag{1.19}$$

With the help of (1.5) $\mathbf{\Delta}$ can be written as a function of deformation gradient $\mathbf{F}$ according to:

$$\mathbf{\Delta} = \frac{1}{2}({}^t\mathbf{F} \cdot \mathbf{F} - \mathbf{1}) \tag{1.20}$$

Being symmetric, tensor $\mathbf{\Delta}$ admits three real eigenvalues Δ_J $(J = I, II, III)$. The latter are the principal strains and are associated with the eigenvectors $\mathbf{e}_J$ $(J = I, II, III)$, which are the principal directions of the deformation such as $\mathbf{\Delta} \cdot \mathbf{e}_J = \Delta_J \mathbf{e}_J$. The orthogonality of the principal directions, writing $\mathbf{e}_I \cdot \mathbf{e}_J = 0$, is preserved in the deformation. Indeed:

$$\left(\mathbf{e}_I \cdot {}^t\mathbf{F}\right) \cdot (\mathbf{F} \cdot \mathbf{e}_J) = 2\mathbf{e}_I \cdot \mathbf{\Delta} \cdot \mathbf{e}_J = 2\Delta_{I \text{ or } J}\mathbf{e}_I \cdot \mathbf{e}_J = 0 \tag{1.21}$$

The gradient $\mathbf{R}$ of the rotation that rigidly transports the set of orthogonal principal directions $\mathbf{e}_J$ to their final directions is an isometry so that the related strain tensor is zero, resulting in ${}^t\mathbf{R} = \mathbf{R}^{-1}$. Therefore the gradient $\mathbf{F}$ of any transformation decomposes as:

$$\mathbf{F} = \mathbf{D} \cdot \mathbf{R} \tag{1.22}$$

that is the rotation $\mathbf{R}$ followed by the actual deformation $\mathbf{D}$, the latter involving no rotation and matching the dilation of the principal directions of deformation. Equivalently, the gradient $\mathbf{F}$ can also decompose as the actual deformation $\mathbf{D}'$ matching the dilation, followed by the rotation $\mathbf{R}'$ relative to that of the eigenvectors. Accordingly, the strain tensor $\mathbf{\Delta}$ accounts entirely for the actual deformation since:

$$\mathbf{\Delta} = \frac{1}{2}(\mathbf{D}^2 - \mathbf{1}) \tag{1.23}$$

By means of (1.8) $\mathbf{\Delta}$ can finally be expressed as a function of the displacement vector $\boldsymbol{\xi}$ according to:

$$\mathbf{\Delta} = \frac{1}{2}\left(\nabla_X \boldsymbol{\xi} + {}^t\nabla_X \boldsymbol{\xi} + {}^t\nabla_X \boldsymbol{\xi} \cdot \nabla_X \boldsymbol{\xi}\right) \tag{1.24a}$$

$$\Delta_{ij} = \frac{1}{2}\left(\frac{\partial \xi_i}{\partial X_j} + \frac{\partial \xi_j}{\partial X_i} + \frac{\partial \xi_k}{\partial X_i}\frac{\partial \xi_k}{\partial X_j}\right) \tag{1.24b}$$

1.2.4 Infinitesimal Transformation and the Linearized Strain Tensor

In many problems a first-order approximation to the finite theory can be carried out under the condition of infinitesimal transformation, that is

$$\|\nabla \boldsymbol{\xi}\| \ll 1 \tag{1.25}$$

where the norm $\|(\cdot)\|$ of $(\cdot)$ has not been specified because of the equivalence of all the norms in a vectorial space of finite three dimensions. Moreover, as far as only spatial derivations are concerned, in the limit of infinitesimal transformation the current and the initial configurations merge, so the nabla operator ∇ can be used with no need for a subscript referring to a particular configuration, that is $\nabla = \nabla_X \equiv \nabla_x$.

Under condition (1.25) the Green–Lagrange strain tensor $\boldsymbol{\Delta}$ reduces to the linearized strain tensor $\boldsymbol{\varepsilon}$:

$$\boldsymbol{\Delta} \simeq \boldsymbol{\varepsilon} = \frac{1}{2}(\nabla \boldsymbol{\xi} + {}^t\nabla \boldsymbol{\xi}); \quad \varepsilon_{ij} = \frac{1}{2}\left(\frac{\partial \xi_i}{\partial x_j} + \frac{\partial \xi_j}{\partial x_i}\right) \tag{1.26}$$

Since $\boldsymbol{\Delta}$ has the same order of magnitude as $\nabla_X \boldsymbol{\xi}$, infinitesimal transformation implies infinitesimal deformation, that is $\|\boldsymbol{\Delta}\| \ll 1$. In contrast, the deformation may be infinitesimal whereas the transformation is not. For instance, in a rigid body motion $\boldsymbol{\Delta}$ is zero whereas $\nabla_X \boldsymbol{\xi}$ can have any order of magnitude.

Under the approximation of infinitesimal transformation, (1.8) gives:

$$(J = \det \mathbf{F}) \simeq \left(1 + \nabla \cdot \boldsymbol{\xi} = 1 + \frac{\partial \xi_i}{\partial x_i} = 1 + \varepsilon_{ii}\right) \tag{1.27}$$

From now on let ϵ be the linearized volume dilation of the skeleton, that is:

$$\epsilon = \varepsilon_{ii} = \nabla \cdot \boldsymbol{\xi} \tag{1.28}$$

so that (1.11) takes the form:

$$d\Omega_t \simeq (1 + \epsilon)\, d\Omega_0 \tag{1.29}$$

The observable macroscopic volume dilation undergone by the skeleton is due both to the change in porosity and to the volume dilation ϵ_s undergone by the solid matrix, although the latter is not accessible from purely macroscopic experiments. Analogously to (1.29) the definition of ϵ_s allows us to write:

$$d\Omega_t^s = (1 + \epsilon_s)\, d\Omega_0^s \tag{1.30}$$

Owing to the respective definition of Eulerian and Lagrangian porosities, n and ϕ (see §1.2.2), the volume occupied by the matrix is linked to the overall volume through the relations:

$$d\Omega_t^s = (1 - n)\, d\Omega_t = d\Omega_t - \phi\, d\Omega_0; \quad d\Omega_0^s = (1 - \phi_0)\, d\Omega_0 \tag{1.31}$$

Combining the above equations, we finally derive the volume balance:

$$\epsilon = (1 - \phi_0)\epsilon_s + \phi - \phi_0 \tag{1.32}$$

In the absence of any occluded porosity, the solid grains forming the matrix generally undergo negligible volume changes so that the matrix can be considered as incompressible. Accordingly we let $\epsilon_s = 0$ in (1.32), giving:

$$\epsilon = \phi - \phi_0 \tag{1.33}$$

As is usually done in soil mechanics, it can be more convenient to use the void ratio e instead of the volumetric dilation ϵ. Combining (1.17), (1.18) and (1.33), we obtain:

$$\epsilon = \frac{e - e_0}{1 + e_0} \tag{1.34}$$

Under the approximation of infinitesimal transformation, the diagonal term ε_{ii} (with no summation) is equal to the linear dilation in the $\mathbf{e}_i$ direction, while twice the non-diagonal term, $\gamma_{ij} = 2\varepsilon_{ij}$ $(i \neq j)$, is equal to the distortion related to directions $\mathbf{e}_i$ and $\mathbf{e}_j$, that is the change undergone by the angle made between the material vectors $\mathbf{e}_i$ and $\mathbf{e}_j$ that were normal prior to the deformation.

1.3 Kinematics

The description of the skeleton deformation by means of the deformation gradient $\mathbf{F}$ is by nature a Lagrangian description. The fields are functions of time t and of position vector $\mathbf{X}$ locating the skeleton particle in the initial configuration. The latter does not vary with time and the kinematics of the skeleton results from a simple time derivation.

In contrast to the Lagrangian approach, the Eulerian approach involves only the current configuration, with no reference to any initial configuration. The approach is carried out by using the velocity field $\mathbf{V}^\pi(\mathbf{x}, t)$ of the particle coinciding at time t with the geometrical point located at $\mathbf{x}$. The particle can be either a skeleton particle, $\pi = s$, or a fluid particle, $\pi = f$.[1] At time t, the same Eulerian approach applies to both particles since the skeleton continuum and the fluid continuum merge in the same current configuration.

1.3.1 Particle Derivative

Definition

The particle derivative $d^\pi \mathcal{G}/dt$ with respect to particle π $(= s$ or $f)$ of some field $\mathcal{G}$ is the time derivative of $\mathcal{G}$ that an observer attached to the particle would derive. This observer records the variation $d^\pi \mathcal{G}$ of quantity $\mathcal{G}$ between times t and $t + dt$. For instance, the

[1] It would have been more rigorous to make a distinction between the index referring to the matter at the macroscopic scale (for instance, sk for the skeleton particle and fl for the fluid particle) and the index referring to the matter at the mesoscopic scale (for instance, s for the solid matrix, as in (1.32), and f for the fluid). However, for the sake of simplicity of notation, we chose not to make this distinction.

origin of the coordinate being fixed, the velocity field $\mathbf{V}^\pi(\mathbf{x}, t)$ of particle π located at $\mathbf{x}$ reads:

$$\frac{d^\pi \mathbf{x}}{dt} = \mathbf{V}^\pi(\mathbf{x}, t); \quad \pi = s \text{ or } f \tag{1.35}$$

Particle derivative of a material vector

Definition (1.35) allows us to write the particle derivative of the material vector $d\mathbf{x}$ in the form:

$$\frac{d^\pi}{dt}(d\mathbf{x}) = \frac{d^\pi}{dt}[(\mathbf{x}+d\mathbf{x}) - \mathbf{x}] = \mathbf{V}^\pi(\mathbf{x}+d\mathbf{x}, t) - \mathbf{V}^\pi(\mathbf{x}, t) \tag{1.36}$$

so that:

$$\frac{d^\pi}{dt}(d\mathbf{x}) = \nabla_x \mathbf{V}^\pi \cdot d\mathbf{x}; \quad \left(\nabla_x \mathbf{V}^\pi\right)_{ij} = \frac{\partial V_i^\pi}{\partial x_j}; \quad \pi = s \text{ or } f \tag{1.37}$$

Particle derivative of a material volume

Starting from (1.9), we express the particle derivative of the material volume $d\Omega_t$ in the form:

$$\frac{d^\pi}{dt}(d\Omega_t) = \frac{d^\pi}{dt}(d\mathbf{x}_1, d\mathbf{x}_2, d\mathbf{x}_3) = \frac{d^\pi}{dt}(d\mathbf{x}_1 \cdot (d\mathbf{x}_2 \times d\mathbf{x}_3)) \tag{1.38}$$

The linearity of the composed product $(\mathbf{v}_1, \mathbf{v}_2, \mathbf{v}_3)$ with regard to vectors $\mathbf{v}_{i=1,2,3}$ allows us to write:

$$\begin{aligned}\frac{d^\pi}{dt}(d\Omega_t) &= \left(\frac{d^\pi}{dt}(d\mathbf{x}_1),\ d\mathbf{x}_2,\ d\mathbf{x}_3\right) \\ &+ \left(d\mathbf{x}_1,\ \frac{d^\pi}{dt}(d\mathbf{x}_2),\ d\mathbf{x}_3\right) + \left(d\mathbf{x}_1,\ d\mathbf{x}_2,\ \frac{d^\pi}{dt}(d\mathbf{x}_3)\right)\end{aligned} \tag{1.39}$$

Use of (1.37) gives:

$$\begin{aligned}\frac{d^\pi}{dt}(d\Omega_t) &= \left(\frac{\partial V_i^\pi}{\partial x_1}\mathbf{e}_i\, dx_1,\ d\mathbf{x}_2,\ d\mathbf{x}_3\right) \\ &+ \left(d\mathbf{x}_1,\ \frac{\partial V_i^\pi}{\partial x_2}\mathbf{e}_i\, dx_2,\ d\mathbf{x}_3\right) + \left(d\mathbf{x}_1,\ d\mathbf{x}_2,\ \frac{\partial V_i^\pi}{\partial x_3}\mathbf{e}_i\, dx_3\right)\end{aligned} \tag{1.40}$$

The product $(\mathbf{v}_1, \mathbf{v}_2, \mathbf{v}_3)$ is zero as soon as two vectors among vectors $\mathbf{v}_{i=1,2,3}$ are colinear. Thus:

$$\frac{d^\pi}{dt}(d\Omega_t) = \left(\frac{\partial V_1^\pi}{\partial x_1} + \frac{\partial V_2^\pi}{\partial x_2} + \frac{\partial V_3^\pi}{\partial x_3}\right)(d\mathbf{x}_1, d\mathbf{x}_2, d\mathbf{x}_3) \tag{1.41}$$

or equivalently:

$$\frac{d^\pi}{dt}(d\Omega_t) = \left(\nabla_x \cdot \mathbf{V}^\pi\right) d\Omega_t; \quad \pi = s \text{ or } f \tag{1.42}$$

Particle derivative of a field

The particle derivative $d^\pi \mathcal{G}/dt$ with respect to particle π $(= s$ or $f)$ of field $\mathcal{G}(\mathbf{x}, t)$ turns out to be the time derivative of $\mathcal{G}$ when letting $\mathbf{x}$ match the successive positions $\mathbf{x}^\pi(t)$ occupied by the particle. We write:

$$\frac{d^\pi \mathcal{G}}{dt} = \frac{\partial \mathcal{G}}{\partial t} + (\nabla_x \mathcal{G}) \cdot \mathbf{V}^\pi \tag{1.43}$$

For instance, the acceleration $\boldsymbol{\gamma}^\pi$ of particle π is the particle derivative of velocity $\mathbf{V}^\pi(\mathbf{x}, t)$:

$$\boldsymbol{\gamma}^\pi = \frac{d^\pi \mathbf{V}^\pi}{dt} = \frac{\partial \mathbf{V}^\pi}{\partial t} + \left(\nabla_x \mathbf{V}^\pi\right) \cdot \mathbf{V}^\pi; \quad \gamma_i^\pi = \frac{\partial V_i^\pi}{\partial t} + \frac{\partial V_i^\pi}{\partial x_j} V_j^\pi \tag{1.44}$$

Particle derivative of a volume integral

The particle derivative applies to the volume integral of any quantity $\mathcal{G}$ according to:

$$\frac{d^\pi}{dt} \int_{\Omega_t} \mathcal{G} d\Omega_t = \int_{\Omega_t} \frac{d^\pi}{dt} (\mathcal{G} d\Omega_t) \tag{1.45}$$

Use of (1.42) and (1.43) allows us to rewrite (1.45) in the form:

$$\frac{d^\pi}{dt} \int_{\Omega_t} \mathcal{G} d\Omega_t = \int_{\Omega_t} \left(\frac{d^\pi \mathcal{G}}{dt} + \mathcal{G} \nabla_x \cdot \mathbf{V}^\pi \right) d\Omega_t \tag{1.46}$$

or, equivalently:

$$\frac{d^\pi}{dt} \int_{\Omega_t} \mathcal{G} d\Omega_t = \int_{\Omega_t} \left(\frac{\partial \mathcal{G}}{\partial t} + \nabla_x \cdot (\mathcal{G} \mathbf{V}^\pi) \right) d\Omega_t \tag{1.47}$$

Use of the divergence theorem finally provides the alternative expression:

$$\frac{d^\pi}{dt} \int_{\Omega_t} \mathcal{G} d\Omega_t = \int_{\Omega_t} \frac{\partial \mathcal{G}}{\partial t} d\Omega_t + \int_{\partial \Omega_t} \mathcal{G} \mathbf{V}^\pi \cdot \mathbf{n}\, da \tag{1.48}$$

where $\partial \Omega_t$ stands for the border of volume Ω_t, while $\mathbf{n}$ is the outward unit normal to surface da.

1.3.2 Strain Rates

The particle derivative allows the Eulerian description of the kinematics of the deformation that refers only to the current configuration. The Eulerian strain rate tensor $\mathbf{d}^\pi$ is defined by the relation:

$$\frac{d^\pi}{dt} (d\mathbf{x} \cdot d\mathbf{y}) = 2 d\mathbf{x} \cdot \mathbf{d}^\pi \cdot d\mathbf{y} \tag{1.49}$$

where $d\mathbf{x}$ and $d\mathbf{y}$ are any infinitesimal skeleton ($\pi = s$) or fluid ($\pi = f$) material vectors. Use of (1.37) into (1.49) gives:

$$\mathbf{d}^{\pi} = \frac{1}{2}\left(\nabla_x \mathbf{V}^{\pi} + {}^t\nabla_x \mathbf{V}^{\pi}\right); \quad d_{ij}^{\pi} = \frac{1}{2}\left(\frac{\partial V_i^{\pi}}{\partial x_j} + \frac{\partial V_j^{\pi}}{\partial x_i}\right) \tag{1.50}$$

Definition (1.50) of $\mathbf{d}^{\pi}$ allows us to decompose the deformation kinematics of material vector $d\mathbf{x}$ in the form:

$$\frac{d^{\pi}}{dt}(d\mathbf{x}) = \boldsymbol{\Omega}^{\pi} \cdot d\mathbf{x} + \mathbf{d}^{\pi} \cdot d\mathbf{x} \tag{1.51}$$

where $\boldsymbol{\Omega}^{\pi}$ is the rotation rate tensor attached to the antisymmetric part of $\nabla_x \mathbf{V}^{\pi}$:

$$\boldsymbol{\Omega}^{\pi} = \frac{1}{2}\left(\nabla_x \mathbf{V}^{\pi} - {}^t\nabla_x \mathbf{V}^{\pi}\right); \quad \Omega_{ij}^{\pi} = \frac{1}{2}\left(\frac{\partial V_i^{\pi}}{\partial x_j} - \frac{\partial V_j^{\pi}}{\partial x_i}\right) \tag{1.52}$$

The term $\boldsymbol{\Omega}^{\pi} \cdot d\mathbf{x}$ in (1.51) induces no strain rate since it accounts for the infinitesimal rotation of material vector $d\mathbf{x}$ according to:

$$\boldsymbol{\Omega}^{\pi} \cdot d\mathbf{x} = 2\boldsymbol{\omega}^{\pi} \times d\mathbf{x} \tag{1.53}$$

where $\boldsymbol{\omega}^{\pi}$ is the vorticity vector:

$$\boldsymbol{\omega}^{\pi} = \nabla_x \times \mathbf{V}^{\pi} \tag{1.54}$$

The Eulerian decomposition (1.51) of the kinematics of deformation is to be compared with the Lagrangian decomposition (1.22) of the deformation.

In contrast to the Eulerian approach to the kinematics of the skeleton deformation, the Lagrangian approach consists in deriving (1.19) with respect to time:[2]

$$\frac{d^s}{dt}(d\mathbf{x} \cdot d\mathbf{y}) = 2d\mathbf{X} \cdot \frac{d\boldsymbol{\Delta}}{dt} \cdot d\mathbf{Y} \tag{1.55}$$

Using transport formulae $d\mathbf{x} = \mathbf{F} \cdot d\mathbf{X}$ and $d\mathbf{y} = \mathbf{F} \cdot d\mathbf{Y}$, the comparison between (1.49) for $\pi = s$ and (1.55) leads to the transport formula:

$$\mathbf{d}^s = {}^t\mathbf{F}^{-1} \cdot \frac{d\boldsymbol{\Delta}}{dt} \cdot \mathbf{F}^{-1}; \quad d_{ij}^s = \frac{\partial X_k}{\partial x_i}\frac{d\Delta_{kl}}{dt}\frac{\partial X_l}{\partial x_j} \tag{1.56}$$

According to (1.26) and (1.56), the infinitesimal transformation approximation turns out to consider $\mathbf{d}^s \simeq d\boldsymbol{\varepsilon}/dt$.

[2]When taking the skeleton particle derivative of Lagrangian quantities, as for instance $d^s\boldsymbol{\Delta}/dt$, we will adopt a standard time derivative notation, such as for instance $d\boldsymbol{\Delta}/dt$ in (1.55). Indeed, there is no ambiguity since $\boldsymbol{\Delta} = \boldsymbol{\Delta}(\mathbf{X}, t)$. Furthermore, the particle derivative with respect to the fluid of a Lagrangian quantity does not generally present any physical interest.

1.4 Mass Balance

1.4.1 Equation of Continuity

Let ρ_s and ρ_f be the mesoscopic or intrinsic matrix and fluid mass densities so that $\rho_s\,(1-n)\,d\Omega_t$ and $\rho_f n\,d\Omega_t$ are respectively the skeleton mass and the fluid mass currently contained in the material volume $d\Omega_t$. Accordingly the macroscopic or apparent skeleton and fluid mass densities are respectively $\rho_s\,(1-n)$ and $\rho_f n$. When no mass change occurs, neither for the skeleton nor the fluid contained in the volume Ω_t, the mass balance can be expressed in the form:

$$\frac{d^s}{dt}\int_{\Omega_t}\rho_s\,(1-n)\;d\Omega_t = 0 \tag{1.57a}$$

$$\frac{d^f}{dt}\int_{\Omega_t}\rho_f n\,d\Omega_t = 0 \tag{1.57b}$$

Applying (1.45) and (1.47) to (1.57) we get:

$$\frac{d^s}{dt}(\rho_s\,(1-n)\,d\Omega_t) = 0 \tag{1.58a}$$

$$\frac{d^f}{dt}(\rho_f n\,d\Omega_t) = 0 \tag{1.58b}$$

and the Eulerian continuity equations:

$$\frac{\partial(\rho_s\,(1-n))}{\partial t} + \nabla_x\cdot\left(\rho_s\,(1-n)\,\mathbf{V}^s\right) = 0 \tag{1.59a}$$

$$\frac{\partial(\rho_f n)}{\partial t} + \nabla_x\cdot\left(\rho_f n\mathbf{V}^f\right) = 0 \tag{1.59b}$$

1.4.2 The Relative Flow Vector of a Fluid Mass. Filtration Vector. Fluid Mass Content

The appropriate formulation of the constitutive equations for the skeleton accounting for the skeleton–fluid couplings will require referring the motion of the fluid to the initial configuration of the skeleton. With that purpose in mind let $J_f\,da$ be the fluid mass flowing between time t and $t+dt$ through the infinitesimal skeleton material surface da oriented by the unit normal $\mathbf{n}$. We write:

$$J_f\,da = \mathbf{w}\cdot\mathbf{n}\,da \tag{1.60}$$

where $\mathbf{w}\,(\mathbf{x},t)$ is the Eulerian relative flow vector of fluid mass. Since the quantity $n(\mathbf{V}^f - \mathbf{V}^s)\cdot\mathbf{n}\,da\,dt$ is the infinitesimal fluid volume flowing through the skeleton surface da during the infinitesimal time dt (see Fig. 1.4), the relative vector of fluid mass $\mathbf{w}$ is consistently defined by:

$$\mathbf{w} = \rho_f\mathcal{V};\quad \mathcal{V} = n(\mathbf{V}^f - \mathbf{V}^s) \tag{1.61}$$

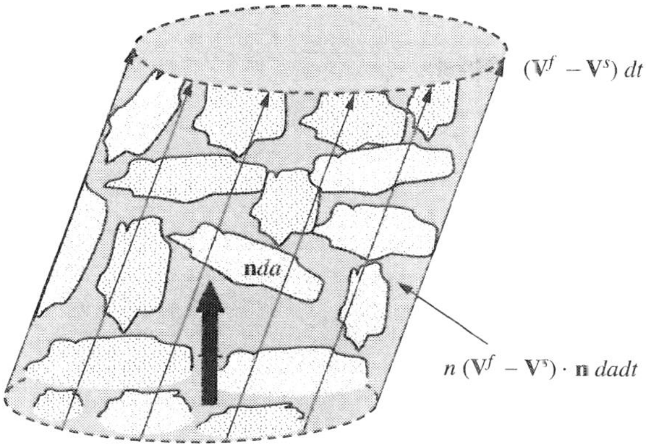

Figure 1.4: The infinitesimal fluid volume flowing through the skeleton surface da during the infinitesimal time dt. Eulerian porosity n is still to be associated with the infinitesimal fluid volume $\left(\mathbf{V}^f - \mathbf{V}^s\right)\cdot \mathbf{n}\, dadt$, and not with the skeleton infinitesimal surface da through which the fluid flows.

where $\mathcal{V}$ is the filtration vector. Use of definition (1.61) allows us to refer the fluid mass balance to the skeleton motion by rearranging the fluid continuity equation (1.59b) in the form:

$$\frac{d^s(\rho_f n)}{dt} + \rho_f n\, \nabla_x \cdot \mathbf{V}^s + \nabla_x \cdot \mathbf{w} = 0 \tag{1.62}$$

The Lagrangian approach to the fluid mass balance can be carried out by introducing the current Lagrangian fluid mass content m_f per unit of initial volume $d\Omega_0$. The latter relates to the current Eulerian fluid mass content $\rho_f n$ per unit of current volume $d\Omega_t$ according to:

$$\rho_f n\, d\Omega_t = m_f\, d\Omega_0 \tag{1.63}$$

Use of (1.17) and (1.63) gives the useful relation:

$$m_f = \rho_f \phi \tag{1.64}$$

where ϕ stands for the Lagrangian porosity (see §1.2.2). Furthermore, let $\mathbf{M}(\mathbf{X}, t)$ be the Lagrangian vector attached to the initial configuration and linked to vector $\mathbf{w}$ through the relation:

$$\mathbf{w}\cdot \mathbf{n}\, da = \mathbf{M}\cdot \mathbf{N}\, dA \tag{1.65}$$

where surfaces da and dA correspond in the skeleton deformation. Accordingly, letting $\mathbf{w} = \mathbf{v}$ and $\mathbf{M} = \mathbf{V}$ in (1.14), Eqs. (1.15) and (1.16) provide the transport formulae:

$$\mathbf{M} = J\mathbf{F}^{-1}\cdot \mathbf{w};\quad \nabla_x \cdot \mathbf{w}\, d\Omega_t = \nabla_X \cdot \mathbf{M}\, d\Omega_0 \tag{1.66a}$$

$$M_i = J\frac{\partial X_i}{\partial x_j} w_j;\quad J\frac{\partial w_i}{\partial x_i} = \frac{\partial M_i}{\partial X_i} \tag{1.66b}$$

Substitution of (1.63) and (1.66) into (1.62) premultiplied by $d\Omega_t$, and use of (1.42) with $\pi = s$, provide the Lagrangian fluid continuity equation in the form:

$$\frac{dm_f}{dt} + \nabla_X \cdot \mathbf{M} = 0; \qquad \frac{\partial m_f\,(\mathbf{X},t)}{\partial t} + \frac{\partial M_i}{\partial X_i} = 0 \tag{1.67}$$

Analogously, the Lagrangian approach to the mass balance of the skeleton turns out to integrate (1.58a) in the form:

$$\rho_s\,(1-n)\,d\Omega_t = \rho_s^0(1-n_0)d\Omega_0 \tag{1.68}$$

where ρ_s^0 and $n_0 = \phi_0$ stand respectively for the initial matrix mass density and for the initial porosity. Use of (1.11) allows us to write:

$$m_s = m_s^0 = \rho_s^0(1-\phi_0) \tag{1.69}$$

where $m_s = J\rho_s\,(1-n)$ denotes the skeleton mass content per unit of initial volume $d\Omega_0$ and remains constantly equal to its initial value m_s^0 representing the skeleton mass density $\rho_s^0(1-\phi_0)$. Equations (1.67) and (1.69) constitute the skeleton Lagrangian alternative to the Eulerian continuity equations (1.59).

1.5 Advanced Analysis

1.5.1 Particle Derivative with a Surface of Discontinuity

Some applications involve propagation fronts across which discontinuities occur. In order to derive the particle derivative of an integral accounting for these discontinuities, let Σ be a surface of discontinuity travelling within the material volume Ω_t and subdividing the latter into two subvolumes Ω_1 and Ω_2. Let $\mathbf{n}$ be the unit normal to Σ oriented in the direction of travel towards the downstream subvolume Ω_2. Finally, let $[[\mathcal{G}]]$ denote the jump across Σ in the direction of $\mathbf{n}$ of the discontinuous quantity $\mathcal{G}$ (see Fig. 1.5):

$$[[\mathcal{G}]] = \mathcal{G}_2 - \mathcal{G}_1 \tag{1.70}$$

The normal speed of displacement $\mathbf{c} = c\,\mathbf{n}$ of the surface of discontinuity Σ is the velocity at which a geometrical point belonging to Σ moves along the normal $\mathbf{n}$. During the time duration dt the infinitesimal surface da belonging to Σ sweeps out the infinitesimal volume $\mathbf{c}\cdot\mathbf{n}\,da\,dt$. During the infinitesimal time dt the volumetric density $\mathcal{G}$ related to the latter undergoes the variation $-\,[[\mathcal{G}]]$. In order to account for this sudden variation, the particle derivative of a volume integral (1.48) has to be modified according to:

$$\frac{d^\pi}{dt}\int_{\Omega_\alpha} \mathcal{G}d\Omega_t = \int_{\Omega_t} \frac{\partial \mathcal{G}}{\partial t} d\Omega_t + \int_{\partial\Omega_t} \mathcal{G}\mathbf{V}^\pi \cdot \mathbf{n}\,da - \int_\Sigma [[\mathcal{G}]]\,\mathbf{c}\cdot\mathbf{n}\,da \tag{1.71}$$

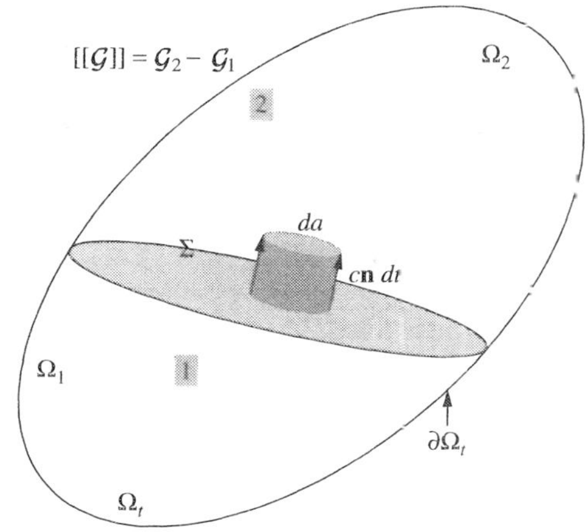

Figure 1.5: Infinitesimal volume swept out during the infinitesimal time dt by an infinitesimal surface da belonging to a surface of discontinuity Σ moving at the normal velocity of displacement $\mathbf{c} = c\,\mathbf{n}$.

Separate application of the divergence theorem to subdomains Ω_α, $\alpha = 1$ or 2, gives:

$$\int_{\Omega_\alpha} \nabla_x \cdot (\mathcal{G}\mathbf{V}^\pi) d\Omega_t$$
$$= \int_{\partial\Omega_\alpha \cap \partial\Omega_t} \mathcal{G}\mathbf{V}^\pi \cdot \mathbf{n}\, da + \int_\Sigma \mathcal{G}\mathbf{V}^\pi \cdot \boldsymbol{\nu}_\alpha\, da; \qquad \boldsymbol{\nu}_1 = -\boldsymbol{\nu}_2 = \mathbf{n} \tag{1.72}$$

Since $\Omega_t = \Omega_1 \cup \Omega_2$ and $\partial\Omega_t = (\partial\Omega_1 \cap \partial\Omega_t) \cup (\partial\Omega_2 \cap \partial\Omega_t)$, the two previous equations lead to:

$$\frac{d^\pi}{dt} \int_{\Omega_t} \mathcal{G} d\Omega_t = \int_{\Omega_t} \left(\frac{\partial \mathcal{G}}{\partial t} + \nabla_x.(\mathcal{G}\mathbf{V}^\pi) \right) d\Omega_t + \int_\Sigma [[\mathcal{G}(\mathbf{V}^\pi - \mathbf{c})]] \cdot \mathbf{n}\, da \tag{1.73}$$

1.5.2 Mass Balance with a Surface of Discontinuity. The Rankine–Hugoniot Jump Condition

Use of (1.73) in (1.57) provides the mass balance equations accounting for a surface of discontinuity. We write:

$$\int_{\Omega_t} \left(\frac{\partial(\rho_s (1-n))}{\partial t} + \nabla_x \cdot \left(\rho_s (1-n)\, \mathbf{V}^s\right) \right) d\Omega_t$$
$$+ \int_\Sigma [[\rho_s (1-n)\, (\mathbf{V}^s - \mathbf{c})]] \cdot \mathbf{n}\, da = 0 \tag{1.74a}$$

$$\int_{\Omega_t} \left(\frac{\partial(\rho_f n)}{\partial t} + \nabla_x \cdot \left(\rho_f n \mathbf{V}^f\right) \right) d\Omega_t + \int_\Sigma [[\rho_f n(\mathbf{V}^f - \mathbf{c})]] \cdot \mathbf{n}\, da = 0 \tag{1.74b}$$

The first term in the above equations allows us to recover continuity equations (1.59), whereas the second term provides the jump or Rankine–Hugoniot conditions:

$$[[\rho_s (1-n)(\mathbf{V}^s - \mathbf{c})]] \cdot \mathbf{n} = 0 \tag{1.75a}$$

$$[[\rho_f n(\mathbf{V}^f - \mathbf{c})]] \cdot \mathbf{n} = 0 \tag{1.75b}$$

The jump condition means that the particles passing across the surface of discontinuity Σ do not undergo any mass change. Referring to the skeleton motion, the jump condition (1.75b) relative to the fluid is conveniently rewritten in the form:

$$[[\mathbf{w} - \rho_f n(\mathbf{c} - \mathbf{V}^s)]] \cdot \mathbf{n} = 0 \tag{1.76}$$

Let us apply the jump condition to the special case of a surface of discontinuity Σ constituted by the interface between two different porous media. Assuming that the two media are perfectly bonded, $\mathbf{V}^s$ remains continuous across their interface so that the displacement speed $c = \mathbf{c} \cdot \mathbf{n}$ equals the normal skeleton velocity $\mathbf{V}^s \cdot \mathbf{n}$. Consequently, at the interface Σ of two bonded porous media, the jump condition (1.76), which eventually ensures that no liquid filtration occurs along the interface, reduces to:

$$[[\mathbf{w}]] \cdot \mathbf{n} = 0 \tag{1.77}$$

During time dt the infinitesimal surface da belonging to the surface of discontinuity Σ sweeps out, in the current configuration, the skeleton material volume $(\mathbf{c} - \mathbf{V}^s) \cdot \mathbf{n}\, da\, dt$. In the meantime the infinitesimal surface dA associated with da through transport formula (1.13) sweeps out, in the initial configuration, the skeleton volume $\mathbf{C} \cdot \mathbf{N}\, dA\, dt$. Lagrangian normal speed $\mathbf{C}$ is the speed at which a geometrical point belonging to the surface of discontinuity moves between times t and $t + dt$ in the reference configuration along the normal $\mathbf{N}$. By using (1.64) and (1.65), the Eulerian jump condition (1.76) can be transported to the initial configuration to furnish the Lagrangian jump condition:

$$[[\mathbf{M} - m_f \mathbf{C}]] \cdot \mathbf{N} = 0 \tag{1.78}$$

With the aim of giving a first illustration of the above Lagrangian approach to the Rankine–Hugoniot jump condition (see §5.4.4 for a second illustration), let us consider a porous material subjected to a dissolution process (e.g. of leaching type, see §7.1), in which the solid matrix (e.g. calcium) progressively dissolves into the interstitial solution filling the porous space. At the microscopic scale of the latter, the problem at hand involves the propagation of a surface of discontinuity in the initial configuration, even though the dissolution front representing the surface of discontinuity in the current configuration remains constantly identified with the current internal walls of the porous space. At the microscopic scale we apply (1.78) in the form:

$$\{(\mathbf{M} - m\mathbf{C}) \cdot \mathbf{N}\}_{\text{intact}} = \{(\mathbf{M} - m\mathbf{C}) \cdot \mathbf{N}\}_{\text{solute}} \tag{1.79}$$

where, for the sake of simplicity, we retain the same notation as that used at the macroscopic scale: $\mathbf{M}$ represents the Lagrangian relative flux vector with respect to the solid matrix; m is the total mass per unit of initial volume of the species subjected to dissolution; $C = \mathbf{C} \cdot \mathbf{N}$ is the Lagrangian speed of propagation front Σ, which separates, in the initial configuration the still intact zone from the already dissolved one. The above equation eventually expresses the mass conservation related to the thin layer of solid matrix currently passing from the intact state, as part of the solid matrix, to the dissolved state, as contributing to the solute. On the intact side we write $\mathbf{M} = 0$ and $m = m_0$ since the still intact material, that is the current solid matrix, has no relative motion with respect to itself. Returning to the macroscopic scale, let $d\phi_{ch}/dt$ be the chemical contribution to the total rate $d\phi/dt$ of Lagrangian porosity, that is the one due only to the dissolution process irrespective of strain effects. According to the above analysis, $d\phi_{ch}/dt$ can be expressed in the form:

$$\frac{d\phi_{ch}}{dt} = \frac{1}{d\Omega_0} \int_{\Sigma} C dA \tag{1.80}$$

1.5.3 Mass Balance and the Double Porosity Network

Some materials such as rocks or concrete sometimes exhibit two very distinct porous networks. Roughly speaking, the first is formed of rounded pores, while the second is formed of penny-shaped cracks. Although these two networks can exchange fluid mass between them, their quite different geometries result in distinct evolution laws of their pore pressure and require a separate analysis. To this end let subscript 1 refer to the porous network formed of cracks and subscript 2 to the one formed of pores. While the continuity equation related to the skeleton remains unchanged, the fluid mass balance now reads:

$$\frac{d^{f_1}}{dt} \int_{\Omega_t} \rho_{f_1} n_1 \, d\Omega_t = \int_{\Omega_t} \overset{\circ}{r}_{2\to 1} \, d\Omega_t \tag{1.81a}$$

$$\frac{d^{f_2}}{dt} \int_{\Omega_t} \rho_{f_2} n_2 \, d\Omega_t = \int_{\Omega_t} \overset{\circ}{r}_{1\to 2} \, d\Omega_t \tag{1.81b}$$

where $\overset{\circ}{r}_{\alpha\to\beta}$ stands for the rate of fluid mass flowing from network α into network β, while n_α is the porosity associated with network α so that $n = n_1 + n_2$.

Mass conservation requires $\overset{\circ}{r}_{\alpha\to\beta} = -\overset{\circ}{r}_{\beta\to\alpha}$ and application of (1.45) to (1.81) gives:

$$\frac{d^{f_1}}{dt}(\rho_{f_1} n_1 \, d\Omega_t) = -\overset{\circ}{r}_{1\to 2} \, d\Omega_t; \quad \frac{d^{f_2}}{dt}(\rho_{f_2} n_2 \, d\Omega_t) = \overset{\circ}{r}_{1\to 2} \, d\Omega_t \tag{1.82}$$

Using (1.46), we obtain two separate continuity equations for the fluid flowing through the crack network and for the one flowing through the pore network:

$$\frac{\partial(\rho_{f_1} n_1)}{\partial t} + \nabla_x \cdot \left(\rho_{f_1} n_1 \mathbf{V}^{f_1}\right) = -\overset{\circ}{r}_{1\to 2} \tag{1.83a}$$

$$\frac{\partial(\rho_{f_2} n_2)}{\partial t} + \nabla_x \cdot \left(\rho_{f_2} n_2 \mathbf{V}^{f_2}\right) = \overset{\circ}{r}_{1\to 2} \tag{1.83b}$$

The Lagrangian alternative to the previous Eulerian fluid continuity equations can be written as:

$$\frac{d^s m_1}{dt} + \nabla_X \cdot \mathbf{M}^{(1)} = -\overset{\circ}{m}_{1\to 2} \tag{1.84a}$$

$$\frac{d^s m_2}{dt} + \nabla_X \cdot \mathbf{M}^{(2)} = \overset{\circ}{m}_{1\to 2} \tag{1.84b}$$

with:

$$m_\alpha = J\rho_{f_\alpha} n_\alpha = \rho_{f_\alpha}\phi_\alpha; \quad \overset{\circ}{m}_{1\to 2} = J\overset{\circ}{r}_{1\to 2} \tag{1.85}$$

The approach can be adapted in order to account for possible long-term exchanges of fluid mass between the occluded porosity, through which no fluid flow significantly occurs in the short-term range, and the connected porous network, through which the fluid flows at any time. This fluid exchange can be viewed as an exchange of mass between the skeleton, including the occluded porosity and its fluid, and still referred to by s, and the fluid saturating the porous network referred to by f. Analogously to (1.83) and (1.84) the Eulerian and Lagrangian approaches to the continuity equation respectively read:

$$\frac{\partial(\rho_s(1-n))}{\partial t} + \nabla_x \cdot \left(\rho_s(1-n)\mathbf{V}^s\right) = -\overset{\circ}{r}_{s\to f} \tag{1.86a}$$

$$\frac{\partial(\rho_f n)}{\partial t} + \nabla_x \cdot \left(\rho_f n \mathbf{V}^f\right) = \overset{\circ}{r}_{s\to f} \tag{1.86b}$$

and:

$$\frac{d^s m_s}{dt} = -\overset{\circ}{m}_{s\to f} \tag{1.87a}$$

$$\frac{d^s m_f}{dt} + \nabla_X \cdot \mathbf{M} = \overset{\circ}{m}_{s\to f} \tag{1.87b}$$

Similarly, consider a material subjected to a dissolution process. Let $\overset{\circ}{m}_{s\to sol}$ be the rate of solid mass (index s) which currently dissolves per unit of initial volume $d\Omega_0$ in solute form (index sol), so that the mass conservation of the solid matrix, the solute and the solvent (index w for liquid water) can be expressed in the form:

$$\frac{d^s m_s}{dt} = -\overset{\circ}{m}_{s\to sol} \tag{1.88a}$$

$$\frac{d^s m_{sol}}{dt} + \nabla_X \cdot \mathbf{M}^{sol} = \overset{\circ}{m}_{s\to sol} \tag{1.88b}$$

$$\frac{d^s m_w}{dt} + \nabla_X \cdot \mathbf{M}^{w} = 0 \tag{1.88c}$$

In addition, from the analysis of §1.5.2 and from (1.80) we derive:

$$\overset{\circ}{m}_{s\to sol} = \rho_s^0 \frac{d\phi_{ch}}{dt} \tag{1.89}$$

Chapter 2

Momentum Balance. Stress Tensor

This chapter is devoted to the formulation of the momentum balance for a porous continuum viewed as the superimposition of two continua in mechanical interaction. As in standard continuum mechanics, the existence of a symmetric total stress tensor and the local momentum equation relative to the porous continuum viewed as a whole can be derived from the overall momentum balance. Nevertheless, the derivation of separate momentum balance equations for the skeleton and for the fluid cannot be carried out from a strictly macroscopic approach. A first step towards the missing momentum equation and an understanding of the balance of mechanical energy involved in the kinetic energy theorem consist in involving the mesoscopic scale through the introduction of partial stress related to each continuum.

2.1 Momentum Balance

2.1.1 The Hypothesis of Local Forces

In continuum mechanics any material domain Ω_t is subjected to two kinds of external forces, namely the external body forces and the external surface forces, as sketched in Fig. 2.1. In most applications, the external body forces, such as the ones due to gravity, are the same for the skeleton and for the fluid. The infinitesimal body force $\delta\mathbf{f}$ acting on the elementary material volume $d\Omega_t$ is defined through a body force density per mass unit $\mathbf{f}$:

$$\delta\mathbf{f} = \rho\mathbf{f}(\mathbf{x}, t)\, d\Omega_t \tag{2.1}$$

where ρ denotes the current mass density of the material volume $d\Omega_t$, here including both the skeleton and the fluid:

$$\rho = \rho_s (1 - n) + \rho_f n \tag{2.2}$$

The body force density $\mathbf{f}$ is assumed to depend only on position vector $\mathbf{x}$ and time t. Accordingly the external body forces applying on the infinitesimal material volume $d\Omega_t$

Poromechanics O. Coussy
 ISBN 0-470-84920-7

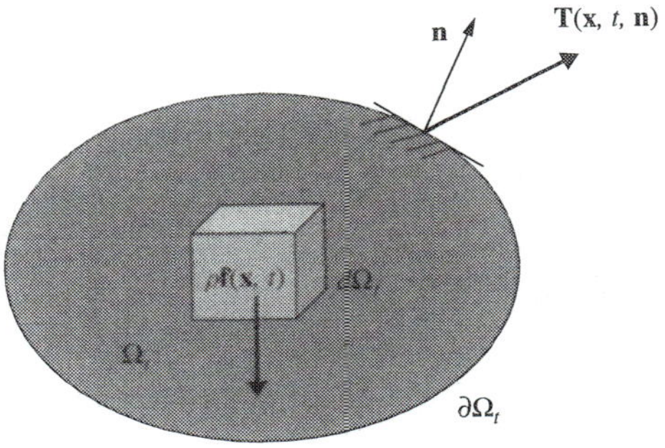

Figure 2.1: Definition of external forces: body forces and surface forces. (from Coussy 2004, reprinted by permission of Pearson Education, Inc.).

express the same thing irrespective of the domain Ω_t the infinitesimal volume $d\Omega_t$ belongs to. In other words, the body forces considered are local forces. Non-local forces such as the forces depending on the distance between particles will not be considered.

The surface forces act on the border $\partial\Omega_t$ of Ω_t. The infinitesimal surface force $\delta\mathbf{T}$ acting on the infinitesimal material surface da is defined through a surface force density $\mathbf{T}$:

$$\delta\mathbf{T} = \mathbf{T}(\mathbf{x}, t, \mathbf{n})\, da \tag{2.3}$$

As with the body force density, the surface force density $\mathbf{T}$ is assumed to depend only on position vector $\mathbf{x}$ and time t and the outward unit normal $\mathbf{n}$ to da. This turns out to assuming that the surface forces result only from local contact forces exerted by the immediately adjacent material points. The action of more distant points, which would at least involve the local curvature of $\partial\Omega_t$ and consequently $\nabla_x\mathbf{n}$, is excluded here. This hypothesis of 'local contact forces', known as Cauchy's hypothesis, will turn out to be essential to define the stress tensor.

2.1.2 The Momentum Balance

In a Galilean referential frame, the instantaneous momentum balance, with respect to all matter included in any porous domain Ω_t, is:

$$\begin{aligned}&\frac{d^s}{dt}\int_{\Omega_t}\rho_s(1-n)\,\mathbf{V}^s d\Omega_t + \frac{d^f}{dt}\int_{\Omega_t}\rho_f n\mathbf{V}^f d\Omega_t\\&\quad = \int_{\Omega_t}\rho\mathbf{f}(\mathbf{x}, t)\,d\Omega_t + \int_{\partial\Omega_t}\mathbf{T}(\mathbf{x}, t, \mathbf{n})\,da\end{aligned} \tag{2.4}$$

$$\begin{aligned}&\frac{d^s}{dt}\int_{\Omega_t}\mathbf{x}\times\rho_s(1-n)\,\mathbf{V}^s d\Omega_t + \frac{d^f}{dt}\int_{\Omega_t}\mathbf{x}\times\rho_f n\mathbf{V}^f d\Omega_t\\&\quad = \int_{\Omega_t}\mathbf{x}\times\rho\mathbf{f}(\mathbf{x}, t)\,d\Omega_t + \int_{\partial\Omega_t}\mathbf{x}\times\mathbf{T}(\mathbf{x}, t, \mathbf{n})\,da\end{aligned} \tag{2.5}$$

where $\rho_s(1-n)\,\mathbf{V}^s d\Omega_t$ and $\rho_s n\mathbf{V}^f d\Omega_t$ represent the linear momentum related respectively to the skeleton particle and to the fluid particle, both particles coinciding with $d\Omega_t$.

Since body and surface forces refer to the whole matter, with no distinction made between the skeleton and the fluid, Eq. (2.4) means that the time derivative of the momentum of the whole matter currently contained in Ω_t is equal to the creation rate of momentum due to the external forces acting on this matter. Equation (2.5) expresses the same thing as it concerns the angular momentum. The use of the particle derivative d^π/dt, with $\pi = s$ or f, accounts for the distinct motion of the solid and the fluid particles forming the volume Ω_t when deriving their change in momentum.

2.1.3 The Dynamic Theorem

Taking the particle derivative of the integrand in (2.4) and (2.5) and using (1.35), (1.44) and (1.58), we get:

$$\int_{\Omega_t} (\rho_s (1-n)\,\boldsymbol{\gamma}^s + \rho_f n \boldsymbol{\gamma}^f) d\Omega_t = \int_{\Omega_t} \rho \mathbf{f}(\mathbf{x}, t)\, d\Omega_t + \int_{\partial\Omega_t} \mathbf{T}(\mathbf{x}, t, \mathbf{n})\, da \tag{2.6}$$

$$\int_{\Omega_t} \mathbf{x} \times (\rho_s (1-n)\,\boldsymbol{\gamma}^s + \rho_f n \boldsymbol{\gamma}^f) d\Omega_t = \int_{\Omega_t} \mathbf{x} \times \rho \mathbf{f}(\mathbf{x}, t)\, d\Omega_t + \int_{\partial\Omega_t} \mathbf{x} \times \mathbf{T}(\mathbf{x}, t, \mathbf{n})\, da \tag{2.7}$$

where $\left(\rho_s (1-n)\,\boldsymbol{\gamma}^s + \rho_f n \boldsymbol{\gamma}^f\right) d\Omega_t$ and $\mathbf{x} \times \left(\rho_s (1-n)\,\boldsymbol{\gamma}^s + \rho_f n \boldsymbol{\gamma}^f\right) d\Omega_t$ represent respectively the dynamic force and the dynamic moment related to the matter currently contained within $d\Omega_t$.

Equation (2.6) represents the theorem of the dynamic resultant, stating that the resultant of the dynamic forces related to the matter contained within Ω_t is equal to the resultant of the forces being exerted on this matter, whereas (2.7) expresses the analogous equality for the dynamic moment.

2.2 The Stress Tensor

2.2.1 Action–Reaction Law

The dynamic resultant theorem (2.6) must hold for any domain Ω_t, whereas the hypothesis of local forces ensures that the body force $\mathbf{f}(\mathbf{x}, t)$ acting on the volume $d\Omega_t$ is independent of the choice of domain Ω_t to which the volume $d\Omega_t$ belongs. Consider the cylinder shown in Fig. 2.2a. The surface $\partial\Omega_t$ is formed by the cylinder wall of surface Σ oriented by the unit outward normal $\mathbf{n}_\Sigma$ and by the two end sections, a^+ and a^-, oriented by the unit outward normal vectors $\mathbf{n}$ and $-\mathbf{n}$. The dynamic resultant theorem (2.6) applied to this cylinder leads to:

$$\int_{\Omega_t} (\rho_s (1-n)\,\boldsymbol{\gamma}^s + \rho_f n \boldsymbol{\gamma}^f - \rho \mathbf{f})\, d\Omega_t = \int_{a^+} \mathbf{T}(\mathbf{n})\, da + \int_{a^-} \mathbf{T}(-\mathbf{n})\, da + \int_{\Sigma} \mathbf{T}(\mathbf{n}_\Sigma)\, da \tag{2.8}$$

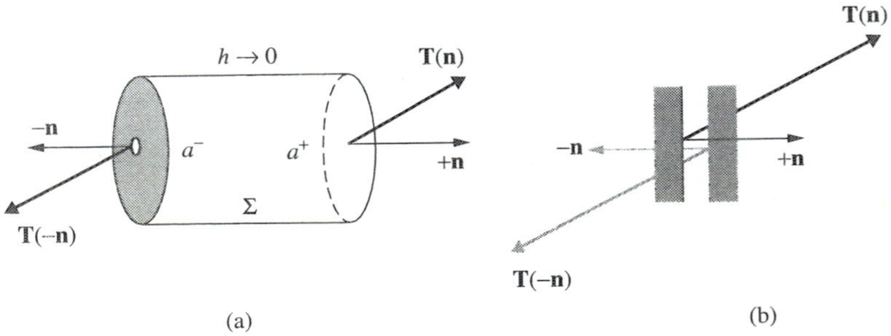

Figure 2.2: Action–reaction law. (from Coussy 2004, reprinted by permission of Pearson Education, Inc.).

Letting the thickness h of the cylinder in the direction normal to a^+ and a^- tend to zero, both the contributions of the surface Σ and volume Ω_t to (2.8) vanish. We write:

$$h \to 0: \quad \int_{a^+} \mathbf{T}(\mathbf{n})\, da + \int_{a^-} \mathbf{T}(-\mathbf{n})\, da = 0 \tag{2.9}$$

Since this identity must hold whatever the value of da, we obtain the 'action–reaction law' sketched in Fig. 2.2b, reading:

$$\mathbf{T}(-\mathbf{n}) = -\mathbf{T}(\mathbf{n}) \tag{2.10}$$

2.2.2 The Tetrahedron Lemma and the Cauchy Stress Tensor

In addition to the action–reaction law let us apply the dynamic resultant theorem to the infinitesimally small tetrahedron whose three facets S_j are parallel to the coordinate planes and oriented by $-\mathbf{e}_j$ (see Fig. 2.3). Surfaces S_j are linked to the base surface S oriented by unit normal $\mathbf{n}$ according to the relation:

$$S_j = S\mathbf{n} \cdot \mathbf{e}_j = Sn_j \tag{2.11}$$

Applying the dynamic resultant theorem (2.6) to this infinitesimally small tetrahedron gives:

$$\frac{hS}{3} O(\rho_s (1-n)\,\boldsymbol{\gamma}^s + \rho_f n \boldsymbol{\gamma}^f - \rho \mathbf{f}) \simeq \mathbf{T}(\mathbf{n})S + \sum_{i=1,2,3} \mathbf{T}(-\mathbf{e}_j)S_j \tag{2.12}$$

where h is the height of the tetrahedron, so that $hS/3$ is its volume, and where $O(\mathcal{G})$ scales as the order of magnitude of quantity $\mathcal{G}$. Substitution of (2.11) into (2.12), together with the use of the action–reaction law (2.10), leads to:

$$\frac{h}{3} O(\rho_s (1-n)\,\boldsymbol{\gamma}^s + \rho_f n \boldsymbol{\gamma}^f - \rho \mathbf{f}) \simeq \mathbf{T}(\mathbf{n}) - \sum_{j=1,2,3} \mathbf{T}(\mathbf{e}_j)\, n_j \tag{2.13}$$

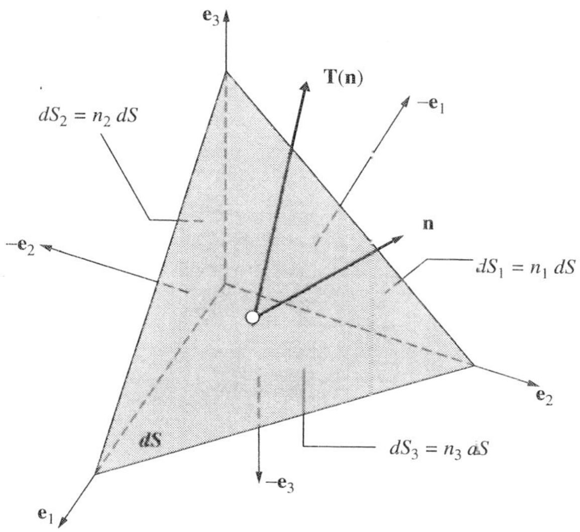

Figure 2.3: Tetrahedron lemma applied to the momentum balance for the existence of the stress as a tensor quantity. (from Coussy 2004, reprinted by permission of Pearson Education, Inc.).

Letting $h \to 0$, that is degenerating the tetrahedron to a point, the left hand side of (2.13) tends to zero, yielding:

$$\mathbf{T}(\mathbf{n} = n_j \mathbf{e}_j) = \sum_{j=1,2,3} \mathbf{T}(\mathbf{e}_j)\, n_j \tag{2.14}$$

This is known as the tetrahedron lemma,[1] here applied to the momentum balance (2.6). Equation (2.14) defines a linear operator relating the vector $\mathbf{T}(\mathbf{x}, t, \mathbf{n})$ to $\mathbf{n}$. This operator is known as the Cauchy (or Euler) stress tensor $\boldsymbol{\sigma} = \boldsymbol{\sigma}$ (x,t) with components σ_{ij}, while vector $\mathbf{T}(\mathbf{x}, t, \mathbf{n})$ is called the stress vector:

$$\mathbf{T}(\mathbf{x}, t, \mathbf{n} = \mathbf{n}_j \mathbf{e}_j) = \boldsymbol{\sigma} \cdot \mathbf{n} = \sigma_{ij} n_j\, \mathbf{e}_i \tag{2.15}$$

The definition (2.15) of stress $\boldsymbol{\sigma}$ as a tensorial quantity is the direct consequence of the hypothesis of local contact forces, that is $\mathbf{T} = \mathbf{T}(\mathbf{x}, t, \mathbf{n})$. Figures 2.4a and 2.4b illustrate

[1]For the purposes of completeness, the tetrahedron lemma states that if a relation of the form

$$\int_{\Omega_t} h(\mathbf{x}, t)\, d\Omega_t + \int_{\partial\Omega_t} f(\mathbf{x}, t, \mathbf{n})\, da = 0$$

holds for any volume Ω_t, then f can be written as a linear function of the n_i components of normal $\mathbf{n}$:

$$f(\mathbf{x}, t, \mathbf{n}) = f_i(\mathbf{x}, t)\, n_i$$

For instance, if f denotes the relative flow J_f of fluid mass with respect to the skeleton motion, the tetrahedron lemma applied to the fluid mass balance requires J_f to have the form of a vector flux, that is $J_f = \mathbf{w} \cdot \mathbf{n}\, da$, as heuristically stated in (1.60).

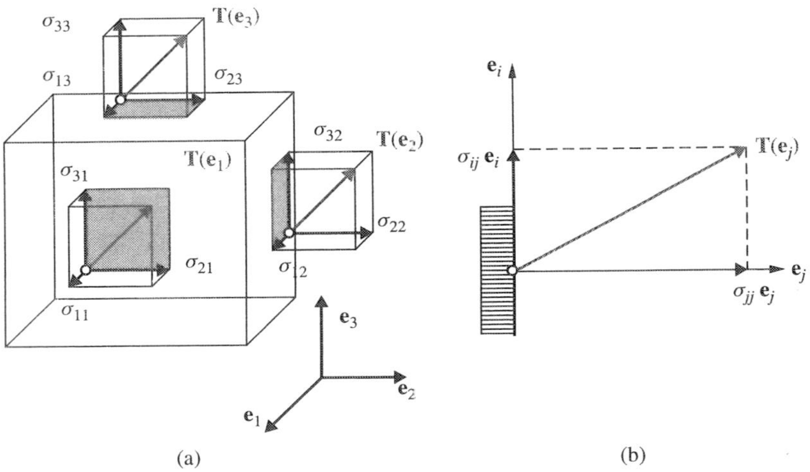

Figure 2.4: Physical significance of the Cauchy stress tensor $\boldsymbol{\sigma}$ relative to the stress vector $\mathbf{T}$. (from Coussy 2004, reprinted by permission of Pearson Education, Inc.).

the physical significance of the Cauchy stress tensor $\boldsymbol{\sigma}$ relative to the stress vector: σ_{ij} $(i = 1, 2, 3)$ are the components in the orthonormal basis $\mathbf{e}_i$ $(i = 1, 2, 3)$ of the surface force density $\mathbf{T}(\mathbf{x}, t, \mathbf{n} = \mathbf{e}_j) = \sigma_{ij}\mathbf{e}_i$ acting per unit of surface area on the material surface oriented by unit normal $\mathbf{n} = \mathbf{e}_j$.

2.3 Equation of Motion

2.3.1 The Local Dynamic Resultant Theorem

Use of (2.15) in the global dynamic resultant theorem (2.6) gives:

$$\int_{\Omega_t} (\rho\mathbf{f} - \rho_s (1 - n)\, \boldsymbol{\gamma}^s - \rho_f n \boldsymbol{\gamma}^f) d\Omega_t + \int_{\partial\Omega_t} \boldsymbol{\sigma} \cdot \mathbf{n}\, da = 0 \tag{2.16}$$

Applying the divergence theorem to the surface integral of (2.16), namely:

$$\int_{\partial\Omega_t} \boldsymbol{\sigma} \cdot \mathbf{n}\, da = \int_{\Omega_t} \nabla_x \cdot \boldsymbol{\sigma}\, d\Omega_t; \quad \int_{\partial\Omega_t} \sigma_{ij} n_j\, da = \int_{\Omega_t} \frac{\partial \sigma_{ij}}{\partial x_j} d\Omega_t \tag{2.17}$$

we can rewrite the dynamic resultant theorem in the form:

$$\int_{\Omega_t} [\nabla_x \cdot \boldsymbol{\sigma} + \rho\mathbf{f} - \rho_s (1 - n)\, \boldsymbol{\gamma}^s - \rho_f n \boldsymbol{\gamma}^f]\, d\Omega_t = 0 \tag{2.18}$$

The dynamic theorem (2.18) must hold irrespective of any particular choice of domain Ω_t. It finally leads to the local equation of motion or equilibrium of the elementary volume $d\Omega_t$:

$$\nabla_x \cdot \boldsymbol{\sigma} + \rho_s (1 - n) (\mathbf{f} - \boldsymbol{\gamma}^s) + \rho_f n(\mathbf{f} - \boldsymbol{\gamma}^f) = 0 \tag{2.19}$$

or in index notation:

$$\frac{\partial \sigma_{ij}}{\partial x_j} + \rho_s (1-n) \left(f_i - \gamma_i^s \right) + \rho_f n (f_i - \gamma_i^f) = 0 \tag{2.20}$$

Equation (2.19) expresses the dynamic resultant theorem applied locally to the matter which constitutes, at time t, any elementary volume $d\Omega_t$, located at material point $\mathbf{x}$. It states that the sum of the elementary body forces $\rho \mathbf{f} d\Omega_t$ and of the surface forces $\mathbf{T}da$ acting on the facets of $d\Omega_t$ is equal to the dynamic force $\left(\rho_s (1-n) \boldsymbol{\gamma}^s + \rho_f n \boldsymbol{\gamma}^f \right) d\Omega_t$. For instance, (2.20) applied for $i = 2$ states that the sum of all forces acting on the material volume $d\Omega_t = dx_1 dx_2 dx_3$ in the $\mathbf{e}_2$ direction is zero.

2.3.2 The Dynamic Moment Theorem and the Symmetry of the Stress Tensor

With the help of (2.15), we can rewrite (2.7) in the form:

$$\int_{\Omega_t} \mathbf{x} \times (\rho \mathbf{f} - \rho_s (1-n) \boldsymbol{\gamma}^s - \rho_f n \boldsymbol{\gamma}^f) \, d\Omega_t + \int_{\partial \Omega_t} \mathbf{x} \times \boldsymbol{\sigma} \cdot \mathbf{n} \, da = 0 \tag{2.21}$$

The divergence theorem applied to the last term of (2.21) yields:

$$\int_{\partial \Omega_t} \mathbf{x} \times \boldsymbol{\sigma} \cdot \mathbf{n} \, da = \int_{\Omega_t} (\mathbf{x} \times \nabla_x \cdot \boldsymbol{\sigma} + 2\boldsymbol{\sigma}^{as}) \, d\Omega_t \tag{2.22}$$

where $\boldsymbol{\sigma}^{as}$ in Cartesian coordinates is the vector defined by:

$$2\boldsymbol{\sigma}^{as} = (\sigma_{23} - \sigma_{32}) \, \mathbf{e}_1 + (\sigma_{13} - \sigma_{31}) \, \mathbf{e}_2 + (\sigma_{12} - \sigma_{21}) \, \mathbf{e}_3 \tag{2.23}$$

Substitution of (2.22) into (2.21) gives:

$$\int_{\Omega_t} \left[\mathbf{x} \times (\rho \mathbf{f} - \rho_s (1-n) \boldsymbol{\gamma}^s - \rho_f n \boldsymbol{\gamma}^f + \nabla_x \cdot \boldsymbol{\sigma}) + 2\boldsymbol{\Sigma}^{as} \right] d\Omega_t = 0 \tag{2.24}$$

Using the local dynamic equation (2.19) in (2.24), the dynamic moment theorem finally gives:

$$\int_{\Omega_t} \boldsymbol{\sigma}^{as} d\Omega_t = 0 \tag{2.25}$$

Since (2.25) must hold for any volume Ω_t, it follows that $\boldsymbol{\sigma}^{as} = 0$ and, from expression (2.23) of Σ^{as}, this eventually implies the symmetry of the strain tensor $\boldsymbol{\sigma}$:

$$\boldsymbol{\sigma} = {}^t\boldsymbol{\sigma}; \quad \sigma_{ij} = \sigma_{ji} \tag{2.26}$$

The symmetry of the stress tensor results from the absence of external moment couples, whether volume or surface related. Indeed, in the absence of external couples, symmetry (2.26) means that the sum of the moments of the surface forces $\mathbf{T}da$ acting on the facets of $d\Omega_t$ is zero. For instance, the symmetry $\sigma_{12} = \sigma_{21}$ expresses the nullity of the dynamic moment in the $\mathbf{e}_3$ direction. The symmetry of the stress tensor eventually ensures the realness of its eigenvalues $\sigma_{J=1,2,3}$:

$$\boldsymbol{\sigma} \cdot \mathbf{u}_J = \sigma_J \mathbf{u}_J \tag{2.27}$$

These eigenvalues are called the principal stresses, and the associated eigenvectors $\mathbf{u}_J$ are the principal stress directions. The equation of motion is finally derived by the combination of local equations (2.19) and (2.26).

2.3.3 Partial Stress Tensor

The stress tensor $\boldsymbol{\sigma}$ does not account separately for the stress related to the skeleton and the one related to the fluid. In order to identify their respective contributions it is tempting to extend the contact force hypothesis (2.3) in the form:

$$\delta\mathbf{T}^s = \mathbf{T}^s(\mathbf{x}, t, \mathbf{n})\, da; \quad \delta\mathbf{T}^f = \mathbf{T}^f(\mathbf{x}, t, \mathbf{n})\, da \tag{2.28}$$

where $\delta\mathbf{T}^s$ and $\delta\mathbf{T}^f$ are the surface forces related respectively to the skeleton continuum and to the fluid continuum. Applying the momentum balance separately to the skeleton and to the fluid, the same analysis as above leads to the separate existence of a partial volumetric stress tensor $\boldsymbol{\sigma}^s$ related to the skeleton and a partial volumetric stress tensor $\boldsymbol{\sigma}^f$ related to the fluid, such as:

$$\mathbf{T}^s(\mathbf{x}, t, \mathbf{n}) = (1-n)\,\boldsymbol{\sigma}^s \cdot \mathbf{n}; \quad \mathbf{T}^f(\mathbf{x}, t, \mathbf{n}) = n\boldsymbol{\sigma}^f \cdot \mathbf{n} \tag{2.29}$$

In addition, partial stresses $\boldsymbol{\sigma}^s$ and $\boldsymbol{\sigma}^f$ must be symmetric, that is $\boldsymbol{\sigma}^s = {}^t\boldsymbol{\sigma}^s$ and $\boldsymbol{\sigma}^f = {}^t\boldsymbol{\sigma}^f$, and satisfy the local equations of motion:

$$\nabla_x \cdot \left[(1-n)\,\boldsymbol{\sigma}^s\right] + \rho_s (1-n)(\mathbf{f} - \boldsymbol{\gamma}^s) + \mathbf{f}_{int}^{\to s} = 0 \tag{2.30a}$$

$$\nabla_x \cdot (n\boldsymbol{\sigma}^f) + \rho_f n(\mathbf{f} - \boldsymbol{\gamma}^f) + \mathbf{f}_{int}^{\to f} = 0 \tag{2.30b}$$

where the volume force $\mathbf{f}_{int}^{\to \pi}$ accounts for the macroscopic interaction force exerted by the other continuum. By virtue of the action–reaction law, each continuum exerts on the other continuum the opposite interaction force, that is:

$$\mathbf{f}_{int}^{\to s} + \mathbf{f}_{int}^{\to f} = 0 \tag{2.31}$$

so that the sum of (2.30) returns (2.19), with:

$$\mathbf{T} = \mathbf{T}^s + \mathbf{T}^f; \quad \boldsymbol{\sigma} = (1-n)\,\boldsymbol{\sigma}^s + n\boldsymbol{\sigma}^f \tag{2.32}$$

Invoking the mesoscopic scale, $\boldsymbol{\sigma}^s$ and $\boldsymbol{\sigma}^f$ can heuristically be identified with the intrinsic averaged stress within the matrix and within the fluid. In a first approach the intrinsic averaged stress within the fluid can be addressed through a spherical tensor:

$$\boldsymbol{\sigma}^f = -p\mathbf{1} \tag{2.33}$$

so that the stress partition (2.32) specializes according to:

$$\boldsymbol{\sigma} = (1-n)\,\boldsymbol{\sigma}^s - np\mathbf{1} \tag{2.34}$$

Accordingly, (2.30b) is rewritten in the form:

$$-\nabla_x\,(np) + \rho_f n(\mathbf{f} - \boldsymbol{\gamma}^f) + \mathbf{f}_{int}^{\rightarrow f} = 0 \tag{2.35}$$

The identification of $\boldsymbol{\sigma}^s$ and $\boldsymbol{\sigma}^f$ as the intrinsic averaged stress cannot be proved using only the macroscopic approach. It requires a micro–macro approach, which is left to the advanced analysis section of this chapter (see §2.5.1). Assumption (2.33) involves the fluid constitutive equation and disregards any shear contribution to the partial stress tensor related to the fluid. The role of the shear stress due to the fluid viscosity will be clarified in the next chapter (see §3.3.1).

2.4 Kinetic Energy Theorem

2.4.1 Strain Work Rates

The strain work rate associated with a velocity field $\mathbf{V}$ is the sum of the work rates developed in this field, both by body and surface forces and by inertia forces, the latter being the opposite of the dynamic forces. Indeed, owing to the dynamic theorem (2.6) this sum is zero for a velocity field related to any rigid body motion of the material domain, which assigns the same velocities to the skeleton and to the fluid. Accordingly, let $\mathbf{V}$ be any velocity field, but let it be the same for the skeleton and the fluid. By the above definition, the strain work rate $\mathcal{P}_{def}(\mathbf{V})$ is:

$$\mathcal{P}_{def}(\mathbf{V}) = \int_{\Omega_t} \rho\mathbf{f}\cdot\mathbf{V}\,d\Omega_t + \int_{\partial\Omega_t} \mathbf{T}\cdot\mathbf{V}\,da - \int_{\Omega_t} \left(\rho_s\,(1-n)\,\boldsymbol{\gamma}^s + \rho_f n\boldsymbol{\gamma}^f\right)\cdot\mathbf{V}\,d\Omega_t \tag{2.36}$$

The divergence theorem and the symmetry of the stress tensor $\boldsymbol{\sigma}$ yield the identity:

$$\int_{\partial\Omega_t} \mathbf{n}\cdot\boldsymbol{\sigma}\cdot\mathbf{V}\,da = \int_{\Omega_t} (\boldsymbol{\sigma}:\mathbf{d} + \mathbf{V}\cdot(\nabla_x\cdot\boldsymbol{\sigma}))\,d\Omega_t \tag{2.37}$$

In (2.37) $\mathbf{d}$ is the strain rate associated with $\mathbf{V}$ (see §1.3.2):

$$\mathbf{d} = \frac{1}{2}\left(\nabla_x\mathbf{V} + {}^t\nabla_x\mathbf{V}\right);\quad d_{ij} = \frac{1}{2}\left(\frac{\partial V_i}{\partial x_j} + \frac{\partial V_j}{\partial x_i}\right) \tag{2.38}$$

whereas $\boldsymbol{\sigma}:\mathbf{d}$ stands for:

$$\boldsymbol{\sigma}:\mathbf{d} = \sigma_{ij}d_{ij} \tag{2.39}$$

so that identity (2.37) can be rewritten in the form:

$$\int_{\partial\Omega_t} V_i \sigma_{ij} n_j \, da = \int_{\Omega_t} \left(\sigma_{ij} d_{ij} + V_i \frac{\partial \sigma_{ij}}{\partial x_j} \right) d\Omega_t \tag{2.40}$$

Using $\mathbf{T} = \boldsymbol{\sigma} \cdot \mathbf{n}$, together with identity (2.37), and local momentum balance equation (2.19), the strain work rate $\mathcal{P}_{def}(\mathbf{V})$ (2.36) is finally expressed in the form:

$$\mathcal{P}_{def}(\mathbf{V}) = \int_{\Omega_t} \boldsymbol{\sigma} : \mathbf{d} \, d\Omega_t = \int_{\Omega_t} \sigma_{ij} d_{ij} \, d\Omega_t \tag{2.41}$$

Choosing for $\mathbf{V}$ the actual velocity $\mathbf{V}^{\pi}$ of particle π, (2.41) gives:

$$\mathcal{P}_{def}(\mathbf{V}^{\pi}) = \int_{\Omega_t} \boldsymbol{\sigma} : \mathbf{d}^{\pi} \, d\Omega_t = \int_{\Omega_t} \sigma_{ij} d_{ij}^{\pi} \, d\Omega_t \tag{2.42}$$

Since the skeleton particles and the fluid particles have distinct velocity fields, the strain work rate, as defined in (2.42) from a unique velocity field, cannot account for the strain work rate in the actual motion of the porous medium. Addressing the skeleton motion and the fluid motion, separately as we did for the stress in §2.3.3, the definition (2.36) can be expanded in the form:

$$\begin{aligned} \mathcal{P}_{def}\left(\mathbf{V}^s, \mathbf{V}^f\right) &= \int_{\Omega_t} (\rho_s (1-n) \mathbf{f} \cdot \mathbf{V}^s + \rho_f n \mathbf{f} \cdot \mathbf{V}^f) d\Omega_t \\ &+ \int_{\partial\Omega_t} (\mathbf{T}^s \cdot \mathbf{V}^s + \mathbf{T}^f \cdot \mathbf{V}^f) da \\ &- \int_{\Omega_t} (\rho_s (1-n) \boldsymbol{\gamma}^s . \mathbf{V}^s + \rho_f n \boldsymbol{\gamma}^f . \mathbf{V}^f) d\Omega_t \end{aligned} \tag{2.43}$$

Using the separate local momentum balance equations (2.30), we now derive:

$$\begin{aligned} \mathcal{P}_{def}(\mathbf{V}^s, \mathbf{V}^{fl}) &= \int_{\Omega_t} (1-n) \boldsymbol{\sigma}^s : \mathbf{d}^s d\Omega_t + \int_{\Omega_t} n \boldsymbol{\sigma}^f : \mathbf{d}^f d\Omega_t \\ &+ \int_{\Omega_t} \mathbf{f}_{int}^{\rightarrow f} \cdot (\mathbf{V}^s - \mathbf{V}^f) d\Omega_t \end{aligned} \tag{2.44}$$

The total strain work rate $\mathcal{P}_{def}\left(\mathbf{V}^s, \mathbf{V}^f\right)$ accounts not only for the strain work rate of the skeleton and the fluid continua considered separately (i.e. the two first terms on the right hand side of (2.44)), but also for the strain work rate associated with the internal interaction force and their relative velocity which acts as a relative strain (i.e. the last term on the right hand side of (2.44)). In view of the forthcoming developments it is, however, more suitable to express the total strain work rate by favouring the skeleton motion and get rid of the unknown interaction force $\mathbf{f}_{int}^{\rightarrow f}$ to the benefit of the fluid pressure p. Use of (2.32)–(2.35) in (2.44) provides the expression:

$$\mathcal{P}_{def}(\mathbf{V}^s, \mathbf{V}^f) = \int_{\Omega_t} \left(\boldsymbol{\sigma} : \mathbf{d}^s - \nabla_x \left(\frac{p}{\rho_f} \mathbf{w} \right) + \left(\mathbf{f} - \boldsymbol{\gamma}^f \right) \cdot \mathbf{w} \right) d\Omega_t \tag{2.45}$$

where $\mathbf{w}$ is the relative flow vector of fluid mass (see (1.60)).

2.4.2 Piola–Kirchhoff Stress Tensor

The Eulerian expression of strain work rates derived in the previous section does not refer to any initial configuration. However, as soon as the skeleton is involved, it is more convenient to relate the quantities to an initial configuration; that is, to adopt a Lagrangian description. Concerning the stress, this is achieved by referring the strain work rate to the initial volume and so introducing the total Piola–Kirchhoff stress tensor $\boldsymbol{\pi}$ according to:

$$\boldsymbol{\sigma}: \mathbf{d}^s d\Omega_t = \boldsymbol{\pi}: \frac{d\boldsymbol{\Delta}}{dt} d\Omega_0 \tag{2.46}$$

Using transport formulae (1.11) and (1.56), the transport formula linking the Piola–Kirchhoff stress tensor $\boldsymbol{\pi}$ to the Cauchy stress tensor is:

$$\boldsymbol{\pi} = J\,\mathbf{F}^{-1} \cdot \boldsymbol{\sigma} \cdot {}^t\mathbf{F}^{-1}; \quad \pi_{ij} = J\,\frac{\partial X_i}{\partial x_k}\sigma_{kl}\frac{\partial X_j}{\partial x_l} \tag{2.47}$$

In the limit of infinitesimal transformations $J \simeq 1$ and $\mathbf{F} \simeq \mathbf{1}$ and the Piola–Kirchhoff stress tensor reduces to the Cauchy stress tensor, namely $\boldsymbol{\pi} \simeq \boldsymbol{\sigma}$.

Using (2.47) and surface transport formula (1.13) we derive:

$$\mathbf{F} \cdot \boldsymbol{\pi} \cdot \mathbf{N}\, dA = \boldsymbol{\sigma} \cdot \mathbf{n}\, da \tag{2.48}$$

so that (1.16) applies, resulting in:

$$\nabla_X \cdot (\mathbf{F} \cdot \boldsymbol{\pi})\, d\Omega_0 = \nabla_x \cdot \boldsymbol{\sigma}\, d\Omega_t \tag{2.49}$$

Using (2.49), together with Lagrangian continuity equations (1.63) and (1.69), the local equation of motion (2.19) is transported from the current configuration to the initial configuration according to:

$$\nabla_X \cdot (\mathbf{F} \cdot \boldsymbol{\pi}) + m_s^0\left(\mathbf{f} - \boldsymbol{\gamma}^s\right) + m_f(\mathbf{f} - \boldsymbol{\gamma}^f) = 0 \tag{2.50}$$

or in index notation:

$$\frac{\partial}{\partial X_j}\left(\frac{\partial x_i}{\partial X_k}\pi_{kj}\right) + m_s^0\left(f_i - \gamma_i^s\right) + m_f\left(f_i - \gamma_i^f\right) = 0 \tag{2.51}$$

2.4.3 Kinetic Energy Theorem

Let $\mathcal{K}_\pi$ be the kinetic energy associated with particles $\pi = s$ and f contained in Ω_t:

$$\mathcal{K}_s = \frac{1}{2}\int_{\Omega_t} \rho_s\,(1-n)\left(\mathbf{V}^s\right)^2 d\Omega_t; \quad \mathcal{K}_f = \frac{1}{2}\int_{\Omega_t} \rho_f n (\mathbf{V}^f)^2 d\Omega_t \tag{2.52}$$

Use of (1.44) and (1.58) leads to the relation:

$$\frac{d^s\mathcal{K}_s}{dt} + \frac{d^f\mathcal{K}_f}{dt} = \int_{\Omega_t} (\rho_s\,(1-n)\,\boldsymbol{\gamma}^s \cdot \mathbf{V}^s + \rho_f n \boldsymbol{\gamma}^f \cdot \mathbf{V}^f) d\Omega_t \tag{2.53}$$

Accordingly (2.43) can be rewritten in the form:

$$\mathcal{P}_{\mathbf{f},\mathbf{T}}(\mathbf{V}^s,\mathbf{V}^f)=\mathcal{P}_{def}(\mathbf{V}^s,\mathbf{V}^f)+\frac{d^s\mathcal{K}_s}{dt}+\frac{d^f\mathcal{K}_f}{dt} \tag{2.54}$$

where $\mathcal{P}_{\mathbf{f},\mathbf{T}}(\mathbf{V}^s,\mathbf{V}^f)$ stands for the work rate of the external body and surface forces:

$$\mathcal{P}_{\mathbf{f},\mathbf{T}}(\mathbf{V}^s,\mathbf{V}^f)=\int_{\Omega_t}(\rho_s(1-n)\,\mathbf{f}\cdot\mathbf{V}^s+\rho_f n\mathbf{f}\cdot\mathbf{V}^f)d\Omega_t$$
$$+\int_{\partial\Omega_t}(\mathbf{T}^s\cdot\mathbf{V}^s+\mathbf{T}^f\cdot\mathbf{V}^f)da \tag{2.55}$$

Equation (2.54) is known as the kinetic energy theorem. It states that the work rate supplied by the external forces to any material domain Ω_t is equal to the sum of the total strain work rate and the particle derivative of the kinetic energy related to both continua contained in Ω_t. The strain work rate $\mathcal{P}_{def}(\mathbf{V}^s,\mathbf{V}^f)$ is not a particle derivative and so the kinetic energy theorem cannot be interpreted as a conservation law. This theorem only expresses a balance of all the mechanical energies involved, without specifying the physical transformations which may affect them. The study of these transformations is the main subject of the next chapter, devoted to the thermodynamics of porous continua.

2.5 Advanced Analysis

2.5.1 The Stress Partition Theorem

In §2.3.3 the stress partition (2.32) has been derived by extending the hypothesis of surface contact forces in a separate way to both the skeleton continuum and the fluid continuum (see (2.28)). The partition of stress can receive better support by using the procedure of micro–macro averaging.[2]

To this end we first introduce the 'sliding' average operator. In the current configuration let $d\Omega_t^0=\omega_0$ be an elementary representative volume centred at the origin of coordinates $\mathbf{x}=0$. Furthermore let $f(\mathbf{z})$ be a weighting function having as argument the position vector $\mathbf{z}$ of points lying in ω_0. We require $f(\mathbf{z})$ to admit continuous derivatives and to satisfy:

$$f(\mathbf{z})=0\quad\text{for }\mathbf{z}\in\partial\omega_0;\quad\frac{1}{\omega_0}\int_{\omega_0}f(\mathbf{z})\,d\omega_0=1 \tag{2.56}$$

where $\partial\omega_0$ stands for the border of ω_0. In practice $f(\mathbf{z})$ can be chosen arbitrarily close to the characteristic function $f_{\omega_0}(\mathbf{z})$ of ω_0. The characteristic function $f_v(\mathbf{z})$ of a volume v is defined by:

$$f_v(\mathbf{z})=1\quad\text{for }\mathbf{z}\in v;\quad f_v(\mathbf{z})=0\quad\text{for }\mathbf{z}\notin\mathbf{v} \tag{2.57}$$

[2]For a general presentation of averaging techniques see Bear J., Bachmat Y. (1990), *Introduction to Modeling of Transport in Porous Media*, Kluwer, Amsterdam.

Any elementary representative volume $d\Omega_t = \omega$ centred at $\mathbf{x}$ can be obtained by the translation of volume ω_0 along vector $\mathbf{x}$. The 'sliding' average $\langle \mathcal{G} \rangle (\mathbf{x})$ related to volume ω centred at point $\mathbf{x}$ is defined by:

$$\langle \mathcal{G} \rangle (\mathbf{x}) = \frac{1}{\omega} \int_{\omega} \mathcal{G} (\mathbf{z}) \, f (\mathbf{z} - \mathbf{x}) \, d\omega \tag{2.58}$$

For instance, letting ρ_Π, with $\Pi = S$ and F, be the macroscopic or apparent skeleton and fluid mass densities, we write:

$$\rho_S = \rho_s (1 - n) = \left\langle f_{\omega_s} (\mathbf{z}) \, \rho_s (\mathbf{z}) \right\rangle; \quad \rho_F = \rho_f n = \left\langle f_{\omega_f} (\mathbf{z}) \, \rho_f (\mathbf{z}) \right\rangle \tag{2.59}$$

In (2.59) ω_π stands for the volume physically occupied by the matrix within ω, $\pi = s$, or by the fluid, $\pi = f$, so that $\omega = \omega_s \cup \omega_f$, while $\rho_\pi (\mathbf{z})$ is the actual mass density within ω_π. Analogously, we define a macroscopic partial stress $\boldsymbol{\sigma}^\Pi$ related to the skeleton particle, $\Pi = S$, and to the fluid particle, $\Pi = F$, and also an intrinsic partial stress $\boldsymbol{\sigma}^\pi$ related to the solid matrix, $\pi = s$, and to the fluid, $\pi = f$. We write:

$$\boldsymbol{\sigma}^S = (1 - n) \, \boldsymbol{\sigma}^s = \left\langle f_{\omega_s} (\mathbf{z}) \, \boldsymbol{\sigma}^s (\mathbf{z}) \right\rangle; \quad \boldsymbol{\sigma}^F = n \boldsymbol{\sigma}^f = \left\langle f_{\omega_f} (\mathbf{z}) \, \boldsymbol{\sigma}^f (\mathbf{z}) \right\rangle \tag{2.60}$$

where $\boldsymbol{\sigma}^s (\mathbf{z})$ and $\boldsymbol{\sigma}^f (\mathbf{z})$ are the actual matrix and fluid stress fields within ω. The stress field $\boldsymbol{\sigma}^\pi (\mathbf{z})$ within volume ω_π must satisfy the equilibrium equation:

$$\nabla_z \cdot \boldsymbol{\sigma}^\pi (\mathbf{z}) + \rho_\pi (\mathbf{z}) \, \mathbf{f} = 0; \quad \frac{\partial \sigma^\pi_{ij} (\mathbf{z})}{\partial z_j} + \rho_\pi (\mathbf{z}) \, f_i = 0 \tag{2.61}$$

where, for the sake of simplicity, we have not considered inertia effects. In addition, on the matrix–fluid interface $\partial \omega_{s,f} = \partial \omega_s \cap \partial \omega_f$ the stress vector, by virtue of the action–reaction law, is subjected to the continuity condition:

$$\boldsymbol{\sigma}^s (\mathbf{z}) \cdot \mathbf{n}^s + \boldsymbol{\sigma}^f (\mathbf{z}) \cdot \mathbf{n}^f = 0; \quad \mathbf{n}^s + \mathbf{n}^f = 0 \tag{2.62}$$

where $\mathbf{n}^\pi$ stands for the unit normal to $\partial \omega_{s,f}$ outwardly oriented relative to ω_π.

With the aim of deriving the macroscopic equilibrium equation from the microscopic one (2.61), we first remark that definitions (2.58) and (2.60) lead to:

$$\frac{\partial \sigma^\Pi_{ij}}{\partial x_j} (\mathbf{x}) = -\frac{1}{\omega} \int_{\omega} f_{\omega_\pi} (\mathbf{z}) \, \sigma^\pi_{ij} (\mathbf{z}) \, \frac{\partial f}{\partial z_j} (\mathbf{z} - \mathbf{x}) \, d\omega \tag{2.63}$$

Integration by parts of the latter gives:

$$\frac{\partial \sigma^\Pi_{ij}}{\partial x_j} (\mathbf{x}) = \frac{1}{\omega} \int_{\omega_\pi} \left[\frac{\partial \sigma^\pi_{ij} (\mathbf{z})}{\partial z_j} f (\mathbf{z} - \mathbf{x}) - \frac{\partial}{\partial z_j} \left(\sigma^\pi_{ij} (\mathbf{z}) \, f (\mathbf{z} - \mathbf{x}) \right) \right] d\omega \tag{2.64}$$

Since f is zero on the border $\partial\omega$ of ω and $\partial\omega_\pi = \partial\omega_{s,f} \cup (\partial\omega_\pi \cap \partial\omega)$, the divergence theorem applied to the last term allows us to rearrange (2.64) in the form:

$$\frac{\partial \sigma_{ij}^{\Pi}}{\partial x_j}(\mathbf{x}) = \frac{1}{\omega}\int_{\omega} f_{\omega_\pi}(\mathbf{z}) \frac{\partial \sigma_{ij}^{\pi}(\mathbf{z})}{\partial z_j} f(\mathbf{z}-\mathbf{x})\, d\omega - \frac{1}{\omega}\int_{\partial\omega_{s,f}} \sigma_{ij}^{\pi}(\mathbf{z})\, n_j^{\pi} f(\mathbf{z}-\mathbf{x})\, da_\omega \tag{2.65}$$

that is:

$$\frac{\partial \sigma_{ij}^{\Pi}}{\partial x_j}(\mathbf{x}) = \left\langle f_{\omega_\pi}(\mathbf{z}) \frac{\partial \sigma_{ij}^{\pi}(\mathbf{z})}{\partial z_j} \right\rangle - \frac{1}{\omega}\int_{\partial\omega_{s,f}} \sigma_{ij}^{\pi}(\mathbf{z})\, n_j^{\pi} f(\mathbf{z}-\mathbf{x})\, da_\omega \tag{2.66}$$

Applying the sliding average operator $\langle\ \rangle$ to (2.61) and using (2.59) and (2.66), we now get:

$$\nabla_x \cdot \boldsymbol{\sigma}^{\Pi} + \rho_\Pi \mathbf{f} + \frac{1}{\omega}\int_{\partial\omega_{s,f}} \boldsymbol{\sigma}^{\pi}(\mathbf{z}) \cdot \mathbf{n}^{\pi} f(\mathbf{z}-\mathbf{x})\, da_\omega = 0 \tag{2.67}$$

Equation (2.67) finally allows us to retrieve the momentum equation (2.30) relative to the π-continuum whereas the interaction force $\mathbf{f}_{int}^{\to\pi}$ is now identified in the form:

$$\mathbf{f}_{int}^{\to\pi} = \frac{1}{\omega}\int_{\partial\omega_{s,f}} \boldsymbol{\sigma}^{\pi}(\mathbf{z}) \cdot \mathbf{n}^{\pi}\, da_\omega \tag{2.68}$$

where we selected the sliding average operator associated with the weighting function $f = f_{\omega_0}$ (see (2.57)). In addition, as required by (2.62), the continuity required for the microscopic stress vector $\boldsymbol{\sigma}^{\pi}(\mathbf{z}) \cdot \mathbf{n}^{\pi}$, together with (2.68), allows us to recover the action–reaction law in its macroscopic form (2.31). Taking into account the action–reaction law and summing the equations obtained by successively letting $\Pi = S$ and F in (2.67), the macroscopic momentum equation (2.19) is recovered, provided that the total stress $\boldsymbol{\sigma}$ is identified as the sum of macroscopic partial stresses $\boldsymbol{\sigma}^S = (1-n)\boldsymbol{\sigma}^s$ and $\boldsymbol{\sigma}^F = n\boldsymbol{\sigma}^f$ resulting in the stress partition theorem (2.32).

2.5.2 Momentum Balance and the Double Porosity Network

The approach to the momentum balance developed throughout this chapter extends to the case of the double porosity network examined in §1.5.3. Using (1.82), the variation of the linear momentum can now be extended in the form:

$$\frac{d^s}{dt}\int_{\Omega_t} \rho_s (1-n)\, \mathbf{V}^s d\Omega_t + \sum_{\alpha=1,2} \frac{d^{f_\alpha}}{dt}\int_{\Omega_t} \rho_{f_\alpha} n_\alpha \mathbf{V}^{f_\alpha} d\Omega_t$$
$$= \int_{\Omega_t}\left(\rho_s (1-n)\, \boldsymbol{\gamma}^s + \sum_{\alpha=1,2} \rho_{f_\alpha} n_\alpha \boldsymbol{\gamma}^{f_\alpha}\right) d\Omega_t + \int_{\Omega_t} \overset{\circ}{r}_{1\to 2}(\mathbf{V}^{f_2} - \mathbf{V}^{f_1}) d\Omega_t \tag{2.69}$$

where the term $\overset{\circ}{r}_{1\to 2}\left(\mathbf{V}^{f_2}-\mathbf{V}^{f_1}\right)$ accounts for the variation in linear momentum related to the mass rate exchanged between the two porous networks. Accordingly, the local equation of motion now reads:

$$\nabla_x \cdot \boldsymbol{\sigma} + \rho_s\,(1-n)\,\boldsymbol{\gamma}^s + \sum_{\alpha=1,2} \rho_{f_\alpha} n_\alpha (\mathbf{f} - \boldsymbol{\gamma}^{f_\alpha}) - \overset{\circ}{r}_{1\to 2}(\mathbf{V}^{f_2} - \mathbf{V}^{f_1}) = 0 \tag{2.70}$$

Favouring the motion of the skeleton by introducing the relative flow vector of fluid mass, $\mathbf{w}^{(\alpha)} = \rho_{f_\alpha} n_\alpha \left(\mathbf{V}^{f_\alpha} - \mathbf{V}^s\right)$, and the fluid pressure p_α related to network α, we express the work rate of the external body and surface forces in the form:

$$\mathcal{P}_{\mathbf{f},\mathbf{T}}(\mathbf{V}^s, \mathbf{V}^{f_\alpha}) = \int_{\Omega_t} \left(\rho \mathbf{f} \cdot \mathbf{V}^s + \mathbf{f} \cdot \sum_{\alpha=1,2} \mathbf{w}^{(\alpha)}\right) d\Omega_t + \int_{\partial\Omega_t} \left(\mathbf{T} \cdot \mathbf{V}^s - \frac{p_\alpha}{\rho_{f_\alpha}} \mathbf{w}^{(\alpha)} \cdot \mathbf{n}\right) da \tag{2.71}$$

where ρ stands for the total apparent mass density, namely $\rho = \rho_s\,(1-n) + \sum_{\alpha=1,2} \rho_{f_\alpha} n_\alpha$. It is instructive to remark that the last integrand of the right hand side integral in (2.71), when multiplied by dt, that is $-(p_\alpha/\rho_{f_\alpha})\mathbf{w}^{(\alpha)} \cdot \mathbf{n}\,da\,dt = -p_\alpha n_\alpha \left(\mathbf{V}^{f_\alpha} - \mathbf{V}^s\right) \cdot \mathbf{n}\,da\,dt$, represents the work supplied by the pressure p_α in the extraction of the fluid volume $n_\alpha \left(\mathbf{V}^{f_\alpha} - \mathbf{V}^s\right) \cdot \mathbf{n}\,da\,dt$ from the domain occupied by the skeleton. In addition, taking into account (1.82) and still favouring the motion of the skeleton, the particle derivative of the kinetic energy reads:

$$\begin{aligned}\frac{d^s \mathcal{K}_s}{dt} + \sum_{\alpha=1,2} \frac{d^{f_\alpha} \mathcal{K}_{f_\alpha}}{dt} = {} & \int_{\Omega_t} \left(\rho_s\,(1-n)\,\boldsymbol{\gamma}^s + \sum_{\alpha=1,2} \rho_{f_\alpha} n_\alpha \boldsymbol{\gamma}^{f_\alpha}\right) \cdot \mathbf{V}^s\, d\Omega_t \\ & + \int_{\Omega_t} \sum_{\alpha=1,2} \boldsymbol{\gamma}^{f_\alpha} \cdot \mathbf{w}^{(\alpha)} d\Omega_t \\ & + \int_{\Omega_t} \frac{1}{2} \overset{\circ}{r}_{1\to 2} \left[(\mathbf{V}^{f_2})^2 - (\mathbf{V}^{f_1})^2\right] d\Omega_t\end{aligned} \tag{2.72}$$

Multiplying (2.70) by $\mathbf{V}^s$ and integrating over the volume Ω_t while using (2.72), we finally extend the kinetic energy theorem in the form:

$$\begin{aligned}\mathcal{P}_{\mathbf{f},\mathbf{T}}(\mathbf{V}^s, \mathbf{V}^{f_\alpha}) = {} & \mathcal{P}_{def}(\mathbf{V}^s, \mathbf{V}^{f_\alpha}) + \frac{d^s \mathcal{K}_s}{dt} + \sum_{\alpha=1,2} \frac{d^{f_\alpha} \mathcal{K}_{f_\alpha}}{dt} \\ & + \int_{\Omega_t} \frac{1}{2} \overset{\circ}{r}_{1\to 2} [(\mathbf{V}^{f_2} - \mathbf{V}^s)^2 - (\mathbf{V}^{f_1} - \mathbf{V}^s)^2] d\Omega_t\end{aligned} \tag{2.73}$$

where $\mathcal{P}_{def}\left(\mathbf{V}^s, \mathbf{V}^{f\alpha}\right)$ can be expanded in the form:

$$\mathcal{P}_{def}\left(\mathbf{V}^s, \mathbf{V}^{f\alpha}\right) = \int_{\Omega_t} \boldsymbol{\sigma} : \mathbf{d}^s d\Omega_t - \int_{\Omega_t} \sum_{\alpha=1,2} \left[\nabla_x \left(\frac{p_\alpha}{\rho_\alpha} \mathbf{w}^{(\alpha)} \right) + \left(\mathbf{f} - \boldsymbol{\gamma}^{f\alpha} \right) \cdot \mathbf{w}^{(\alpha)} \right] d\Omega_t \tag{2.74}$$

Expression (2.74) extends expression (2.45) to the case of the double porous network. The approach here used to derive the kinetic energy theorem constitutes an alternative to the approach we used in §2.4. Indeed, to derive (2.73) and (2.74) we did not involve interaction forces such as $\mathbf{f}_{int}^{\rightarrow \pi}$. In fact, with regard to the macroscopic approach, the kinetic energy theorem is based only on the expression retained for the work rate of the surface forces, that is the surface integral on the right hand side of (2.71).

2.5.3 The Tortuosity Effect

Expression (2.52) of the kinetic energy does not account for the 'tortuosity' effect.[3] This effect cannot be captured by the macroscopic approach. Indeed, the actual fluid kinetic energy reads:

$$\mathcal{K}_f = \frac{1}{2} \int_{\Omega_t} \left\langle f_{\omega_f}(\mathbf{z})\, \rho_f(\mathbf{z})\, (\mathbf{v}^f(\mathbf{z}))^2 \right\rangle d\Omega_t \tag{2.75}$$

where we adopt the notation of §2.5.1, $\mathbf{v}^f(\mathbf{z})$ being the velocity field of the fluid particles within the elementary volume ω_f. When identifying the macroscopic fluid velocity $\mathbf{V}^f$ as the barycentric average of $\mathbf{v}^f(\mathbf{z})$, that is:

$$\rho_f n \mathbf{V}^f = \left\langle f_{\omega_f}(\mathbf{z})\, \rho_f(\mathbf{z})\, \mathbf{v}^f(\mathbf{z}) \right\rangle \tag{2.76}$$

some calculations allow us to rewrite (2.75) in the form:

$$\mathcal{K}_f = \frac{1}{2} \int_{\Omega_t} \rho_f n (\mathbf{V}^f)^2 d\Omega_t + \frac{1}{2} \int_{\Omega_t} \left\langle f_{\omega_f}(\mathbf{z})\, \rho_f(\mathbf{z}) \left((\mathbf{v}^f(\mathbf{z}) - \mathbf{V}^s)^2 - (\mathbf{V}^f - \mathbf{V}^s)^2 \right) \right\rangle d\Omega_t \tag{2.77}$$

In order to evaluate the second term we firstly note that:

$$\left\langle f_{\omega_f}(\mathbf{z})\, \rho_f(\mathbf{z})\, (\mathbf{v}^f(\mathbf{z}) - \mathbf{V}^s)^2 \right\rangle \geq \rho_f n (\mathbf{V}^f - \mathbf{V}^s)^2 \tag{2.78}$$

[3] The tortuosity effect, related to the dynamics of porous media, was introduced in Biot M.A. (1956), 'Theory of propagation of elastic waves in a fluid-saturated porous solid', *Journal of the Acoustic Society of America*, **28**, 168–178. See also Biot M.A. (1962), 'Mechanics of deformation and acoustic propagation in porous media', *Journal of Applied Physics*, **27**, 459–467.

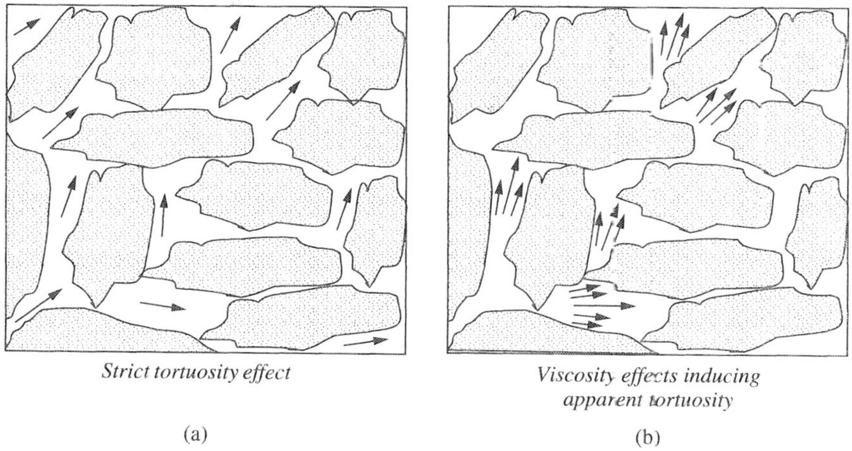

Figure 2.5: Tortuosity effect.

The above inequality holds for two reasons. The first reason is that the direction of the microscopic relative velocity $\mathbf{v}^f(\mathbf{z}) - \mathbf{V}^s$ does not remain constant within volume ω_f. It is the strict tortuosity effect as sketched in Fig. 2.5a. The second reason is that the modulus of $\mathbf{v}^f(\mathbf{z}) - \mathbf{V}^s$ does not remain constant too. Owing to the fluid viscosity, the modulus of $\mathbf{v}^f(\mathbf{z}) - \mathbf{V}^s$ is zero on the internal walls of the skeleton delimiting the porous network through which the fluid flows. It reaches its maximal value far from this matrix–fluid interface (Fig. 2.5b). In the isotropic case these effects are usually captured through a dynamic tortuosity factor a, depending on both the current geometry of the porous network and the fluid viscosity, such as:[4]

$$a = \frac{\left\langle f_{\omega_f}(\mathbf{z})\,\rho_f(\mathbf{z})\left(\mathbf{v}^f(\mathbf{z}) - \mathbf{V}^s\right)^2\right\rangle}{\rho_f n\left(\mathbf{V}^f - \mathbf{V}^s\right)^2} \geq 1 \tag{2.79}$$

Substituting (2.79) into (2.77), the expression of the kinetic energy taking account into the tortuosity effect reads:

$$\mathcal{K}_f = \frac{1}{2}\int_{\Omega_t} \rho_f n\left[(\mathbf{V}^f)^2 + (a-1)(\mathbf{V}^f - \mathbf{V}^s)^2\right] d\Omega_t \tag{2.80}$$

Accordingly we now obtain:

$$\frac{d^f \mathcal{K}_f}{dt} = \int_{\Omega_t} \rho_f n\left(\boldsymbol{\gamma}^f \cdot \mathbf{V}^f + (\mathbf{V}^f - \mathbf{V}^s)\cdot \mathbf{a}\right) d\Omega_t \tag{2.81}$$

[4]The tortuosity factor a admits the limit $a \to 1$ as $n \to 1$, that is for a fluid, and the limit $a \to \infty$ as $n \to 0$, that is for a solid. In the latter case, as $n \to 0$ the relative velocity $\mathbf{V}^f - \mathbf{V}^s$ tends to zero with regard to important viscosity effects. For a matrix formed of spherical solid grains, the expression $a = \frac{1}{2}\left(1 + \frac{1}{n}\right)$ is sometimes proposed. Although fitting the limiting values, this expression does not involve the fluid viscosity and can only account for the strict 'tortuosity' effect (Fig. 2.5a).

where **a** is the dynamic tortuosity vector. When assuming that the dynamic tortuosity factor a is constant with respect to both space and time, **a** can be expressed as:

$$\mathbf{a} = (a-1)\left(\boldsymbol{\gamma}^f - \boldsymbol{\gamma}^s - \nabla_x \mathbf{V}^s \cdot (\mathbf{V}^f - \mathbf{V}^s)\right) \tag{2.82}$$

The tortuosity adds a coupling inertia force $-\rho_f n\,\mathbf{a}$ between the two superimposed continua. As an interaction force, eventually included in $\mathbf{f}_{int}^{\rightarrow f}$, this coupling force does not figure in the overall equation of motion (2.19) and is not associated with any kind of strain. Indeed the kinetic energy theorem (2.54) remains formally unchanged provided that (2.44) and (2.45) are replaced by:

$$\begin{aligned}\mathcal{P}_{def}\left(\mathbf{V}^s, \mathbf{V}^f\right) &= \int_{\Omega_t} (1-n)\,\boldsymbol{\sigma}^s : \mathbf{d}^s d\Omega_t + \int_{\Omega_t} n\boldsymbol{\sigma}^f : \mathbf{d}^f d\Omega_t \\ &+ \int_{\Omega_t} \left(\mathbf{f}_{int}^{\rightarrow f} - \mathbf{a}\right) \cdot (\mathbf{V}^s - \mathbf{V}^f) d\Omega_t\end{aligned} \tag{2.83}$$

and:

$$\begin{aligned}\mathcal{P}_{def}(\mathbf{V}^s, \mathbf{V}^f) &= \int_{\Omega_t} \left(\boldsymbol{\sigma} : \mathbf{d}^s - \nabla_x \left(\frac{p}{\rho_f}\mathbf{w}\right)\right) d\Omega_t \\ &+ \int_{\Omega_t} (\mathbf{f} - \boldsymbol{\gamma}^f - \mathbf{a}) \cdot \mathbf{w}\, d\Omega_t\end{aligned} \tag{2.84}$$

Chapter 3

Thermodynamics

Thermodynamics analyses the transformations affecting all the forms of energy involved in the evolution of a system. Thermodynamics constitutes the natural extension of thermostatics. Thermostatics is restricted to reversible and infinitely slow evolutions between successive states of equilibrium of homogeneous systems. Based upon the postulate of the local state, thermodynamics extends thermostatics to any system, discrete or continuous, and to any evolution, reversible or irreversible, irrespective of its time scale. Thermodynamics is based upon two laws. The first law expresses the conservation of energy when considering all its possible forms. The second law is of a quite different kind. It expresses that the quality of energy can only deteriorate in what concerns its transformability into efficient mechanical work. Applied to a particular system, these laws involve the variables characterizing its state of internal energy and eventually provide the appropriate framework to formulate the constitutive equations governing its evolution. This chapter applies the laws of thermodynamics to a porous continuum with the aim of capturing the various possible couplings in the formulation of its constitutive equations.[1]

3.1 Thermostatics of Homogeneous Fluids

We start by recalling the two fundamental laws of thermostatics and their application to homogeneous fluids.

3.1.1 Energy Conservation and Entropy Balance

The first law expresses the conservation of energy when considering all its possible forms. Restricting consideration here to mechanical work and to heat as forms of energy, the first law of thermostatics can be stated as follows.

[1] A founding paper on the energy approach to poromechanics is Biot M.A. (1972), 'Theory of finite deformation of porous solids', *Indiana University Mathematical Journal*, **21**, 579–620. See also Biot M.A. (1977), 'Variational Lagrangian-thermodynamics of non isothermal finite strain. Mechanics of porous solid and thermomolecular diffusion', *International Journal of Solids and Structures*, **13**, 579–597.

Poromechanics O. Coussy
 ISBN 0-470-84920-7

First law. In any infinitely slow evolution from one homogeneous equilibrium state of a material system to another, the variation of the internal energy of the latter is the sum of the mechanical work performed by the external forces on the system and the external heat supply.

The first law applied to a fluid between two infinitely close homogeneous equilibrium states can be expressed in the form:

$$de_f = -p\,d\left(\frac{1}{\rho_f}\right) + \delta Q \tag{3.1}$$

In (3.1) e_f is the fluid-specific (i.e. per mass unit) *internal* energy; $-p\,d(1/\rho_f)$ represents the infinitesimal mechanical work supplied to the fluid by the pressure p in the infinitesimal volume change $d(1/\rho_f)$ of its specific material volume $1/\rho_f$, while δQ is the infinitesimal heat supply.

The second law of thermostatics, takes that the quality of thermal energy can only deteriorate with regard to its ulterior transformation into efficient mechanical work. In its explicit formulation the second law introduces two new conjugate quantities, the entropy and the absolute temperature. An amount of heat supplied at high temperature has a 'mechanical quality' superior to the same amount of heat supplied at a lower temperature. Indeed it will be easier eventually to extract efficient mechanical work from the former than from the latter by finding a cold source. Hence, if the entropy somehow has to be a measure of the deterioration of the 'mechanical quality' of heat given the same heat supply, the higher the temperature, the lower the related supply of entropy. More formally the second law of thermostatics can be stated as follows.

Second law. There exists an additive function called entropy such that the variation of entropy of a material system in any infinitely slow evolution from one of its homogeneous equilibrium states to another, is equal to the external supply of entropy. It exists on a universal scale of positive absolute temperature so that the external infinitesimal supply of entropy is defined as the external infinitesimal supply of heat divided by the absolute temperature.

Excluding irreversible transformations, the second law of thermostatics applied to the fluid-specific material volume $1/\rho_f$ between two infinitely close homogeneous equilibrium states can be expressed in the form:

$$ds_f = \frac{\delta Q}{T} \tag{3.2}$$

where s_f is the fluid-specific entropy and T the absolute temperature.[2]

[2]Such as defined here, temperature T is actually only relative. Indeed T relates to the association of two specific systems, the material volume under consideration and the exterior. The definition of the absolute temperature requires the addition of the so-called zero law stating that two systems in thermal equilibrium with the same third system (no heat exchange between the systems) are themselves in thermal equilibrium. A temperature scale T being then defined, any other scale of temperature Θ now becomes linked to the former by a relation of the form $\Theta = \vartheta\,(T)$ irrespective of the systems considered.

3.1.2 Fluid State Equations. Gibbs Potential

Eliminating the infinitesimal heat supply, (3.1) and (3.2) combine to give the following energy balance:

$$de_f = -p\,d\left(\frac{1}{\rho_f}\right) + T ds_f \tag{3.3}$$

Equation (3.3) must hold for any infinitely slow evolution of the fluid from one of its homogeneous equilibrium states to another. As a consequence $1/\rho_f$ and s_f constitute a complete set of independent thermodynamical state variables such as:

$$e_f = e_f\left(\frac{1}{\rho_f}, s_f\right): \quad p = -\frac{\partial e_f}{\partial\left(\frac{1}{\rho_f}\right)}; \quad T = \frac{\partial e_f}{\partial s_f} \tag{3.4}$$

Equations (3.4) are the fluid state equations where the internal energy e_f acts as a potential which links the set of thermodynamical state variables $(1/\rho_f, T)$ to the conjugate set $(-p, s_f)$.

The state equations can be partially inverted with respect to the couple of conjugate variables $(1/\rho_f, -p)$ by introducing the fluid-specific enthalpy h_f:

$$h_f = e_f + \frac{p}{\rho_f} \tag{3.5}$$

so that the state equations take the alternative form:

$$h_f = h_f(p, s_f): \quad \frac{1}{\rho_f} = \frac{\partial h_f}{\partial p}; \quad T = \frac{\partial h_f}{\partial s_f} \tag{3.6}$$

Equations (3.4) can also be partially inverted with respect to the couple of conjugate variables (s_f, T). Introducing the fluid-specific Helmholtz free energy ψ_f:

$$\psi_f = e_f - T s_f \tag{3.7}$$

we get:

$$\psi_f = \psi_f\left(\frac{1}{\rho_f}, T\right): \quad p = -\frac{\partial \psi_f}{\partial\left(\frac{1}{\rho_f}\right)}; \quad s_f = -\frac{\partial \psi_f}{\partial T} \tag{3.8}$$

Equation (3.4) can finally be totally inverted. Introducing the fluid-specific free enthalpy g_f:

$$g_f = \psi_f + \frac{p}{\rho_f} = h_f - T s_f \tag{3.9}$$

we obtain:

$$g_f = g_f(p, T): \quad \frac{1}{\rho_f} = \frac{\partial g_f}{\partial p}; \quad s_f = -\frac{\partial g_f}{\partial T} \tag{3.10}$$

The specific free enthalpy g_f is also called the Gibbs potential. The work δW required to make the infinitesimal fluid mass δm_f enter a given fluid volume is the work needed to make room for the new volume $\delta m_f/\rho_f$, which is:

$$\delta W = \frac{p}{\rho_f}\delta m_f \tag{3.11}$$

Consequently, the free energy supplied to the fluid volume considered as an open thermodynamic system is not $\psi_f\,\delta m_f$, but $g_f\,\delta m_f = (\psi_f + p/\rho_f)\delta m_f$, indicating the actual meaning of the Gibbs potential.

3.2 Thermodynamics of Porous Continua

3.2.1 Postulate of Local State

The extension of thermostatics to thermodynamics of continua is based upon the postulate of local state with regard to both time and space.

With regard to time, the postulate of local state stipulates that the current state of internal energy of a homogeneous system in any evolution is irrespective of the rate of evolution and can be characterized by the same state variables as the ones characterizing equilibrium states. For instance, the postulate of local state applied to the fluid of the previous section eventually comes down to considering that Eq. (3.3) holds irrespective of the evolution rate, so that it can be divided by the infinitesimal time dt and consequently can be written in terms of time rates.

With regard to space, the postulate of local state addresses the thermodynamics of the continuum by considering that the thermodynamics of any material volume Ω_t results from the thermodynamics of the juxtaposed material elementary volumes $d\Omega_t$ forming Ω_t and exchanging heat and mechanical work between them. As a consequence the laws of thermodynamics can be applied in an integral form to additive quantities such as the energy and the entropy.

The postulate of local state is extended to a porous continuum by considering that the thermodynamics of such a continuum results from those of the superimposed interacting continua forming it, that is the skeleton continuum and the fluid continuum.

3.2.2 The First Law

Energy conservation

Based on the postulate of local state, the first law of thermodynamics, when applied to any material volume Ω_t of a porous continuum, expresses the energy conservation in the form:

$$\frac{d^s}{dt}\int_{\Omega_t}\rho_s(1-n)\left(e_s+\frac{1}{2}(\mathbf{V}^s)^2\right)d\Omega_t + \frac{d^f}{dt}\int_{\Omega_t}\rho_f n\left(e_f+\frac{1}{2}(\mathbf{V}^f)^2\right)d\Omega_t$$
$$= \mathcal{P}_{\mathbf{f,T}}(\mathbf{V}^s,\mathbf{V}^f) + \overset{\circ}{Q} \tag{3.12}$$

In (3.12) e_s stands for the specific internal energy of the matrix (or equivalently the intrinsic specific energy of the skeleton). Accordingly the left hand term accounts for the time rate of the energy attached to the whole matter currently contained within the volume Ω_t, including its kinetic form. This energy time rate is produced by the right hand side term, which accounts for the total rate of energy externally supplied to the matter currently contained within the volume Ω_t: $\mathcal{P}_{\mathbf{f},\mathbf{T}}(\mathbf{V}^s, \mathbf{V}^f)$ accounts for the mechanical work rate of the external forces (see §2.4.3), while $\overset{\circ}{Q}$ accounts for the rate of heat externally supplied. As it concerns the thermal term, we write:[3]

$$\overset{\circ}{Q} = \int_{\partial\Omega_t} J_Q(\mathbf{x}, \mathbf{n}, t)\, da \tag{3.13}$$

where J_Q is the surface rate of heat supplied by conduction. This heat supply is due to contact effects and therefore, as it concerns space, depends only on the position vector $\mathbf{x}$ and on the outward unit normal $\mathbf{n}$ to surface da.

The energy equation

Use of the kinetic energy theorem (see §2.4.3) allows us to rewrite (3.12) in the form:

$$\frac{d^s}{dt}\int_{\Omega_t} \rho_s (1-n)\, e_s\, d\Omega_t + \frac{d^f}{dt}\int_{\Omega_t} \rho_f n e_f\, d\Omega_t = \mathcal{P}_{def}(\mathbf{V}^s, \mathbf{V}^f) + \overset{\circ}{Q} \tag{3.14}$$

Use of expression (1.46) or (1.47) for the particle derivative of an integral and of definition (1.61) for the relative flow vector of fluid mass furnishes:

$$\frac{d^s}{dt}\int_{\Omega_t} \rho_s (1-n)\, e_s\, d\Omega_t + \frac{d^f}{dt}\int_{\Omega_t} \rho_f n e_f\, d\Omega_t$$
$$= \int_{\Omega_t} \left(\frac{d^s e}{dt} + e\nabla_x \cdot \mathbf{V}^s + \nabla_x \cdot (e_f \mathbf{w})\right) d\Omega_t \tag{3.15}$$

where e is the overall density of internal energy per unit of volume $d\Omega_t$, that is:

$$e = \rho_s (1-n)\, e_s + \rho_f n e_f \tag{3.16}$$

Expression (2.45) for the strain work rate $\mathcal{P}_{def}(\mathbf{V}^s, \mathbf{V}^f)$, together with equations (3.13)–(3.15), yields:

$$\int_{\Omega_t} \left(\frac{d^s e}{dt} + e\nabla_x \cdot \mathbf{V}^s + \nabla_x \cdot (h_f \mathbf{w}) - \boldsymbol{\sigma} : \mathbf{d}^s - (\mathbf{f} - \boldsymbol{\gamma}^f) \cdot \mathbf{w}\right) d\Omega_t$$
$$= \int_{\partial\Omega_t} J_Q(\mathbf{x}, \mathbf{n}, t)\, da \tag{3.17}$$

[3]We do not consider here possible volume sources of heat which would add a volume integral term.

The tetrahedron lemma (see §2.2.2, including the footnote) applied to (3.17) shows the existence of an outgoing heat flow vector $\mathbf{q}$, such that the surface rate J_Q relies linearly on $\mathbf{n}$ and reads:

$$J_Q = -\mathbf{q} \cdot \mathbf{n} \tag{3.18}$$

Substitution of (3.18) into (3.17) provides the Euler energy equation:

$$\frac{d^s e}{dt} + e\nabla_x \cdot \mathbf{V}^s = \boldsymbol{\sigma} : \mathbf{d}^s - \nabla_x \cdot (h_f \mathbf{w} + \mathbf{q}) + (\mathbf{f} - \boldsymbol{\gamma}^f) \cdot \mathbf{w} \tag{3.19}$$

eventually expressing the first law applied to the infinitesimal volume $d\Omega_t$. With the aim of transporting energy equation (3.19) to the skeleton initial configuration let E and $\mathbf{Q}$ be respectively the overall Lagrangian density of internal energy per unit of initial volume $d\Omega_0$ and the Lagrangian heat flow vector such that:

$$E\, d\Omega_0 = e\, d\Omega_t; \quad \mathbf{Q} \cdot \mathbf{N}\, dA = \mathbf{q} \cdot \mathbf{n}\, da \tag{3.20}$$

Use of (1.42), (3.20) and the transport formulae derived in the previous chapters (see §1.2.1, §1.4 and §2.4.2) finally delivers the Lagrangian energy equation:

$$\frac{dE}{dt} = \boldsymbol{\pi} : \frac{d\boldsymbol{\Delta}}{dt} - \nabla_X \cdot (h_f \mathbf{M} + \mathbf{Q}) + (\mathbf{f} - \boldsymbol{\gamma}^f) \cdot \mathbf{F} \cdot \mathbf{M} \tag{3.21}$$

3.2.3 The Second Law

The balance of entropy

Based on the postulate of local state, the second law of thermodynamics expresses the entropy balance of any material volume Ω_t of a porous continuum in the form:

$$\frac{d^s}{dt} \int_{\Omega_t} \rho_s (1-n)\, s_s\, d\Omega_t + \frac{d^f}{dt} \int_{\Omega_t} \rho_f n\, s_f\, d\Omega_t \geq \int_{\partial\Omega_t} -\frac{\mathbf{q} \cdot \mathbf{n}}{T}\, da \tag{3.22}$$

where s_s stands for the specific entropy of the matrix (or equivalently the intrinsic specific entropy of the skeleton). As formulated in (3.22) the second law assumes that the fluid and the matrix included in $d\Omega_t$ are in thermal equilibrium, that is at the same temperature T.

Compared with thermostatics, the second law of thermodynamics states that the time rate of the entropy associated with the matter contained in Ω_t (left hand side of (3.22)) cannot be less than the rate of entropy externally supplied (right hand side of (3.22)). In contrast to the conservation laws which involve equalities, the second law of thermodynamics breaks the time symmetry; the flow of time cannot be reversed since a time inversion $t \to -t$ would change (3.22) into its opposite. Indeed, the $\geq$ sign in (3.22), instead of the $=$ sign in (3.2), accounts for the spontaneous production of entropy associated with the irreversible processes at work.

Starting from (3.22), similar calculations to those which led to (3.21) yield the local Lagrangian equation of entropy balance:

$$\frac{dS}{dt} \geq -\nabla_X \cdot \left(s_f \mathbf{M} + \frac{\mathbf{Q}}{T} \right) \tag{3.23}$$

where S stands for the overall Lagrangian density of entropy per unit of initial volume $d\Omega_0$:

$$S d\Omega_0 = (\rho_s (1-n) s_s + \rho_f n s_f) d\Omega_t \tag{3.24}$$

Identification of dissipation. Thermal equation

Let Ψ denote the overall Lagrangian density of Helmholtz free energy:

$$\Psi = E - TS \tag{3.25}$$

Getting rid of E, $\mathbf{Q}$ and s_f from (3.21), (3.23), (3.25) and using (3.9) and (3.10), we derive the Clausius–Duhem inequality related to deformable porous continua:

$$\Phi = \Phi_s + \Phi_f + \Phi_{th} \geq 0 \tag{3.26}$$

with:

$$\Phi_s = \boldsymbol{\pi} : \frac{d\boldsymbol{\Delta}}{dt} - g_f\, (\nabla_X \cdot \mathbf{M}) - S\frac{dT}{dt} - \frac{d\Psi}{dt} \tag{3.27a}$$

$$\Phi_f = \left[-(\nabla_X g_f)_T + (\mathbf{f} - \boldsymbol{\gamma}^f) \cdot \mathbf{F}\right] \cdot \mathbf{M} \tag{3.27b}$$

$$\Phi_{th} = -\frac{\mathbf{Q}}{T} \cdot \nabla_X T \tag{3.27c}$$

where $(\nabla_X g_f)_T$ stands for the gradient of g_f taken at temperature T held constant.

According to (3.26) the overall dissipation Φ (or the spontaneous production of entropy Φ/T) results from three distinct sources of dissipation.

Skeleton dissipation. The first source of dissipation, Φ_s, eventually accounts for the dissipation related to the sole skeleton. In order to achieve this identification, the use of mass balance equation (1.67) firstly allows us to rewrite Φ_s in the form:

$$\Phi_s = \boldsymbol{\pi} : \frac{d\boldsymbol{\Delta}}{dt} + g_f \frac{dm_f}{dt} - S\frac{dT}{dt} - \frac{d\Psi}{dt} \tag{3.28}$$

where m_f is the Lagrangian fluid mass content (see §1.4). Owing to the additive character of energy and entropy, the skeleton Lagrangian densities Ψ_s and S_s of free energy and entropy per unit of initial volume $d\Omega_0$ are given by:

$$\Psi_s = \Psi - m_f \psi_f; \qquad S_s = S - m_f s_f \tag{3.29}$$

The fluid state equations recalled in §3.1.2, when combined with the above definitions, together with relation (1.64), that is $m_f = \rho_f \phi$, allow us to express Φ_s in the form:

$$\Phi_s = \boldsymbol{\pi} : \frac{d\boldsymbol{\Delta}}{dt} + p\frac{d\phi}{dt} - S_s\frac{dT}{dt} - \frac{d\Psi_s}{dt} \tag{3.30}$$

Dissipation Φ_s such as given by (3.30) matches the standard expression of the dissipation for a solid, here the skeleton. Indeed, the strain work rate for an ordinary solid would reduce to the term $\boldsymbol{\pi} : d\boldsymbol{\Delta}/dt$. In the case of a porous solid, the strain work rate related to the skeleton is obtained by adding $p d\phi/dt$ to $\boldsymbol{\pi} : d\boldsymbol{\Delta}/dt$ in order to account for the action of the pore pressure on the skeleton through the internal walls of the porous network.

Fluid dissipation. The second source of dissipation, Φ_f, involves a gradient related to the fluid, $(\nabla_X g_f)_T$, and consequently the state of energy of the fluid particles contiguous to $d\Omega_t$. Hence, as debated more lengthily in §3.3.1, Φ_f accounts for the viscous dissipation due to the relative motion of the fluid with respect to the skeleton. It can be expressed more conveniently in the Eulerian form:

$$\varphi_f = (-\nabla_x p + \rho^f(\mathbf{f} - \boldsymbol{\gamma}^f)) \cdot \mathcal{V} \tag{3.31}$$

where we let $\varphi_f \, d\Omega_t = \Phi_f \, d\Omega_0$ and where we used fluid state equations (3.10), while $\mathcal{V}$ is the filtration vector defined by (1.61).

Thermal dissipation. The third and last source of dissipation, Φ_{th}, involves the temperature gradient $\nabla_X T$ and therefore the thermal state of the elementary systems contiguous to $d\Omega_t$. It is related to the dissipation due to heat conduction and can be written more naturally in the Eulerian form:

$$\varphi_{th} = -\frac{\mathbf{q}}{T} \cdot \nabla_x T \tag{3.32}$$

where we let $\varphi_{th} \, d\Omega_t = \Phi_{th} \, d\Omega_0$.

Owing to the very distinct nature of the dissipations so identified, the decoupling hypothesis consists of substituting the unique inequality (3.26) by the three separate inequalities:

$$\Phi_s = \boldsymbol{\pi} : \frac{d\boldsymbol{\Delta}}{dt} + p\frac{d\phi}{dt} - S_s\frac{dT}{dt} - \frac{d\Psi_s}{dt} \geq 0 \tag{3.33}$$

$$\varphi_f = (-\nabla_x p + \rho_f(\mathbf{f} - \boldsymbol{\gamma}^f)) \cdot \mathcal{V} \geq 0 \tag{3.34}$$

$$\varphi_{th} = -\frac{\mathbf{q}}{T} \cdot \nabla_x T \geq 0 \tag{3.35}$$

Inequality (3.33) states that only the part $d\Psi_s$ of the infinitesimal strain work supplied to the skeleton, that is $\pi_{ij} d\Delta_{ij} + p\, d\phi$ minus the corrective thermal term $S_s\, dT$, can be stored by the skeleton in the form of free energy, that is in a form eventually recoverable into efficient mechanical work. Indeed, owing to possible irreversible sliding of the matter forming the matrix, the non-stored part of the strain work is spontaneously transformed into the form of heat. Equation (3.34) states the same for the fluid. In addition, (3.35) stipulates that heat spontaneously flows from high temperatures to low temperatures. The associated spontaneous increase of entropy, $\varphi_{th}/T \geq 0$, expresses the harder task of extracting efficient mechanical work from the heat after its flow (i.e. by finding a cold source).

Once the spontaneous production of entropy Φ/T is identified, inequality (3.23) finally provides the entropy balance in the form of the Lagrangian thermal equation:

$$T\left(\frac{dS}{dt} + \nabla_X \cdot (s_f \mathbf{M})\right) = -\nabla_X \cdot \mathbf{Q} + \Phi_M \tag{3.36}$$

where $\Phi_M = \Phi_s + \Phi_f$ stands for the Lagrangian mechanical dissipation density and acts as a spontaneous heat source term.

3.3 Conduction Laws

3.3.1 Darcy's Law

Formulation of the law. Expression (3.34) for dissipation φ_f associated with the viscous flow of the fluid through the porous continuum can be written in the form:

$$\varphi_f = \mathcal{L} \cdot \mathcal{V} \geq 0 \tag{3.37}$$

where:

$$\mathcal{V} = n(\mathbf{V}^f - \mathbf{V}^s); \quad \mathcal{L} = -\nabla_x p + \rho_f(\mathbf{f} - \boldsymbol{\gamma}^f) \tag{3.38}$$

Dissipation φ_f is the product of the filtration vector $\mathcal{V} = n(\mathbf{V}^f - \mathbf{V}^s)$ and the force $\mathcal{L}$ producing the filtration. The law governing the fluid conduction or filtration has therefore to relate $\mathcal{V}$ to $\mathcal{L}$. Its simplest form is Darcy's law, which linearly links the flux $\mathcal{V}$ to the force $\mathcal{L}$. In the isotropic case Darcy's law reads:[4]

$$\mathcal{V} = k\mathcal{L}: \quad n(\mathbf{V}^f - \mathbf{V}^s) = k(-\nabla_x p + \rho_f(\mathbf{f} - \boldsymbol{\gamma}^f)) \tag{3.39}$$

where k is the permeability of the fluid. It has to be positive in order to ensure the positiveness of the dissipation associated with the fluid flow:

$$\varphi_f = \frac{n^2}{k}(\mathbf{V}^f - \mathbf{V}^s)^2 = \frac{\mathcal{V}^2}{k} \geq 0 \tag{3.40}$$

Fluid viscosity and the intrinsic permeability. When the macroscopic law of fluid conduction results from the microscopic viscous flow of the fluid through the porous network, Darcy's law can receive some support from a straightforward dimensional analysis. Force $\mathcal{L}$ already includes the pressure effect through the term $-\nabla_x p$, as does the dynamic inertia and body forces through the term $\rho_f(\mathbf{f} - \boldsymbol{\gamma}^f)$. In order to account for the missing effect of the viscous force resisting the flow, the isotropic fluid conduction law relating component $\mathcal{V}_i$ of $\mathcal{V}$ to component $\mathcal{L}_i$ of $\mathcal{L}$ can be formally written:

$$\mathcal{V}_i = f(\mathcal{L}_i, \eta_f, \ell, n) \tag{3.41}$$

where η_f is the dynamic fluid (shear) viscosity and ℓ the characteristic length of the porous network through which the fluid flow occurs. The previous relation will be physically relevant only if the relation is dimensionally consistent, that is if $\mathcal{V}_i$ and f have the same physical dimension, namely the dimension LT^{-1} of a velocity in the LMT base

[4]The more general form of Darcy's law is:

$$n(\mathbf{V}^f - \mathbf{V}^s) = \mathbf{k} \cdot (-\nabla_x p + \rho_f(\mathbf{f} - \boldsymbol{\gamma}^f - \mathbf{a}))$$

where $\mathbf{k}$ stands for the anisotropic permeability tensor and where $\mathbf{a}$ accounts for the tortuosity effect (see §2.5.3).

dimension, with L for length, M for mass and T for time. In this base any physical quantity Q can be expressed through its dimension function $[Q] = L^{\alpha} M^{\beta} T^{\gamma}$, which is a power function of the fundamental dimensions. For instance, since viscosity η_f linearly links a (shear) stress to a strain rate, its dimension function is $[\eta_f] = L^{-1} M T^{-1}$. The physical dimension of function f results from the physical dimensions of its arguments, $\mathcal{L}_i$, η_f, ℓ and n. The dimension functions are conveniently summarized in the form of an exponent matrix of dimensions, in which exponents α, β and γ form the columns:

	$\mathcal{V}_i$	$\mathcal{L}_i$	η_f	ℓ	n
L	1	−2	−1	1	0
M	0	1	1	0	0
T	−1	−2	−1	0	0

(3.42)

The number of dimensionally independent quantities among $\mathcal{V}_i$, $\mathcal{L}_i$, η_f, ℓ and n is three, which turns out to be the rank of their exponent matrix of dimensions, that is the maximum number of linearly independent columns in (3.42). Excluding the porosity n which is dimensionless, only one dimensionless quantity Π based on the independent set $\mathcal{L}_i$, η_f and ℓ remains to be formed from $\mathcal{V}_i$:

$$\Pi = \frac{\eta_f \mathcal{V}_i}{\ell^2 \mathcal{L}_i} \tag{3.43}$$

Consequently, dimensional analysis requires the relation linking $\mathcal{V}_i$ to $\mathcal{L}_i$, η_f, ℓ and n to be in the form:

$$\Pi = \Lambda\,(n) \tag{3.44}$$

resulting in:

$$\mathcal{V}_i = \frac{\varkappa}{\eta_f}\mathcal{L}_i; \qquad \varkappa = \ell^2 \delta\,(n) \tag{3.45}$$

and providing isotropic Darcy's law in the form:

$$n(\mathbf{V}^f - \mathbf{V}^s) = \frac{\varkappa}{\eta_f}(-\nabla_x p + \rho_f(\mathbf{f} - \boldsymbol{\gamma}^f)) \tag{3.46}$$

where permeability k is now identified as:

$$k = \frac{\varkappa}{\eta_f} = \frac{\ell^2}{\eta_f}\,\delta\,(n) \tag{3.47}$$

Expression (3.47) for k lies essentially on the physical dimension of η_f and eventually on the assumed linear viscous or Newtonian behaviour of the fluid. The material property $\varkappa = \ell^2 \delta\,(n)$ represents the square of a length scaling the geometry of the flow and is called the *intrinsic* permeability of the skeleton since it depends only on the geometry of the porous network irrespective of the fluid. In addition to porosity n, a unique length ℓ is here supposed to characterize the porous network geometry as far as only the fluid flow is concerned. This assumption holds only for simple geometries for which various

Table 3.1: Order of magnitude of intrinsic permeability for different materials (after de Marsily (1986) and Cowin (1998), see footnote for references).

Material	$\varkappa$ [m^2(= 10^{12} Darcy)]
Concrete	10^{-16}–10^{-21}
Clays	10^{-16}–10^{-20}
Bone	10^{-20}
Granites, gneiss, compact basalts	10^{-16}–10^{-20}
Marble	10^{-19}
Sandstones	10^{-11}–10^{-17}
Limestone	10^{-12}–10^{-16}
Fine sands, silts and loess	10^{-12}–10^{-16}
Gravels and sands	10^{-9}–10^{-12}

expressions of $\delta(n)$ have been derived in the literature. An expression often referred to is Kozeny–Carman's formula, which relates to a solid matrix formed by the packing of regular spheres. It reads:

$$\delta(n) = \frac{n^3}{1-n^2} \tag{3.48}$$

For more complex geometries, or when the filtration process does not result from a viscous flow (see §3.6.3), an experimental determination of intrinsic permeability $\varkappa$ must be preferred. Typical values of $\varkappa$ are given in Table 3.1.[5] Finally, through the dependence of η_f upon temperature T, expression (3.47) identifies the possible temperature dependence of permeability k.

The combination of (2.35) and Darcy's law (3.39) eventually allows us to identify the macroscopic interaction force $\mathbf{f}_{int}^{\rightarrow f}$ exerted by the skeleton on the fluid in the form:

$$\mathbf{f}_{int}^{\rightarrow f} = p\nabla_x n - \frac{n^2}{k}(\mathbf{V}^f - \mathbf{V}^s) \tag{3.49}$$

The first term, $p\nabla_x n$, accounts for the pressure effect resulting from the variation of the section offered to the fluid flow; the second term, $-\frac{n^2}{k}(\mathbf{V}^f - \mathbf{V}^s)$, accounts for the viscous resistance opposed by the shear stress to the fluid flow from the drag at the internal walls of the porous network.[6]

Deviations from Darcy's law. Darcy's law (3.46) assumes laminar flows where, by opposing turbulent flows, the viscous forces opposing the flow prevail over the forces related to the fluid acceleration. Neglecting the skeleton velocity, the strength of the fluid

[5]See de Marsily J. (1986), *Quantitative Hydrogeology. Groundwater Hydrology for Engineers*, Academic Press, New York, and Cowin S.C. (1998), 'Bone fluid poroelasticity', *Poromechanics, A tribute to M.A. Biot*, Proceedings of the Biot Conference on Poromechanics, ed. J.F. Thymus *et al.*, Balkema, Rotterdam. For concrete see Černy R., Rovnanikova P. (2002), *Transport Processes in Concrete*, Spon Press, London.

[6]Substitution of (3.49) into (2.68) for $\pi = f$ allows us to rewrite (2.67) for $\Pi = F$ in the form:

$$-\nabla_x \cdot (np) + \nabla_x \cdot \boldsymbol{\tau}^f + \rho_f n\mathbf{f} - \frac{n^2}{k}(\mathbf{V}^f - \mathbf{V}^s) = 0$$

acceleration forces is comparable with that of the viscous forces through the Reynolds number, which we express in the form:

$$\mathrm{Re} = \frac{\rho_f \mathcal{V}_\infty \sqrt{\varkappa}}{\eta_f} \tag{3.50}$$

where $\mathcal{V}_\infty$ quantifies the strength of the upstream flow. Owing to the low values of the characteristic length $\sqrt{\varkappa}$ of the usual porous materials (see Table 3.1), the Reynolds number is generally low enough for the usual flows to be laminar.[7]

Although the intrinsic permeability $\varkappa$ would have to depend only on the current porous network geometry, the experimental values of $\varkappa$ estimated from gaseous flows are generally found to be higher than those estimated from liquid flows. This difference is commonly attributed to the possible sliding of the gas molecules along the internal walls of the porous network. The sliding occurs when the mean free path λ of a molecule has the same order of magnitude as the diameter d of the pore. The kinetic theory of ideal gases gives the expression for λ:

$$\lambda = \frac{2\eta_g}{\rho_g \langle v \rangle} \tag{3.51}$$

where the index g refers to the gas and where $\langle v \rangle$ is the mean thermal molecular velocity:

$$\langle v \rangle = \sqrt{\frac{8RT}{\pi \mathcal{M}_g}} \tag{3.52}$$

---

where $\boldsymbol{\tau}^f(\mathbf{x})$ stands for the averaged fluid viscous stress, that is $\boldsymbol{\tau}^f(\mathbf{x}) = \langle f_{\omega_f}(\mathbf{z})\, \boldsymbol{\tau}^f(\mathbf{z}) \rangle$, when adopting the notations of §2.5.1 and §2.5.3. $\boldsymbol{\tau}^f(\mathbf{z})$ is the microscopic viscous stress, namely:

$$\boldsymbol{\tau}^f(\mathbf{z}) = 2\eta_f\, \mathbf{d}^f(\mathbf{z}) = \eta_f \left(\nabla_z \mathbf{v}^f(\mathbf{z}) + {}^t\nabla_z \mathbf{v}^f(\mathbf{z}) \right)$$

Let L_c be a characteristic length scaling the macroscopic position vector $\mathbf{x}$ as far as the macroscopic flow is concerned. The microscopic fluid velocity has to satisfy the fluid–matrix adherence condition $\mathbf{v}^f(\mathbf{z}) = \mathbf{V}^s(\mathbf{x})$ on the internal walls of the porous network. Consequently, length $\ell \equiv \|\mathbf{z} - \mathbf{x}\|$ appropriately scales as the microscopic position vector $\mathbf{z}$ as far as the microscopic flow is concerned. The order of magnitude of $\|\nabla_x \cdot \boldsymbol{\tau}^f\|$ can eventually be evaluated according to:

$$\left\| \nabla_x \cdot \boldsymbol{\tau}^f \right\| \equiv \frac{n\eta_f}{\ell L_c} \left\| \mathbf{V}^f - \mathbf{V}^s \right\|$$

Using (3.47) the latter can be compared with the order of magnitude of $\left\| \frac{n^2}{k} \left(\mathbf{V}^f - \mathbf{V}^s \right) \right\|$ according to:

$$\left\| \nabla_x \cdot \boldsymbol{\tau}^f \right\| / \left\| \frac{n^2}{k} \left(\mathbf{V}^f - \mathbf{V}^s \right) \right\| \equiv \frac{\ell}{nL_c}$$

In more general cases $\ell / nL_c \ll 1$ and the averaged viscous fluid shear stress $\boldsymbol{\tau}^f(\mathbf{x})$ can be neglected. Accordingly the macroscopic partial stress related to the fluid can be approximated by $n\boldsymbol{\sigma}^f \equiv -np\mathbf{1}$, as we assumed from the outset (see §2.3.3).

[7]The order of magnitude for the Reynolds number below which there is no doubt that the flow is mainly governed by the viscous forces and is therefore laminar is about 100, the possibility of a turbulent flow generally appearing for an order of magnitude 10 times greater.

where R (= 8.314 J/(mol K)) and $\mathcal{M}_g$ are respectively the universal constant of ideal gases and the molar mass of the gas considered. For ideal gases the fluid pressure p is:

$$p = \frac{RT}{\mathcal{M}_g}\rho_g \tag{3.53}$$

so that the mean free path λ can be rewritten in the form:

$$\lambda = \frac{\eta_g}{p}\sqrt{\frac{\pi RT}{2\mathcal{M}_g}} \tag{3.54}$$

The lower the gas pressure p, the more rarefied the gas and the higher the mean free path λ as, consequently, the filtration vector. The sliding phenomenon becomes significant when the value of the Knudsen number $\mathrm{Kn} = \lambda/d$ is close to unity. Provided that the macroscopic fluid acceleration $\boldsymbol{\gamma}^f$ can be neglected in (3.39), the resulting 'slip flow' can be captured by considering in (3.39) a permeability k of the gas modified according to the expression:

$$k = \frac{\varkappa}{\eta_g} + \frac{4}{3}l\frac{\langle v\rangle}{p} \tag{3.55}$$

where l is a characteristic length depending only on the geometry of the porous network. Expression (3.55) is conveniently rewritten according to Klinkenberg's formula:[8]

$$k = \frac{\varkappa}{\eta_g}\left(1 + \frac{\varpi}{p}\right) \tag{3.56}$$

where ϖ is a characteristic pressure depending on both the gas and the porous network geometry. The actual permeability $\varkappa/\eta_g$ and, consequently, the intrinsic permeability $\varkappa$ are then obtained by looking in the plane $(1/p, k)$ for the value of k at $1/p = 0$ along the line extrapolating the experimental results obtained for different values of $1/p$. For $\mathrm{Kn} \ll 1$ the viscosity vanishes: the flow reduces to free molecular diffusion ('Knudsen flow') and the first term in (3.55) does not even have to be considered.[9]

3.3.2 Fourier's Law

Expression (3.35) for thermal dissipation φ_{th} can be written as the product of the entropy efflux vector $\mathbf{q}/T$ multiplied by the force $-\nabla_x T$ producing the heat conduction. The law of heat conduction has therefore to relate $\mathbf{q}$ to $-\nabla_x T$. Its simplest form is Fourier's law, which linearly links $\mathbf{q}$ to $-\nabla_x T$. In the isotropic case Fourier's Law reads:

$$\mathbf{q} = -\kappa\nabla_x T \tag{3.57}$$

[8] See Klinkenberg L.J. (1941), 'The permeability of porous media to liquids and gases', *Drilling and production practices*, American Petroleum Institute, New York, 200–214.

[9] For a comprehensive presentation see Dullien F.A.L. (1979), *Porous media: fluid transport and pore structure*, Academic Press, New York. See also Carman P.C. (1956), *Flow of gases through porous media*, Butterworth Scientific London.

Table 3.2: Order of magnitude of thermal conductivities (see references in the footnote related to the permeability).

Material	κ (W/(m K))
Concrete	1.5–2.1
Rock	3
Dry sand	0.4–0.8
Wet sand	2.5–3.5
Dry clay	0.8–2
Wet clay	1.2–1.7
Granite	2.5–3.8
Sandstone	1.5–4.3
Water (0°C)	0.598
Air	0.026

where κ is the thermal conductivity and has to be replaced by a tensor of thermal conductivities in the general anisotropic case. Thermal conductivity κ has to be positive in order to ensure the positiveness of the dissipation associated with heat conduction:

$$\varphi_{th} = \frac{\mathbf{q}^2}{\kappa T} \geq 0 \tag{3.58}$$

Typical values of κ are given in Table 3.2.

Fourier's law (3.57) looks the same at the macroscopic scale and the microscopic scale and the overall thermal conductivity κ mainly depends on the matrix and fluid thermal conductivities, κ_s and κ_f, and on porosity n. Explicit expressions can be derived for simple geometries. For instance, consider the porous material made up of the progressive filling of volume $d\Omega_t$ by fluid-saturated hollow spheres formed by the matrix material. In addition, for any sphere let the ratio of its internal volume ω_f to its overall volume ω be equal to the porosity, that is $\omega_f/\omega = n$, irrespective of the sphere's outer radius. The overall thermal conductivity $\kappa_{sph}^{f\subset s}$ of such a material can be shown to be equal to:

$$\kappa_{sph}^{f\subset s} = \kappa_s \frac{2+\lambda-2n(1-\lambda)}{2+\lambda+n(1-\lambda)}; \quad \lambda = \frac{\kappa_f}{\kappa_s} \tag{3.59}$$

The thermal conductivity $\kappa_{sph}^{s\subset f}$ of the porous material, which is obtained by reversing the role of the matrix and the fluid in the filling process, reads:

$$\kappa_{sph}^{s\subset f} = \kappa_f \frac{2\lambda+1+2(1-n)(1-\lambda)}{2\lambda+1-(1-n)(1-\lambda)} \tag{3.60}$$

Conductivities $\kappa_{sph}^{s\subset f}$ and $\kappa_{sph}^{s\subset f}$ are associated with the most ideal isotropic porous material which can ever be encountered. As a consequence, they can be shown to constitute the

Hashin–Shtrikman[10] lower and upper bounds for conductivity κ, and for any isotropic material we write:

$$\kappa_f \leq \kappa_s : \quad \kappa_{sph}^{s \subset f} \leq \kappa \leq \kappa_{sph}^{s \subset f} \tag{3.61}$$

3.4 Constitutive Equations of the Skeleton

3.4.1 State Equations of the Skeleton

State variables and state equations

Skeleton free energy Ψ_s admits Δ_{ij}, ϕ and T as natural arguments since the rates of the latter three explicitly appear in expression (3.33) for the dissipation Φ_s related to the skeleton. In addition, on account of the postulate of local state, the free energy of the skeleton is rate independent and can finally be expressed in the form:

$$\Psi_s = \Psi_s(\Delta_{ij}, \phi, T;\ \chi_J) \tag{3.62}$$

Variables Δ_{ij}, ϕ, T and χ_J ($J = 1$ to N) form a set of state variables for the skeleton. Indeed, when given their values, the state of free energy of the skeleton is known irrespective of its past evolution. The variables Δ_{ij}, ϕ and T form a subset of *external* state variables since their variations can be externally controlled. By contrast, variables χ_J, which cannot be externally controlled, form the subset of *internal* state variables.

Considering evolutions where the internal variables do not vary:

$$\frac{d\chi_J}{dt} = 0 \tag{3.63}$$

and using (3.62) in (3.33), we get:

$$\left(\pi_{ij} - \frac{\partial \Psi_s}{\partial \Delta_{ij}}\right)\frac{d\Delta_{ij}}{dt} + \left(p - \frac{\partial \Psi_s}{\partial \phi}\right)\frac{d\phi}{dt} - \left(S_s + \frac{\partial \Psi_s}{\partial T}\right)\frac{dT}{dt} \geq 0 \tag{3.64}$$

Variations of any variable among the set of external variables Δ_{ij}, ϕ and T can actually occur irrespective of the variations of the other variables. In addition inequality (3.64) must hold whatever the sign of time variations $d\Delta_{ij}$, $d\phi$ and dT. Assuming also that π_{ij}, p and S_s are rate independent, we conclude that:

$$\pi_{ij} = \frac{\partial \Psi_s}{\partial \Delta_{ij}}; \qquad p = \frac{\partial \Psi_s}{\partial \phi}; \qquad S_s = -\frac{\partial \Psi_s}{\partial T} \tag{3.65}$$

Equations (3.65) associate the state variables Δ_{ij}, ϕ and T to their conjugate thermodynamic state variables π_{ij}, p and $-S$. They are the state equations relative to the skeleton. According to the postulate of local state, since they hold at equilibrium, they still hold in any evolution occurring at any rate, in particular in evolutions where $d\ \chi_J/dt \neq 0$.

[10] Hashin Z. (1983), 'Analysis of composite materials—A survey', *Journal of Applied Mechanics*, **50**, 481–504.

State equations (3.65) can be inverted, partially or totally, following the same procedure as in §3.1.2. For instance, defining G_s as:

$$G_s = \Psi_s - p\phi \tag{3.66}$$

we obtain:

$$\pi_{ij} = \frac{\partial G_s}{\partial \Delta_{ij}}; \quad \phi = -\frac{\partial G_s}{\partial p}; \quad S_s = -\frac{\partial G_s}{\partial T} \tag{3.67}$$

Finally, let us stress the important role played by an approach to the physical laws governing the evolution of a medium which is both tensorial and Lagrangian. The tensorial character ensures the spatial objectivity of these laws, that is their independence with regard to any particular reference frame adopted to express them by means of vector and tensor components. For its part, the Lagrangian formulation (i.e. referring to a fixed initial configuration both in (3.65) and in (3.67)), without being the only way to do so, ensures the time objectivity of these laws, that is their independence with respect to movement of the observer. This twofold objectivity, with regard to both space and time, is often called the law of material indifference.

State equations of the skeleton and porous material

When expressing the skeleton dissipation, we can also start from (3.28) instead of (3.30). Accordingly we write:

$$\Psi = \Psi\left(\Delta_{ij}, m_f, T;\ \chi_J\right) \tag{3.68}$$

Instead of (3.65) we derive the alternative state equations:

$$\pi_{ij} = \frac{\partial \Psi}{\partial \Delta_{ij}}; \quad g_f = \frac{\partial \Psi}{\partial m_f}; \quad S = -\frac{\partial \Psi}{\partial T} \tag{3.69}$$

Equations (3.69) are the state equations relative to the porous material formed by the open system $d\Omega_0$ which is followed in the motion of the skeleton and exchanges fluid mass with the outside. Indeed, for this open system $g_f\, dm_f = (\psi_f + p/\rho_f)\, dm_f$ accounts for the free energy supplied by the entering fluid, so that g_f and m_f become conjugate thermodynamic state variables. The consideration of the open system $d\Omega_0$, instead of the skeleton system, is the starting point to introduce the chemical potential and to address chemically active porous materials (see §3.6.3). When addressing ordinary porous materials, the point of view as developed in the previous section is more convenient and there is eventually no need to consider explicitly the fluid Gibbs potential g_f. Indeed the fluid mass content m_f can be reintroduced in (3.67) by combining the saturation condition, $\phi = m_f/\rho_f$, together with the fluid state equations expressed in the form $\rho_f = \rho_f(p, T)$, $s_f = s_f(p, T)$, in order finally to produce:

$$\pi_{ij} = \frac{\partial G_s}{\partial \Delta_{ij}}; \quad \frac{m_f}{\rho_f(p, T)} = -\frac{\partial G_s}{\partial p}; \quad S = -\frac{\partial G_s}{\partial T} + s_f(p, T)\, m_f \tag{3.70}$$

Matrix incompressibility and the effective stress

In the absence of any occluded porosity, the solid grains forming the matrix generally undergo negligible volume changes and the matrix can be assumed to be incompressible as is commonly done in soil mechanics. The matrix volume remains unchanged in the deformation and we write:

$$(1-n)\,d\Omega_t = (1-n_0)\,d\Omega_0 \tag{3.71}$$

Use of transport formulae (1.11) and (1.17) gives:

$$J = 1 + \phi - \phi_0 = \frac{1-n_0}{1-n} \tag{3.72}$$

Using identity:[11]

$$\frac{dJ}{dt} = J(\mathbf{F}^{-1}\cdot{}^t\mathbf{F}^{-1}) : \frac{d\mathbf{\Delta}}{dt} \tag{3.73}$$

together with (3.72), we rewrite (3.33) in the form:

$$\Phi_s = (\boldsymbol{\pi} + pJ(\mathbf{F}^{-1}\cdot{}^t\mathbf{F}^{-1})) : \frac{d\mathbf{\Delta}}{dt} - S_s\frac{dT}{dt} - \frac{d\Psi_s}{dt} \geq 0 \tag{3.74}$$

Using the same procedure as the one we used to derive state equations (3.65), we now obtain:

$$\pi'_{ij} = \frac{\partial \Psi_s}{\partial \Delta_{ij}}; \qquad S_s = -\frac{\partial \Psi_s}{\partial T} \tag{3.75}$$

where we note:

$$\pi'_{ij} = \pi_{ij} + pJ\frac{\partial X_i}{\partial x_k}\frac{\partial X_j}{\partial x_k} \tag{3.76}$$

The stress tensor $\boldsymbol{\pi}'$ of components π'_{ij} is therefore defined by:

$$\boldsymbol{\pi}' = \boldsymbol{\pi} + pJ(\mathbf{F}^{-1}\cdot{}^t\mathbf{F}^{-1}) \tag{3.77}$$

---

[11] Starting from:

$$J = \prod_{J=I,II,III} (1+2\Delta_J)^{1/2}$$

where $\Delta_{J=I,II,III}$ stands for the principal strains, we derive:

$$\frac{1}{J}\frac{dJ}{dt} = \sum_{J=I,II,III} (1+2\Delta_J)^{-1}\frac{d\Delta_J}{dt}$$

which results in (3.73) when returning to the definition of $\mathbf{\Delta}$ as a function of the gradient of the deformation $\mathbf{F}$ (see §1.2.3).

and is called the *effective* stress tensor. Indeed, irrespective of the pore pressure p, the state equations of an ordinary solid subjected to the stress $\boldsymbol{\pi}'$ would read the same as (3.75) with regard to the strain tensor $\boldsymbol{\Delta}$. Use of (2.47) to transport $\boldsymbol{\pi}'$ to the current configuration allows to us recover the Eulerian form of Terzaghi's effective stress widely used in soil mechanics, that is:

$$\boldsymbol{\sigma} = \frac{1}{J}\mathbf{F}\cdot\boldsymbol{\pi}'\cdot{}^t\mathbf{F} = \boldsymbol{\sigma} + p\,\mathbf{1};\qquad \sigma'_{ij} = \sigma_{ij} + p\,\delta_{ij} \tag{3.78}$$

A standard problem in soil mechanics and in geophysics is the unidimensional consolidation or sedimentation of a layer, where body and surface forces (the gravity forces and the applied consolidation pressure) act only in the $\mathbf{e}_3$ direction. As soon as the layer is transversally homogeneous, the displacement is non-zero only in the $\mathbf{e}_3$ direction, resulting in:

$$F_{33} = J = \frac{\partial x_3}{\partial X_3};\qquad \Delta_{33} = \frac{1}{2}\left[J^2 - 1\right];\quad \text{other } \Delta_{ij} = 0 \tag{3.79}$$

In isothermal evolutions and in the absence of internal variables χ_J, the free energy and consequence the effective stress $\boldsymbol{\pi}'$ depend only on J and therefore on the change in porosity due to (3.72). Owing to (3.78) and (3.79) it allows us to recover the state equation widely use in unidimensional problems in the geosciences, reading:

$$\sigma'_{33} = \sigma'_{33}\,(n) \tag{3.80}$$

3.4.2 Complementary Evolution Laws

Internal variables and dissipation

Substitution of state equations (3.65) into (3.33) leads to an expression of the skeleton dissipation in the form:

$$\Phi_s = \mathcal{L}_{\dot{\chi}}\cdot\dot{\chi} \geq 0 \tag{3.81}$$

where we note:

$$\mathcal{L}_{\dot{\chi}} = -\frac{\partial\Psi_s}{\partial\chi};\qquad \dot{\chi} = \frac{d\chi}{dt} \tag{3.82}$$

Skeleton dissipation Φ_s involves only the set of yet unspecified internal state variables $\chi = (\chi_J)$. Variables χ_J are associated with irreversible relative material movements occurring within the skeleton and eventually the matrix, each variable χ_J being associated with a particular dissipative mechanism. Unlike external variables, Δ_{ij}, ϕ and T, internal variables χ_J are not accessible to direct observation; they do not explicitly appear in the balance equations and thus in the Clausius–Duhem inequality $\Phi_s \geq 0$. When addressing irreversible material behaviour, state equations must be completed by laws governing the evolution of hidden variables χ_J and the associated dissipative mechanisms, that is the laws linking the components of the thermodynamic force $\mathcal{L}_{\dot{\chi}}$ to the rate $\dot{\chi} = \left(\dot{\chi}_J\right)$ that the force produces. These laws are called the complementary evolution laws.

Normal dissipative mechanisms

The complementary evolution laws are only required to satisfy the positiveness of dissipation (3.81). With the final aim of formulating the latter let $\mathcal{D}(\dot{\chi})$ be a non-negative convex function of its arguments $\dot{\chi}_J$ ($J = 1$ to N) with:

$$\mathcal{D}(\dot{\chi}) \geq \mathcal{D}(0) = 0 \tag{3.83}$$

Let us briefly recall the general properties of a convex function. A scalar function $f(x)$ is convex if the tangent to the curve $y = f(x)$ at any point M of coordinates $(x, y = f(x))$ is always below the secant connecting M to another point M' of coordinates $(x', y = f(x'))$. It results in the inequality:

$$\frac{df}{dx}\Big|_x (x' - x) \leq f(x') - f(x) \tag{3.84}$$

In addition, for a twice differentiable function, the previous convexity condition equivalently reads (see Fig. 3.1):

$$\frac{df^2}{dx^2} \geq 0 \tag{3.85}$$

The convexity property extends to a vectorial function $\mathcal{D}(\dot{\chi})$ in the form:

$$\frac{\partial \mathcal{D}}{\partial \dot{\chi}}\Big|_{\dot{\chi}} \cdot (\dot{\chi}' - \dot{\chi}) \leq \mathcal{D}(\dot{\chi}') - \mathcal{D}(\dot{\chi}) \tag{3.86a}$$

$$\dot{\chi}_J \frac{\partial^2 \mathcal{D}}{\partial \dot{\chi}_J \partial \dot{\chi}_K} \dot{\chi}_K \geq 0 \tag{3.86b}$$

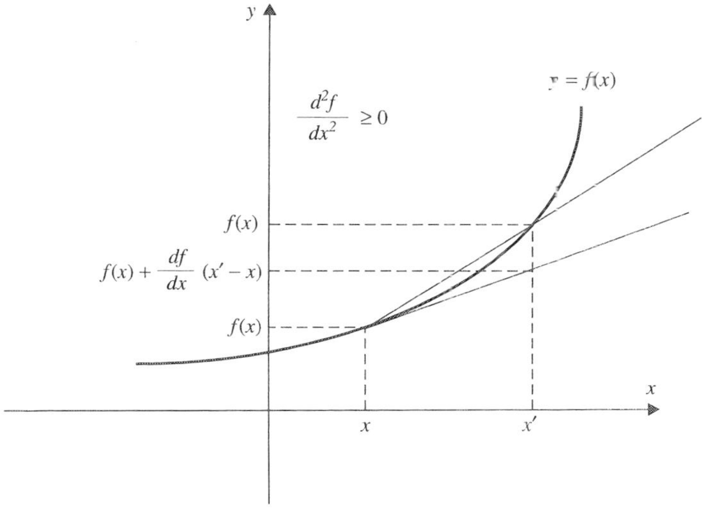

Figure 3.1: Properties of a convex function.

Returning now to the formulation of the complementary evolution laws, the dissipative mechanism associated with internal variables χ_J is said to be 'normal' if there exists a convex function $\mathcal{D}$, called the dissipation potential, such that the complementary evolution laws can be expressed in the form:

$$\mathcal{L}_{\dot{\chi}} = \frac{\partial \mathcal{D}}{\partial \dot{\chi}} \quad \leftrightarrow \quad \mathcal{L}_{\dot{\chi}_J} = \frac{\partial \mathcal{D}}{\partial \dot{\chi}_J} \tag{3.87}$$

The terminology of 'normal' dissipative mechanism refers to the normality of the thermodynamic force $\mathcal{L}_{\dot{\chi}}$ to surfaces defined by isovalues of $\mathcal{D}$ in the $\dot{\chi}_J$ space. Use of (3.87) in (3.86a) gives:

$$\mathcal{L}_{\dot{\chi}} \cdot (\dot{\chi}' - \dot{\chi}) \leq \mathcal{D}(\dot{\chi}') - \mathcal{D}(\dot{\chi}) \tag{3.88}$$

A useful inequality is provided by reversing the role of $\dot{\chi}$ and $\dot{\chi}'$ in (3.88) and adding the resulting inequality to (3.88). The procedure eventually gives:

$$\left(\mathcal{L}_{\dot{\chi}'} - \mathcal{L}_{\dot{\chi}}\right) \cdot (\dot{\chi}' - \dot{\chi}) \geq 0 \tag{3.89}$$

Moreover, use of (3.87) provides the symmetry relation:

$$\frac{\partial \mathcal{L}_{\dot{\chi}_J}}{\partial \dot{\chi}_K} = \frac{\partial \mathcal{L}_{\dot{\chi}_K}}{\partial \dot{\chi}_J} \tag{3.90}$$

When in addition $\mathcal{D}$ is a quadratic form of its arguments, it follows that $\Phi_s = 2\mathcal{D}$, while the relation linking $\mathcal{L}_{\dot{\chi}}$ and $\dot{\chi}$ is linear. Accordingly the irreversible process is said to be linear and (3.90) reduces to the celebrated Onsager symmetry relations between the coefficients of the quadratic form $\mathcal{D}$.

Darcy's law (3.39) and Fourier's law (3.57) constitute examples of such normal linear irreversible processes with:

$$\mathcal{D}_{Darcy}(\mathcal{V}) = \frac{1}{2k}\mathcal{V}^2; \quad \mathcal{D}_{Fourier}\left(\frac{\mathbf{q}}{T}\right) = \frac{T}{2\kappa}\left(\frac{\mathbf{q}}{T}\right)^2 \tag{3.91}$$

where the positiveness of permeability k and of thermal conductivity κ ensures the fulfilment of condition (3.86b). As a first example of a normal non-linear irreversible process, a filtration law possibly accounting for the flow of a non-linear viscous fluid can be obtained by extending the potential $\mathcal{D}_{Darcy}(\mathcal{V})$ in the form:

$$\mathcal{D}(\mathcal{V}) = \frac{1}{2nk}(\mathcal{V}^2)^n; \quad k > 0; \quad \frac{1}{2} \leq n \leq 1 \tag{3.92}$$

Letting $\dot{\chi} \equiv \mathcal{V}$ and $\mathcal{L}_{\dot{\chi}} \equiv -\nabla_x p$, that is not considering body nor dynamic forces in (3.38), the non-linear filtration law is derived from (3.87) and (3.92) in the form:

$$(\mathcal{V}^2)^{n-1}\mathcal{V} = -k\nabla_x p \tag{3.93}$$

so that:

$$\|\nabla_x p\| = \frac{1}{k} \|\mathcal{V}\|^{2n-1} \tag{3.94}$$

where $\| \ \|$ is the norm $\sqrt{()^2}$. It is worthwhile to note that according to the filtration law relative to the limit case $n \to 1/2$, the fluid is actually set in motion if only the strength $\|\nabla_x p\|$ of the fluid pressure gradient reaches the threshold value $1/k$. The resulting flow then occurs along the same direction as $-\nabla_x p$ at a rate remaining undetermined (pseudo plastic flow, see §8.2).

Substitution of (3.87) into (3.81) leads to an expression for the dissipation Φ_s in the form:

$$\Phi_s = \Phi_s\left(\dot{\chi}_J\right) = \frac{\partial \mathcal{D}}{\partial \dot{\chi}} \cdot \dot{\chi} \tag{3.95}$$

Applying (3.86a) for $\dot{\chi}' = 0$ and using (3.83), we derive:

$$\frac{\partial \mathcal{D}}{\partial \dot{\chi}} \cdot \dot{\chi} \geq 0 \tag{3.96}$$

Comparison of (3.95) and (3.96) shows that the normality of the dissipation mechanism allows us automatically to satisfy the positiveness of the dissipation as stated by the second law of thermodynamics. This does not mean that the positiveness of the dissipation requires any dissipation mechanism to be normal. When the dissipation mechanism is not normal, the positiveness condition (3.81) only requires the material properties involved in the complementary evolution laws to satisfy some specific conditions (e.g. (8.64)).

3.5 Recapitulating the Laws

The equations introduced up to now represent a closed set of equations: The number of equations in the set is equal to the number of unknown functions which have to be determined. A detailed account of unknown functions and equations is given below when using the Lagrangian formulation, so that the unknown functions admit $\mathbf{X}$ and t as arguments (see Table 3.3). It can be checked that the number of governing equations equals the number of unknown functions (see Table 3.4).

In addition to these equations the boundary conditions fix the values of the variables on the border of the porous media. These boundary conditions concern the stress vector, $\mathbf{T} = \boldsymbol{\sigma} \cdot \mathbf{n}$, and, consequently, vector $\mathbf{F} \cdot \boldsymbol{\pi} \cdot \mathbf{N}$, since $\mathbf{F} \cdot \boldsymbol{\pi} \cdot \mathbf{N} dA = \mathbf{T} da$, or alternatively the components of the skeleton displacement vector $\boldsymbol{\xi}$. The boundary conditions also fix the values of the fluid pressure p, or alternatively the fluid flow $\mathbf{M} \cdot \mathbf{N}$, together with the absolute temperature T, or alternatively the heat flow $\mathbf{Q} \cdot \mathbf{N}$. The system of equations is then complete, once the expressions of both the skeleton free energy and the specific fluid enthalpy, Ψ_s and g_f, have been specified and once the complementary evolution laws governing the evolution of internal variables χ have been given. This is the aim of the next chapter, devoted to poroelasticity, and Chapters 8 and 9, devoted to poroplasticity and poroviscoelasticity, respectively.

Table 3.3: Recapitulating the unknown functions.

Unknown	Notation	Number
Skeleton displacement	$\boldsymbol{\xi}$	3
Strain tensor	$\boldsymbol{\Delta}$	6
Fluid mass content	m_f	1
Lagrangian porosity	ϕ	1
Stress tensor	$\boldsymbol{\pi}$	6
Temperature	T	1
Skeleton entropy	S_s	1
Internal variables	χ_J	N
Fluid pressure	p	1
Fluid-specific density	ρ_f	1
Fluid-specific entropy	s_f	1
Fluid mass flow vector	$\mathbf{M}$	3
Heat flow vector	$\mathbf{Q}$	3
Total	—	$28 + N$

Table 3.4: Recapitulating the governing equations.

Nature	Equation
Strain and displacement, (1.24a)	$\boldsymbol{\Delta} = \frac{1}{2}\left(\nabla_X\boldsymbol{\xi} +^t \nabla_X\boldsymbol{\xi} +^t \nabla_X\boldsymbol{\xi} \cdot \nabla_X\boldsymbol{\xi}\right)$
Fluid continuity, (1.67)	$\dfrac{dm_f}{dt} + \nabla_x \cdot \mathbf{M} = 0$
Filling condition, (1.64)	$m_f = \rho_f \phi$
Motion equation, (2.50)	$\nabla_X \cdot (\mathbf{F} \cdot \boldsymbol{\pi}) + m_s^0 (\mathbf{f} - \boldsymbol{\gamma}^s) + m_f \left(\mathbf{f} - \boldsymbol{\gamma}^f\right) = 0$
Thermal equation, (3.36)	$T\dfrac{d\left(S_s + m_f s_f\right)}{dt} = -T\nabla_X \cdot \left(s_f\,\mathbf{M} + \mathbf{Q}\right) + \Phi_M$
Skeleton state equations, (3.67)	$\pi_{ij} = \dfrac{\partial \Psi_s}{\partial \Delta_{ij}};\quad p = \dfrac{\partial \Psi_s}{\partial \phi};\quad S_s = -\dfrac{\partial \Psi_s}{\partial T}$
Complementary equations, (3.87)	$\mathcal{L}_X = \dfrac{\partial \mathcal{D}}{\partial \dot{\chi}}$
Darcy's law, (3.39)	$\dfrac{\mathbf{F} \cdot \mathbf{M}}{\rho_f} = Jk^t\mathbf{F}^{-1}\left(-\nabla_X p + \rho_f\,{}^t\mathbf{F} \cdot \left(\mathbf{f} - \boldsymbol{\gamma}^f\right)\right)$
Fourier's law, (3.57)	$\mathbf{F} \cdot \mathbf{Q} = J\kappa^t\mathbf{F}^{-1}\nabla_X T$
Fluid state equations, (3.10)	$\dfrac{1}{\rho_f} = \dfrac{\partial g_f}{\partial p};\quad s_f = -\dfrac{\partial g_f}{\partial T}$
Number of equations	$28 + N$

3.6 Advanced Analysis

3.6.1 Fluid Particle Head. Bernoulli Theorem

In the usual applications the body forces are the gravity forces. Denoting by z the vertical ascendant coordinate, we write:

$$\mathbf{f} = -\nabla_x (gz) \tag{3.97}$$

where g stands for the gravity intensity. Considering quasistatic fluid flows, that is $\boldsymbol{\gamma}^f \sim 0$, (3.97) allows us to recover Darcy's law (3.46) in the form commonly used in hydrology and soil mechanics:

$$n(\mathbf{V}^f - \mathbf{V}^s) = -\lambda \nabla_x H; \quad \lambda = \frac{\rho_f g \varkappa}{\eta_f} \tag{3.98}$$

where H is the fluid particle head as defined in soil mechanics:

$$H = \frac{p}{\rho_f g} + z \tag{3.99}$$

At zero pressure H is the height of the liquid whose field determination can be achieved by means of piezometric measurements. Property λ is a characteristic filtration velocity and is the ('hydraulic') permeability commonly defined in soil mechanics.[12] It is the vertical filtration velocity which would result from gravity forces in the absence of any pressure gradient whatever its origin.

Consider now steady state flows such as:

$$\frac{\partial \mathcal{G}}{\partial t} = 0; \qquad \frac{d^f \mathcal{G}}{dt} = \nabla_x \mathcal{G} \cdot \mathbf{V}^f \tag{3.100}$$

for any quantity $\mathcal{G}$. Let us assume in addition that the skeleton is at rest, reading $\mathbf{V}^s = 0$. Multiplying (3.39) by $\mathbf{V}^f dt$ and using (3.40), (3.97) and (3.100), we get:

$$\frac{d^f p}{\rho_f} + d^f \left(gz + \frac{1}{2} (\mathbf{V}^f)^2 \right) = -\frac{\varphi_f}{\rho_f n} dt \tag{3.101}$$

In the absence of an external heat supply, the usual flows are sufficiently slow to be considered as isothermal. Using (3.10), (3.101) can be rewritten in the form:

$$\rho_f n \frac{d^f \mathcal{H}}{dt} = -\varphi_f \leq 0 \tag{3.102}$$

[12] For liquid water $\eta_f = 1 \times 10^{-3}$ kg/ (ms) and a useful numerical relation between $\varkappa$ and λ is found to be:

$$\varkappa \left(\mathrm{m}^2 \right) \simeq 10^{-7} \lambda \ (\mathrm{m/s})$$

while the usual unit used for intrinsic permeability $\varkappa$ is the Darcy, which is equal to 10^{-12} m^2 (see Table 3.1 for the usual values).

where $\mathcal{H}$ is the specific fluid head as defined in fluid mechanics:

$$\mathcal{H} = g_f + \mathrm{g}z + \frac{1}{2}(\mathbf{V}^f)^2 \tag{3.103}$$

Equation (3.102) constitutes an extension to poromechanics of the usual Bernoulli theorem of fluid mechanics. It means that the fluid flow along a flow line necessarily occurs in the direction of the head loss due to viscous dissipation.

3.6.2 Thermodynamics and the Double Porosity Network

Let us consider a porous continuum with a double porosity network such as examined in §1.5.3 and §2.5.2. Using the kinetic energy theorem as expressed by (2.73), together with (1.85) and (2.74), we extend (3.26) in the form:

$$\begin{aligned}\Phi = \boldsymbol{\pi} : \frac{d\boldsymbol{\Delta}}{dt} &- \sum_{\alpha=1,2} g_{f_\alpha}(\nabla_X \cdot \mathbf{M}^{(\alpha)}) - S\frac{dT}{dt} - \frac{d\Psi}{dt} \\ &+ \frac{1}{2}\overset{\circ}{m}_{1\to 2}((\mathbf{V}^{f_1} - \mathbf{V}^s)^2 - (\mathbf{V}^{f_2} - \mathbf{V}^s)^2) \\ &+ \sum_{\alpha=1,2} \left[-(\nabla_X g_{f_\alpha})_T + (\mathbf{f} - \boldsymbol{\gamma}^f) \cdot \mathbf{F}\right] \cdot \mathbf{M}^{(\alpha)} \\ &- \frac{\mathbf{Q}}{T} \cdot \nabla_X T \geq 0\end{aligned} \tag{3.104}$$

Mass balance equation (1.84) relative to the double porosity network allows us to split the overall dissipation into the form:

$$\Phi = \Phi_s + \Phi_{1\to 2} + \sum_{\alpha=1,2} \Phi_{f_\alpha} + \Phi_{th} \geq 0 \tag{3.105}$$

Using (3.29), the skeleton dissipation Φ_s can now be written:

$$\Phi_s = \boldsymbol{\pi} : \frac{d\boldsymbol{\Delta}}{dt} + \sum_{\alpha=1,2} p_\alpha \frac{d\phi_\alpha}{dt} - S_s \frac{dT}{dt} - \frac{d\Psi_s}{dt} \geq 0 \tag{3.106}$$

which leads to the state equations:

$$\boldsymbol{\pi} = \frac{\partial \Psi_s}{\partial \boldsymbol{\Delta}}; \quad p_\alpha = \frac{\partial \Psi_s}{\partial \phi_\alpha}; \quad S_s = -\frac{\partial \Psi_s}{\partial T} \tag{3.107}$$

The Eulerian form φ_{f_α} of the fluid dissipation Φ_{f_α} is:

$$\varphi_{f_\alpha} = \left[-\nabla_x p_\alpha + \rho_{f_\alpha}(\mathbf{f} - \boldsymbol{\gamma}^{f_\alpha})\right] \cdot \mathcal{V}_\alpha \geq 0 \tag{3.108}$$

where $\mathcal{V}_\alpha = n_\alpha \left(\mathbf{V}^{f_\alpha} - \mathbf{V}^s\right)$ is the filtration vector related to the porous network α and which leads us to formulate Darcy's law separately for each network. The thermal dissipation Φ_{th} is still given by (3.27c) and Fourier's law remains unchanged.

Dissipation $\Phi_{1\to 2}$ is the dissipation related to the fluid mass exchange between the two networks. Its Eulerian expression is:

$$\varphi_{1\to 2} = \overset{\circ}{r}_{1\to 2}\Delta\mathcal{H} \geq 0 \tag{3.109}$$

where $\Delta\mathcal{H}$ is:

$$\Delta\mathcal{H} = \left(g_{f_1} + \frac{1}{2}(\mathbf{V}^{f_1} - \mathbf{V}^s)^2\right) - \left(g_{f_2} + \frac{1}{2}(\mathbf{V}^{f_2} - \mathbf{V}^s)^2\right) \tag{3.110}$$

Comparing (3.109) and (3.110) with (3.102) and (3.103), $\Delta\mathcal{H}$ is identified as the specific head loss undergone by the fluid mass flowing from network 2 to network 1.

Assuming quasistatic incompressible fluid flow, the kinetic energy can be neglected, while $\rho_{f_\alpha} = \rho_f$ remains constant and $g_{f_1} - g_{f_2} = p_1/\rho_f - p_2/\rho_f$, so that $\varphi_{1\to 2}$ reduces to:

$$\varphi_{1\to 2} = \frac{\overset{\circ}{r}_{1\to 2}}{\rho_f}(p_1 - p_2) \geq 0 \tag{3.111}$$

In order to identify the relation linking $\overset{\circ}{r}_{1\to 2}/\rho_f$ and $p_1 - p_2$ we can proceed in the same way as we did in §3.3.1 when introducing the intrinsic permeability $\varkappa$. Among the set of the three relevant quantities involved here, that is $\overset{\circ}{r}_{1\to 2}/\rho_f$, $p_1 - p_2$ and η_f, only two are dimensionally independent. Dimensional analysis requires that the unique dimensionless quantity which can be formed from the set, namely $(\overset{\circ}{r}_{1\to 2}/\rho_f)(\eta_f/(p_1 - p_2))$, is a dimensionless constant υ so that:

$$\overset{\circ}{r}_{1\to 2} = \upsilon\frac{\rho_f}{\eta_f}(p_1 - p_2) \tag{3.112}$$

3.6.3 Chemically Active Porous Continua

Chemical potential and activity. Effective porosity and permeability

In the absence of skeleton dissipation, that is in thermoporoelasticity (see Chapter 4), (3.28) yields:

$$\Phi_s = \pi : \frac{d\mathbf{\Delta}}{dt} + g_f\frac{dm_f}{dt} - S\frac{dT}{dt} - \frac{d\Psi}{dt} = 0 \tag{3.113}$$

In (3.113), $g_f\,dm_f = (\psi_f + p/\rho_f)dm_f$ accounts for the free energy supplied to the open thermodynamic elementary system volume $d\Omega_0$ followed in the skeleton deformation. However, the energy required to make the fluid mass dm_f enter the volume $d\Omega_t$ does not always reduce to the mechanical work $p\,dm_f/\rho_f$ required against the internal fluid pressure p to make room for the entering volume dm_f/ρ_f. For instance, non-local interaction forces can exist between the saturating solution and the solid internal walls of the porous space. Let μ_f be the specific chemical potential of the saturating solution so that $\mu_f \, dm_f$ accounts for the free energy supply associated with the introduction of mass dm_f, whatever the non-local interaction forces at work. Instead of (3.113) we now write:

$$\Phi_s = \boldsymbol{\pi} : \frac{d\boldsymbol{\Delta}}{dt} + \mu_f \frac{dm_f}{dt} - S\frac{dT}{dt} - \frac{d\Psi}{dt} = 0 \tag{3.114}$$

resulting in state equations:

$$\boldsymbol{\pi} = \frac{\partial \Psi}{\partial \boldsymbol{\Delta}}; \quad \mu_f = \frac{\partial \Psi}{\partial m_f}; \quad S = -\frac{\partial \Psi}{\partial T} \tag{3.115}$$

Consider now an effective solution, possibly fictitious and usually referred to as the bulk solution, which is defined as the solution formed from the same components as those of the actual solution saturating volume $d\Omega_t$, which is in thermodynamic equilibrium with the latter. As a consequence of this equilibrium the immersion of material volume $d\Omega_t$ into the effective solution would not produce any effect. Denoting g_f^{eff} as the specific Gibbs potential of the effective solution, we write:

$$\mu_f = g_f^{eff}(p, T) \tag{3.116}$$

The effective solution serves as a 'chemical thermometer' to determine the thermodynamic pressure p of the actual solution. However, owing to the interaction energy Ψ_{int} of the forces of non-local type acting between the actual solution and the matrix, the strength $\mu_f = g_f^{eff}$ for the chemical potentials is achieved for a density ρ_f of the actual solution lower than the density ρ_f^{eff} of the effective solution. The chemical activity A of the saturating solution is then defined as the ratio of the two densities according to:

$$\rho_f^{eff} = A\rho_f; \quad A \geq 1 \tag{3.117}$$

Use of fluid state equations (3.10) where $g_f = g_f^{eff}(p, T)$, together with (3.116) and (3.117), yields:

$$\frac{\partial \mu_f}{\partial p} = \frac{1}{A\rho_f}; \quad s_f = -\frac{\partial \mu_f}{\partial T} \tag{3.118}$$

In addition, since the interaction energy Ψ_{int} represents the work required for the mass m_f to change from the specific volume of the effective solution to that of the saturating solution, we write:

$$\Psi_{int} = p\left(\frac{m_f}{\rho_f} - \frac{m_f}{\rho_f^{eff}}\right) = m_f\left(\psi_f^{eff} - \psi_f\right) \tag{3.119}$$

where ψ_f and ψ_f^{eff} stand for the actual and effective specific free energies, respectively. Including the interaction energy Ψ_{int} in the free energy Ψ_s of the skeleton, we extend (3.29) in the form:

$$\Psi_s = \Psi - m_f \psi_f^{eff} \tag{3.120}$$

Use of (3.116)–(3.120) allows us to rewrite (3.114) in the form:

$$\Phi_s = \boldsymbol{\pi} : \frac{d\boldsymbol{\Delta}}{dt} + p\frac{d}{dt}\left(\frac{\phi}{A}\right) - S_s\frac{dT}{dt} - \frac{d\Psi_s}{dt} = 0 \tag{3.121}$$

With regard to the skeleton, the chemical activity of the saturating solution eventually acts the same as if the effective porosity ϕ/A were substituted for the actual porosity ϕ according to:

$$\boldsymbol{\pi} = \frac{\partial\Psi_s}{\partial\boldsymbol{\Delta}}; \quad p = \frac{\partial\Psi_s}{\partial\left(\frac{\phi}{A}\right)}; \quad S_s = -\frac{\partial\Psi_s}{\partial T} \tag{3.122}$$

Accordingly, substitution of the chemical potential μ_f instead of g_f into (3.27b) leads us to extend expression (3.34) for the dissipation φ_f associated with the filtration process in the form:

$$\varphi_f = \left[-\frac{1}{A}\nabla_x p + \rho_f(\mathbf{f} - \boldsymbol{\gamma}^f)\right] \cdot \mathcal{V} \geq 0 \tag{3.123}$$

Assuming that the fluid filtration is a linear irreversible process and neglecting body and inertia forces, we write:

$$\mathcal{V} = n(\mathbf{V}^f - \mathbf{V}^s) = -k^{eff}\nabla_x p \tag{3.124}$$

In comparison with the original formulation (3.39) of Darcy's law, k^{eff} now includes the activity effect and plays the role of an effective permeability even though the filtration process through the porous medium does not result from a viscous flow owing to the interaction forces existing between the solution and the matrix. By analogy to (3.47) k^{eff} can be set in the form:

$$k^{eff} = \frac{\varkappa^{eff}}{\eta_f} \tag{3.125}$$

where the effective intrinsic permeability $\varkappa^{eff}$ is conveniently expressed in m^2 for comparison with the usual values.

From poromechanics to chemomechanics

Reactive saturating mixture. Consider now the case where the saturating fluid is a mixture formed of M components referred to by index α ($\alpha = 1$ to M). In addition a chemical reaction occurs between these components. Analogously to the case of the double porosity network examined in §1.5.3, the mass conservation related to the αth component is expressed in the form:

$$\frac{dm_\alpha}{dt} = -\nabla_X \cdot \mathbf{M}^{(\alpha)} + \overset{\circ}{m}_{\to\alpha} \tag{3.126}$$

where $\overset{\circ}{m}_{\to\alpha}$ is the rate of mass formation (> 0) or consumption (< 0) of the αth component during the chemical reaction. Expression (3.26) then extends to:

$$\Phi_{s,\to} + \sum_{\alpha=1}^{\alpha=M} \Phi_\alpha + \Phi_{th} \geq 0 \tag{3.127}$$

where:[13]

$$\Phi_{s,\to} = \pi : \frac{d\mathbf{\Delta}}{dt} - \sum_{\alpha=1}^{\alpha=M} \mu_\alpha \left(\nabla_X \cdot \mathbf{M}^{(\alpha)}\right) - S\frac{dT}{dt} - \frac{d\Psi}{dt} \tag{3.128}$$

Owing to activity effects the use of the chemical potential μ_α is now preferred to that of the Gibbs potential g_α (see previous section). In (3.127), Φ_{th} and Φ_α stand respectively for the thermal dissipation and for the dissipation associated with the transport of the αth component. They do not need further analysis. The mass balance equation (3.126) allows us to split dissipation $\Phi_{s,\to}$ according to:

$$\Phi_{s,\to} = \Phi_s + \Phi_\to \geq 0 \tag{3.129}$$

Dissipation Φ_s is the skeleton dissipation:

$$\Phi_s = \pi : \frac{d\mathbf{\Delta}}{dt} + \sum_{\alpha=1}^{\alpha=M} \mu_\alpha \frac{dm_\alpha}{dt} - S\frac{dT}{dt} - \frac{d\Psi}{dt} \geq 0 \tag{3.130}$$

while $\Phi_\to$ is the dissipation associated with the reaction and expressed in the form:

$$\Phi_\to = - \sum_{\alpha=1}^{\alpha=M} \mu_\alpha \overset{\circ}{m}_{\to\alpha} \geq 0 \tag{3.131}$$

Analogously to (3.68) we write:

$$\Psi = \Psi\left(\Delta_{ij}, m_\alpha, T; \chi\right) \tag{3.132}$$

resulting in the state equations:

$$\pi = \frac{\partial\Psi}{\partial\mathbf{\Delta}}; \qquad \mu_\alpha = \frac{\partial\Psi}{\partial m_\alpha}; \qquad S = -\frac{\partial\Psi}{\partial T} \tag{3.133}$$

so that the skeleton dissipation Φ_s can still be expressed in the form:

$$\Phi_s = -\frac{\partial\Psi}{\partial\chi} \cdot \frac{d\chi}{dt} \tag{3.134}$$

[13]We did not consider the kinetic energy term associated with $\overset{\circ}{m}_{\to\alpha}$ such as the one associated with $\overset{\circ}{m}_{\to\alpha}$ in (3.104).

Equations (3.133) extend the standard state equations of physical chemistry[14] to a mixture saturating a deformable porous material.

Chemical affinity and reaction rate. Since there is no overall mass creation, mass formation rates $\overset{\circ}{m}_{\to\alpha}$ satisfy:

$$\sum_{\alpha=1}^{\alpha=M} \overset{\circ}{m}_{\to\alpha} = 0 \tag{3.135}$$

In addition, rates $\overset{\circ}{m}_{\to\alpha}$ are subjected to match the molar ratios according to the way in which the components are involved in the chemical reaction. More precisely, a reaction can be written in the general form:

$$\sum_{\alpha=1}^{\alpha=L} \mathrm{r}_\alpha R_\alpha \to \sum_{\alpha=L+1}^{\alpha=M} \mathrm{p}_\alpha P_\alpha \tag{3.136}$$

where R_α are the L reacting moles, while P_α are the $M - L$ produced moles. The above equation indicates that r_α moles of reactant R_α are needed to form p_α moles of product P_α. Let ν_α be the stoichiometric coefficient relative to P_α or R_α, with the convention $\nu_\alpha = \mathrm{p}_\alpha > 0$ for a produced mole, and $\nu_\alpha = -\mathrm{r}_\alpha < 0$ for a reacting mole. If $\mathcal{M}_\alpha$ denotes the molar mass of component R_α or P_α, the stoichiometry of reaction (3.136) requires that:

$$\frac{\overset{\circ}{m}_{\to 1}}{\nu_1 \mathcal{M}_1} = \frac{\overset{\circ}{m}_{\to 2}}{\nu_2 \mathcal{M}_2} = \cdots = \frac{\overset{\circ}{m}_{\to M}}{\nu_M \mathcal{M}_M} \tag{3.137}$$

while (3.135) implies in addition:

$$\sum_{\alpha=1}^{\alpha=M} \nu_\alpha \mathcal{M}_\alpha = 0 \tag{3.138}$$

The reaction rate $\overset{\circ}{\xi}$ is defined as the common value of ratios involved in (3.137):

$$\overset{\circ}{\xi} = \frac{\overset{\circ}{m}_{\to\alpha}}{\nu_\alpha \mathcal{M}_\alpha} \tag{3.139}$$

Use of (3.139) in (3.131) allows us to express $\Phi_{\to}$ in the form:

$$\Phi_{\to} = \mathcal{A}\overset{\circ}{\xi} \geq 0 \tag{3.140}$$

where $\mathcal{A}$ is de Donder's chemical affinity relative to the reaction, that is:

$$\mathcal{A} = -\sum_{\alpha=1}^{\alpha=M} \nu_\alpha \mathcal{M}_\alpha \mu_\alpha \tag{3.141}$$

---

[14] See for instance Atkins P.W. (1990), *Physical Chemistry*, Fourth Edition, Oxford University Press.

Kinetics of reaction. According to (3.141), the law governing the reaction kinetics is the law linking the reaction rate $\overset{\circ}{\xi}$ to the chemical affinity $\mathcal{A}$, that is the potential difference or thermodynamic force producing the reaction. In a diffusion-controlled reaction the macroscopic rate results from the velocity at which the reactants diffuse at the microscopic scale through the mixture. When they meet, they instantaneously combine to form the products without dissipation. Since diffusion can be often considered as a linear irreversible process, a first approximation consists in adopting the linear law:

$$\mathcal{A} = \eta \overset{\circ}{\xi} \exp\left(\frac{E_a}{RT}\right) \tag{3.142}$$

By analogy to the Arrhenius law, the factor E_a/RT accounts for the final formation of the products obeying an activation-controlled reaction; E_a stands for the activation energy and R is the universal constant for an ideal gas.

Thermodynamic equilibrium is achieved at a zero reaction rate obtained when the potential difference between the reactants and the products, as captured by affinity $\mathcal{A}$, vanishes. Using state equations (3.133), the equilibrium condition can be expressed in the form:

$$\mathcal{A} = -\sum_{\alpha=1}^{\alpha=M} \nu_\alpha \mathcal{M}_\alpha \frac{\partial \Psi}{\partial m_\alpha} = 0 \tag{3.143}$$

The characteristic time τ_η associated with (3.142) is:

$$\tau_\eta \equiv \frac{\eta}{\mathcal{A}_0} \tag{3.144}$$

where $\mathcal{A}_0$ scales as the order of magnitude of the chemical affinity. When τ_η is much smaller than the characteristic time of the other processes, the equilibrium condition (3.143) can be conveniently used instead of kinetics law (3.142), so that the number of independent state variables relative to the mixture decreases from M to $M-1$.

Closed systems. The extent of the reaction as an internal variable. For closed systems, there is no matter supplied by an external flow, that is $-\nabla_X \cdot \mathbf{M}^{(\alpha)} = 0$, and (3.126) and (3.139) give:

$$\frac{dm_\alpha}{dt} = \overset{\circ}{m}_{\rightarrow\alpha} = \nu_\alpha \mathcal{M}_\alpha \overset{\circ}{\xi} \tag{3.145}$$

Since dm_α/dt is an actual time derivative, the chemical reaction rate $\overset{\circ}{\xi}$ turns out to be a time derivative too. We write:

$$\overset{\circ}{\xi} = \frac{d\xi}{dt} \tag{3.146}$$

The variable ξ so introduced is a measure of the degree of advancement of the reaction and is called the extent of the reaction.

Using (3.134), (3.140) and (3.146), the dissipation associated with a closed porous material, where an internal chemical reaction occurs, can be identified as:

$$\Phi = \Phi_s + \Phi_{\rightarrow} + \Phi_{th} = -\frac{\partial \Psi}{\partial \chi} \cdot \frac{d\chi}{dt} + \mathcal{A}\frac{d\xi}{dt} + \Phi_{th} \geq 0 \tag{3.147}$$

Let $\psi = \psi\left(\Delta_{ij}, T; \chi, \xi\right)$ be the free energy of any closed system, where ξ is a yet unspecified internal variable, and let Φ_{clo} be the dissipation relative to this system not including the thermal dissipation. The first and second laws of thermodynamics applied to any closed system yield:

$$\Phi = \Phi_{clo} + \Phi_{th} = -\frac{\partial \psi}{\partial \chi} \cdot \frac{d\chi}{dt} - \frac{\partial \psi}{\partial \xi}\frac{d\xi}{dt} + \Phi_{th} \geq 0 \tag{3.148}$$

A chemical reaction occurring in a closed system is an internal process, so that its extent ξ eventually is the internal state variable characterizing its progress. Expressions (3.147) and (3.148) of dissipation Φ must coincide and we get:

$$\psi\left(\Delta_{ij}, T; \chi, \xi\right) = \Psi\left(\Delta_{ij}, m_\alpha = \nu_\alpha \mathcal{M}_\alpha \xi, T; \chi\right); \quad -\frac{\partial \psi}{\partial \xi} = \mathcal{A} = -\nu_\alpha \mathcal{M}_\alpha \mu_\alpha \tag{3.149}$$

The above sequence of identification is quite general:

- First, consider an open system and identify the explicit expression of the intensive thermodynamic force (as here chemical affinity $\mathcal{A} = -\nu_\alpha \mathcal{M}_\alpha \mu_\alpha$) to be linked to a given rate (such as chemical reaction rate $\overset{\circ}{\xi}$ here).
- Second, apply the closure conditions (such as $\mathbf{M}^{(\alpha)} = 0$ here resulting in (3.146)).
- Last, provided that the complementary evolution law (or equilibrium condition) of the irreversible process at work is known irrespective of the porous medium, use it without further investigation.

For instance, if the kinetics of the reaction obeys (3.142), the laws governing the evolution of a closed solid system subjected to an internal chemical reaction are the state equations:

$$\psi = \psi\left(\Delta_{ij}, T; \chi, \xi\right); \quad \pi = \frac{\partial \psi}{\partial \mathbf{\Delta}}; \quad S = -\frac{\partial \psi}{\partial T}; \quad \mathcal{L}_\chi = -\frac{\partial \psi}{\partial \chi}; \quad \mathcal{A} = -\frac{\partial \psi}{\partial \xi} \tag{3.150}$$

and the complementary evolution laws:

$$\mathcal{L}_\chi = \frac{\partial \mathcal{D}}{\partial \dot{\chi}}; \quad \mathcal{A} = \eta \frac{d\xi}{dt} \exp\left(\frac{E_a}{RT}\right) \tag{3.151}$$

where we assumed that the internal variables χ are associated with a normal dissipative mechanism (see §3.4.2). Equations (3.150) and (3.151) achieve the transition from poromechanics to chemomechanics.

Thermodynamics of a poroelastic material subjected to dissolution. Consider a material whose solid matrix can dissolve in solute form (index *sol*) within a solvent, for example liquid water (index w), forming the solution saturating the porous space. For isothermal evolutions expression (3.128) of the dissipation $\Phi_{s,\rightarrow}$ attached to both the skeleton and the dissolution process is specified in the form:

$$\Phi_{s,\rightarrow} = \pi : \frac{d\mathbf{\Delta}}{dt} - \mu_w\left(\nabla_X \cdot \mathbf{M}^w\right) - \mu_{sol}\left(\nabla_X \cdot \mathbf{M}^{sol}\right) - \frac{d\Psi}{dt} \geq 0 \tag{3.152}$$

Use of continuity equations (1.88) allows us to express $\Phi_{s,\rightarrow}$ in the alternative form:

$$\Phi_{s,\rightarrow} = \boldsymbol{\pi} : \frac{d\boldsymbol{\Delta}}{dt} + \mu_w \frac{dm_w}{dt} + \mu_{sol}\frac{dm_{sol}}{dt} - \mu_{sol}\overset{\circ}{m}_{s\rightarrow sol} - \frac{d\Psi}{dt} \geq 0 \tag{3.153}$$

Similarly to (3.29), let Ψ_s be the free energy of the skeleton:

$$\Psi_s = \Psi - m_w\psi_w - m_{sol}\psi_{sol} \tag{3.154}$$

Substitution of (3.154) into (3.153) and using a similar procedure to the one that led from (3.28) to (3.30) give:

$$\Phi_{s,\rightarrow} = \boldsymbol{\pi} : \frac{d\boldsymbol{\Delta}}{dt} + p\frac{d\phi}{dt} - \mu_{sol}\overset{\circ}{m}_{s\rightarrow sol} - \frac{d\Psi_s}{dt} \geq 0 \tag{3.155}$$

where $p = p_w + p_{sol}$ is the solution pressure. Considering an elastic matrix, the only source of dissipation is the dissolution process so that $\Phi_{s,\rightarrow}$ must be zero when the dissolution process is not active, that is for $\overset{\circ}{m}_{s\rightarrow sol} = 0$. We conclude that state equations (3.65) still remain valid and that the chemical porosity ϕ_{ch} defined by (1.89) is the relevant internal state variable related to the dissipative dissolution process. We write:

$$\Psi_s = \Psi_s\left(\Delta_{ij}, \phi, \phi_{ch}\right) : \quad \pi_{ij} = \frac{\partial \Psi_s}{\partial \Delta_{ij}}; \quad p = \frac{\partial \Psi_s}{\partial \phi} \tag{3.156}$$

A combination of (1.89), (3.155) and (3.156) finally gives the dissipation $\Phi_{\rightarrow} = \Phi_{s,\rightarrow}$ occurring in the dissolution process in the form:

$$\Phi_{\rightarrow} = \left(-\frac{\partial \Psi_s}{\partial \phi_{ch}} - \rho_s^0\mu_{sol}\right)\frac{d\phi_{ch}}{dt} \geq 0 \tag{3.157}$$

A further identification requires an analysis of the dissipation at the microscopic scale where the dissolution process does occur. The dissipation is the one attached to the thin layer of the solid matrix currently dissolving in solute form. Let $\overset{\circ}{r}_{s\rightarrow sol}$ be the Eulerian rate of mass dissolving per unit of microscopic surface of the current internal walls of the porous space. According to the analysis we performed in §1.5.2, $\overset{\circ}{r}_{s\rightarrow sol}$ can be expressed in the form:

$$\overset{\circ}{r}_{s\rightarrow sol} = \rho_s(\mathbf{c} - \mathbf{V}^s)\cdot\mathbf{n} = \rho_{sol}(\mathbf{c} - \mathbf{V}^{sol})\cdot\mathbf{n} \tag{3.158}$$

where, for the sake of simplicity, we maintain the same notation as that used at the macroscopic scale: $\mathbf{c}$ is the Eulerian speed of displacement of the dissolution front, which coincides at any time with the current internal walls of the porous space; $\mathbf{n}$ is the outward unit normal to the surface enclosing the porous volume. The momentum balance along the direction $\mathbf{n}$ related to the mass dissolving between times t and $t + dt$ is:

$$\overset{\circ}{r}_{s\rightarrow sol}\left(v_{sol} - v_s\right) = p_s - p_{sol} \tag{3.159}$$

where $v_\alpha = \mathbf{V}^{(\alpha)}\cdot\mathbf{n}$. Along the direction normal to $\mathbf{n}$ there is no shear stress at the contact between the solution and the internal solid walls of the porous space. Accordingly the momentum balance implies that the tangential velocities of the solid and the solute are the same. Letting $\overset{\circ}{q}$ be the heat supply rate and e_α the specific energy of constituent $\alpha = sol$ or s, the first law of thermodynamics applied to the thin layer dissolving between times t and $t + dt$ gives:

$$\overset{\circ}{r}_{s\to sol}\left(e_{sol} + \frac{1}{2}v_{sol}^2 - e_s - \frac{1}{2}v_s^2\right) = p_s v_s - p_{sol}v_{sol} + \overset{\circ}{q} \tag{3.160}$$

Combining (3.158)–(3.160), we get:

$$\overset{\circ}{r}_{s\to sol}\left(h_{sol} + \frac{1}{2}(v_{sol} - c)^2 - h_s - \frac{1}{2}(v_s - c)^2\right) = \overset{\circ}{q} \tag{3.161}$$

where $h_\alpha = e_\alpha + p_\alpha/\rho_\alpha$ is the specific enthalpy. In addition, letting s_α be the specific entropy, the second law of thermodynamics provides the inequality:

$$\overset{\circ}{r}_{s\to sol}(s_{sol} - s_s) \geq \frac{\overset{\circ}{q}}{T} \tag{3.162}$$

The dissolution process is a slow process so that the evolutions can be considered as isothermal while the kinetic energy terms $\frac{1}{2}(v_\alpha - c)^2$ can be neglected as we actually did in deriving (3.157). Combining (3.161) and (3.162), we obtain:

$$\mu_\alpha = h_\alpha - Ts_\alpha : \quad \overset{\circ}{r}_{s\to sol}(\mu_s - \mu_{sol}) \geq 0 \tag{3.163}$$

The mass $\overset{\circ}{r}_{s\to sol}da\,dt$ which dissolves between time t and $t + dt$ refers to the current surface da across which the dissolution physically takes place. It can be alternatively expressed in the Lagrangian form according to:

$$\overset{\circ}{r}_{s\to sol}da\,dt = \rho_s^0 C\,dA\,dt \tag{3.164}$$

where dA is the initial material surface corresponding to da in the initial non-deformed reference configuration, whereas C stands for the Lagrangian normal speed as defined in §1.5.2. Substituting (3.164) into (3.163) and integrating over the dissolution front Σ formed by the internal walls dA of the porous network, we finally derive:

$$\Phi_{\to} = \frac{1}{d\Omega_0}\int_\Sigma \rho_s^0 C(\mu_s - \mu_{sol})\,dA \geq 0 \tag{3.165}$$

Except in the particular case of the pressure-dissolution process (see §4.4.4), chemical potential μ_s can be assumed to be almost the same along the dissolution front so that (1.80) and (3.165) give:

$$\Phi_{\to} = \rho_s^0(\mu_s - \mu_{sol})\frac{d\phi_{ch}}{dt} \geq 0 \tag{3.166}$$

A comparison of (3.157) and (3.166) finally leads to the identification:

$$-\frac{\partial \Psi_s}{\partial \phi_{ch}} = \rho_s^0 \mu_s \tag{3.167}$$

The chemical potential μ_s can be expressed in the form:

$$\mu_s = \mu_s^0 + \Delta\mu_s \tag{3.168}$$

where μ_s^0 stands for the chemical energy associated with the chemical bonds in the non-deformed state, while $\Delta\mu_s$ accounts for the elastic energy stored during the solid deformation. Dissolution is a sufficiently slow process so that the solid and the solute can be assumed to remain constantly in thermodynamic equilibrium, resulting in a zero dissipation, $\Phi_{\rightarrow} = 0$, so that $\mu_{sol} = \mu_s$ at any time. With $\Delta\mu_s \ll \mu_s^0$ (see §4.4.4) and $\mu_{sol} = \mu_{sol}(\rho_{sol})$, the solid–solute equilibrium eventually requires the solute mass density ρ_{sol} to remain constantly equal to some equilibrium value ρ^{Eq}.

Chapter 4

Thermoporoelasticity

Thermoporoelasticity extends the theory of thermoelasticity to porous continua. This extension is achieved by considering an underlying thermoelastic skeleton. The dissipation related to the skeleton is zero and there are no internal variables. The constitutive equations reduce to state equations (see §3.4.1). Their operational formulation needs an explicit expression for the skeleton free energy Ψ_s. Except for the stability conditions examined later on (see §5.4.1), this expression is not restricted by any particular constraint and the determination of the thermoporoelastic properties involved by the state equations is finally left to experiments. Nevertheless these macroscopic properties depend on the microscopic properties and the compatibility relations between both properties can be derived. This chapter progressively explores these different issues.[1]

4.1 Non-linear Thermoporoelastic Skeleton

4.1.1 Infinitesimal Transformation and State Equations

The dissipation related to a thermoelastic skeleton is zero, that is:

$$\pi_{ij}\, d\Delta_{ij} + p\, d\phi - S_s\, dT - d\Psi_s = 0 \tag{4.1}$$

The transformation will be infinitesimal as soon as condition (1.25) is satisfied. The fulfilment of condition (1.25) depends on the strength of the applied stress compared with the material stiffness and can only be checked a posteriori. Indeed, the order of magnitude of the displacement gradient $\nabla\boldsymbol{\xi}$ can be compared with one only when the material properties are known and the field displacement $\boldsymbol{\xi}$ derived for the problem at hand (see Chapter 5).

When condition (1.25) is fulfilled, Green strain components Δ_{ij} and Piola–Kirchhoff stress components π_{ij} can be replaced by linearized strain components ε_{ij} and Cauchy

[1] A founding paper on poroelasticity, although not the first on the topic, is without doubt, Biot M. A. (1941), 'General theory of three dimensional consolidation', *Journal of Applied Physics*, **12**, 155–164.

Poromechanics O. Coussy
© 2004 John Wiley & Sons, Ltd ISBN 0-470-84920-7

stress components σ_{ij}. We write:

$$\sigma_{ij}\, d\varepsilon_{ij} + p\, d\phi - S_s\, dT - d\Psi_s = 0 \tag{4.2}$$

Alternatively, use of energy G_s defined by:

$$G_s = \Psi_s - p\,\phi \tag{4.3}$$

leads to:

$$\sigma_{ij}\, d\varepsilon_{ij} - \phi\, dp - S_s\, dT - dG_s = 0 \tag{4.4}$$

From (4.4) we finally derive the state equations in the form:

$$G_s = G_s(\varepsilon_{ij}, p, T): \quad \sigma_{ij} = \frac{\partial G_s}{\partial \varepsilon_{ij}}; \quad \phi = -\frac{\partial G_s}{\partial p}; \quad S_s = -\frac{\partial G_s}{\partial T} \tag{4.5}$$

Owing to (4.5) it is worthwhile to note Maxwell's symmetry relations:

$$\frac{\partial \sigma_{ij}}{\partial \varepsilon_{kl}} = \frac{\partial \sigma_{kl}}{\partial \varepsilon_{ij}}; \quad \frac{\partial \sigma_{ij}}{\partial p} = -\frac{\partial \phi}{\partial \varepsilon_{ij}}; \quad \frac{\partial \sigma_{ij}}{\partial T} = -\frac{\partial S_s}{\partial \varepsilon_{ij}}; \quad \frac{\partial \phi}{\partial T} = \frac{\partial S_s}{\partial p} \tag{4.6}$$

Let σ and s_{ij} be the hydrostatic and deviatoric components of the stress tensor, and let ϵ and e_{ij} be the volumetric dilation and the deviatoric components of the strain tensor:

$$\sigma = \frac{1}{3}\sigma_{ii}; \quad s_{ij} = \sigma_{ij} - \sigma\, \delta_{ij} \tag{4.7a}$$

$$\epsilon = \varepsilon_{ii}; \quad e_{ij} = \varepsilon_{ij} - \frac{1}{3}\epsilon\, \delta_{ij} \tag{4.7b}$$

Use of (4.7) allows us to rewrite (4.4) in the form:

$$\sigma\, d\epsilon + s_{ij}\, de_{ij} - \phi\, dp - S_s\, dT - dG_s = 0 \tag{4.8}$$

so that (4.5) can be written equivalently:

$$G_s = G_s(\epsilon, e_{ij}, p, T): \\ \sigma = \frac{\partial G_s}{\partial \epsilon}; \quad s_{ij} = \frac{\partial G_s}{\partial e_{ij}}; \quad \phi = -\frac{\partial G_s}{\partial p}; \quad S_s = -\frac{\partial G_s}{\partial T} \tag{4.9}$$

In addition, let H_s be defined by:

$$H_s = \sigma\epsilon + s_{ij}e_{ij} - G_s \tag{4.10}$$

We get:

$$\epsilon\, d\sigma + e_{ij}\, ds_{ij} + \phi\, dp + S_s\, dT - dH_s = 0 \tag{4.11}$$

providing the inversion of state equations in the form:

$$H_s = H_s(\sigma, s_{ij}, p, T):$$
$$\epsilon = \frac{\partial H_s}{\partial \sigma}; \quad e_{ij} = \frac{\partial H_s}{\partial s_{ij}}; \quad \phi = \frac{\partial H_s}{\partial p}; \quad S_s = \frac{\partial H_s}{\partial T} \tag{4.12}$$

4.1.2 Tangent Thermoporoelastic Properties

Differentiating state equations (4.5) and taking into account Maxwell's symmetry relations (4.6), we obtain:

$$d\sigma_{ij} = C_{ijkl}\, d\varepsilon_{kl} - b_{ij}\, dp - C_{ijkl}\alpha_{kl}\, dT \tag{4.13a}$$

$$d\phi = b_{ij}\, d\varepsilon_{ij} + \frac{dp}{N} - 3\alpha_\phi\, dT \tag{4.13b}$$

$$dS_s = C_{ijkl}\alpha_{kl}\, d\varepsilon_{ij} - 3\alpha_\phi\, dp + C\,\frac{dT}{T} \tag{4.13c}$$

In (4.13) C_{ijkl}, b_{ij}, $C_{ijkl}\alpha_{kl}$, $1/N$, $3\alpha_\phi$ and C are the thermoporoelastic tangent properties. They are functions of state variables ε_{ij}, p and T and must satisfy the relations expressing the integrability of (4.13). For instance, the relation expressing the integrability of (4.13a) is:[2]

$$\frac{\partial C_{ijkl}}{\partial \varepsilon_{mn}} = \frac{\partial C_{ijmn}}{\partial \varepsilon_{kl}}; \quad \frac{\partial C_{ijkl}}{\partial p} = -\frac{\partial b_{ij}}{\partial \varepsilon_{kl}}; \quad \frac{\partial C_{ijkl}}{\partial T} = -\frac{\partial (C_{ijmn}\alpha_{mn})}{\partial \varepsilon_{kl}} \tag{4.14}$$

Letting $dp = 0$ in (4.13a) and (4.13c), we recover the standard equations of incremental thermoelasticity:

- $C_{ijkl} = \partial^2 G_s / \partial \varepsilon_{ij} \partial \varepsilon_{kl}$ is the $ijkl^{th}$ component of the tensor of skeleton tangent elastic stiffness moduli. Owing to Maxwell's symmetry relations (4.6) and to symmetry conditions $\sigma_{ij} = \sigma_{ji}$ and $\varepsilon_{kl} = \varepsilon_{lk}$, the tangent elastic moduli C_{ijkl} admit the following symmetries:

$$C_{ijkl} = C_{klij}; \quad C_{ijkl} = C_{ijlk}; \quad C_{ijkl} = C_{jikl} \tag{4.15}$$

  Owing to symmetries (4.15), among the 81 components of C_{ijkl} only 21 are eventually independent.
- α_{kl} is the kl^{th} component of the tensor of skeleton tangent thermal dilation coefficients, with symmetry $\alpha_{kl} = \alpha_{lk}$. Owing to Maxwell's symmetry relations (4.6), the term $TC_{ijkl}\alpha_{kl} = -T\partial^2 G_s / \partial T \partial \varepsilon_{ij}$ also represents the skeleton tangent strain latent heat, that is the heat per unit of strain that the skeleton exchanges with the outside in an evolution when both temperature and pressure are held constant ($dT = dp = 0$).

[2] The other relations are: $\frac{\partial B_{ij}}{\partial p} = \frac{\partial (1/N)}{\partial \varepsilon_{ij}}$; $\frac{\partial B_{ij}}{\partial T} = -3\frac{\partial \alpha_\phi}{\partial \varepsilon_{ij}}$; $\frac{\partial (1/N)}{\partial T} = -3\frac{\partial \alpha_\phi}{\partial p}$; $\frac{\partial (C/T)}{\partial p} = -3\frac{\partial \alpha_\phi}{\partial T}$; $\frac{\partial (C/T)}{\partial \varepsilon_{ij}} = \frac{\partial \left(C_{ijkl}\alpha_{kl}\right)}{\partial T}$; $-3\frac{\partial \alpha_\phi}{\partial \varepsilon_{ij}} = \frac{\partial \left(C_{ijkl}\alpha_{kl}\right)}{\partial p}$.

- $C = -T\partial^2 G_s/\partial T^2$ is the skeleton tangent volumetric heat capacity, when strain ε_{ij} and pressure p are held constant ($d\varepsilon_{ij} = dp = 0$).

With regard to thermoelasticity, thermoporoelasticity includes the incremental state equation (4.13b) related to the change in porosity involving new thermoporoelastic properties:

- $b_{ij} = -\partial^2 G_s/\partial\varepsilon_{ij}\partial p$ is the ij^{th} component of Biot's tangent tensor with symmetry $b_{ij} = b_{ji}$. It linearly relates the change in porosity to the strain variation when both pressure and temperature are held constant ($dp = dT = 0$). Owing to Maxwell's symmetry relations (4.6), $-b_{ij}$ also linearly relates the stress increment to the pressure increment in an evolution when both strain and temperature are held constant ($d\varepsilon_{ij} = dT = 0$).
- $1/N = -\partial^2 G_s/\partial p^2$ is the inverse of Biot's tangent modulus linking the pressure variation dp and the porosity variation $d\phi$ in an evolution when both strain and temperature are held constant ($d\varepsilon_{ij} = dT = 0$).
- $3\alpha_\phi = \partial^2 G_s/\partial p\partial T$ stands for a volumetric thermal dilation coefficient related to the porosity. Owing to Maxwell's symmetry relations (4.6), the term $-3T\alpha_\phi$ also represents the skeleton tangent pressure latent heat, that is the heat per unit of pressure p that the skeleton exchanges with the outside in an evolution when both temperature and strain are held constant ($d\varepsilon_{ij} = dT = 0$).

4.1.3 The Incompressible Matrix and the Effective Stress

In the absence of any occluded porosity, the solid grains forming the matrix generally undergo negligible volume changes so that the overall skeleton volumetric change $\epsilon = \varepsilon_{ii}$ reduces to the change in porosity ϕ. Differentiation of (1.33) finally gives:

$$d\phi = d\epsilon \tag{4.16}$$

State equation (4.13b) must reduce to the incompressibility condition (4.16) whatever the values of increments dp and dT, yielding:

$$b_{ij} = \delta_{ij}; \quad \frac{1}{N} = \alpha_\phi = 0 \tag{4.17}$$

In the case of an incompressible matrix, incremental state equations (4.13) eventually simplify according to:

$$d(\sigma_{ij} + p\,\delta_{ij}) = C_{ijkl}\,d\varepsilon_{kl} - C_{ijkl}\alpha_{kl}\,dT \tag{4.18a}$$

$$d\phi = d\varepsilon_{ii} \tag{4.18b}$$

$$dS_s = C\,\frac{dT}{T} + C_{ijkl}\alpha_{kl}\,d\varepsilon_{ij} \tag{4.18c}$$

In agreement with (3.78) we determine that $\sigma'_{ij} = \sigma_{ij} + p\,\delta_{ij}$ plays the role of an effective stress for the skeleton: irrespective of pore pressure p, the state equations of an ordinary solid subjected to stress σ'_{ij} would be written in the same form as (4.18a) and (4.18c) with regard to the linearized strain ε_{ij}.

4.2 Linear Thermoporoelastic Skeleton

4.2.1 Linear Thermoporoelasticity

Linear thermoporoelasticity consists in setting the tangent properties constant. Constitutive equations (4.13) can be integrated in the form:

$$\sigma_{ij} - \sigma_{ij}^0 = C_{ijkl}\,\varepsilon_{kl} - b_{ij}(p - p_0) - C_{ijkl}\alpha_{kl}(T - T_0) \tag{4.19a}$$

$$\phi - \phi_0 = b_{ij}\,\varepsilon_{ij} + \frac{p - p_0}{N} - 3\alpha_\phi(T - T_0) \tag{4.19b}$$

$$S_s - S_s^0 = C_{ijkl}\alpha_{kl}\,\varepsilon_{ij} - 3\alpha_\phi(p - p_0) + \frac{C}{T_0}(T - T_0) \tag{4.19c}$$

where σ_{ij}^0, p_0 and T_0 stand respectively for the initial stress, pressure and temperature, whereas (4.19c) assumes small variations of temperature, that is $(T - T_0)/T_0 \ll 1$.

4.2.2 Isotropic Linear Thermoporoelasticity

A material is isotropic when no material frame is preferred to formulate its constitutive equations. As a consequence energy functions can depend only on the scalar invariants of the involved tensors. Linear thermoporoelasticity consists in choosing a quadratic expression with regard to those invariants for the energy functions. Accordingly, for an isotropic linear thermoporoelastic material, only the first invariant of the strain tensor, namely $\epsilon = \varepsilon_{ii}$, and the second invariant of its deviator, namely $e_{ij}e_{ji}$, are to be considered in the expression of the energy function G_s involved in (4.5), that is:

$$\begin{aligned} G_s = {} & \sigma_0\epsilon + s_{ij}^0\,e_{ij} - \phi_0\,p - S_s^0 T \\ & + \frac{1}{2}K\epsilon^2 - b(p - p_0)\epsilon - 3\alpha K(T - T_0)\epsilon \\ & + 3\alpha_\phi(p - p_0)(T - T_0) - \frac{1}{2}\frac{(p - p_0)^2}{N} - \frac{1}{2}\frac{C}{T_0}(T - T_0)^2 + \mu\, e_{ij}e_{ji} \end{aligned} \tag{4.20}$$

Isotropic state equations are derived from (4.9) and (4.20) in the form:

$$\sigma - \sigma_0 = K\epsilon - b(p - p_0) - 3\alpha K(T - T_0) \tag{4.21a}$$

$$s_{ij} - s_{ij}^0 = 2\mu\, e_{ij} \tag{4.21b}$$

$$\phi - \phi_0 = b\epsilon + \frac{p - p_0}{N} - 3\alpha_\phi(T - T_0) \tag{4.21c}$$

$$S_s - S_s^0 = 3\alpha K\epsilon - 3\alpha_\phi(p - p_0) + \frac{C}{T_0}(T - T_0) \tag{4.21d}$$

In (4.21) K and μ are the skeleton bulk and shear moduli while 3α is the volumetric skeleton thermal dilation coefficient. Equations (4.21a) and (4.21b) can be conveniently condensed according to:

$$\sigma_{ij} - \sigma_{ij}^0 = \left(K - \frac{2}{3}\mu\right)\epsilon\,\delta_{ij} + 2\mu\,\varepsilon_{ij} - b(p - p_0)\delta_{ij} - 3\alpha K(T - T_0)\delta_{ij} \tag{4.22}$$

Owing to the form (4.22) of the constitutive equations, it can sometimes be more convenient to consider the couple of Lamé coefficients, that is $\lambda = K - \frac{2}{3}\mu$ and μ, instead of the couple of properties K and μ.

Conditions that are often encountered in the geosciences (see §3.4.1 and §5.4.4) are the 'oedometric' conditions where the only non-zero strain component is ε_{zz} in the vertical direction $0z$. Accordingly constitutive equation (4.22) gives:

$$\sigma_{zz} - \sigma_{zz}^0 = \left(K + \frac{4}{3}\mu\right)\varepsilon_{zz} - b(p - p_0) - 3\alpha K(T - T_0) \tag{4.23}$$

Referring to the strain conditions that led to (4.23), the modulus $K + \frac{4}{3}\mu$ is called the oedometric modulus. Thermoelastic properties K (or $K + \frac{4}{3}\mu$), μ and 3α can eventually be determined by the same experiments as for a standard solid, provided that the pressure p is maintained at its initial value p_0. Indeed these properties are those of the 'dry' porous material.

Coefficient b is Biot's coefficient. According to (4.21c) it quantifies the part $b\epsilon$ of the volumetric strain ϵ caused by the change in porosity, when both pressure and temperature are held constant ($dp = dT = 0$). In the context of linear thermoporoelasticity Biot's coefficient allows us to extend Terzaghi's effective stress to the case of a compressible matrix. Indeed (4.21a) can be rewritten in the form:

$$\sigma'' - \sigma_0'' = K\epsilon - 3\alpha K(T - T_0) \tag{4.24}$$

and (4.22) inverted in the form:

$$\varepsilon_{ij} = \frac{1+\nu}{E}(\sigma_{ij}'' - \sigma_{ij}''^0) - \frac{\nu}{E}(\sigma_{kk}'' - \sigma_{kk}''^0)\delta_{ij} - \alpha(T - T_0)\delta_{ij} \tag{4.25}$$

where σ_{ij}'' denotes Biot's effective stress defined by:

$$\sigma_{ij}'' = \sigma_{ij} + bp\,\delta_{ij}; \quad \sigma'' = \frac{1}{3}\sigma_{kk}'' \tag{4.26}$$

and where E and ν are the skeleton Young modulus and the skeleton Poisson coefficient with:

$$E = \mu\frac{9K}{3K + \mu}; \quad \nu = \frac{3K - 2\mu}{2(3K + \mu)} \tag{4.27}$$

According to (4.24), a variation $d\sigma''$ of the mean effective stress results in the same variation $d\epsilon$ of the volumetric strain, whether the variation $d\sigma''$ is achieved through a mean stress variation, namely $d\sigma'' = d\sigma$, or through a pressure variation, namely $d\sigma'' = b\,dp$. This remark provides the experimental means to determine Biot's coefficient b through a loading path constituted by successive two-step incremental loadings—a stress loading ($d\sigma = d\omega, dp = 0$) followed by a pore pressure loading ($d\sigma = 0, dp = d\omega$), and so on (see Fig. 4.1a). The ratio between the successive strain responses is identified as coefficient b. With b so determined between two successive incremental loadings, Biot's effective mean pressure $-(\sigma'' - \sigma_0'') = -(\sigma - \sigma_0 + b(p - p_0))$ plotted against the volumetric contraction $\varepsilon = -\epsilon$ must be a straight line. Figure 4.1b represents such a plot for a limestone sample.

A comparison between (4.19) and (4.21) shows that Biot's modulus N, coefficient α_ϕ and heat capacity C have the same meaning whether the material is isotropic or not. Equations (4.21c) and (4.21d) indicate that their measurement requires experiments where the strain is held constant. Experiments where the stress is held constant (at atmospheric pressure) are obviously easier to carry out. The combination of (4.21a) and (4.21d) allows us to express the entropy variation as a function of the volumetric stress σ instead of the

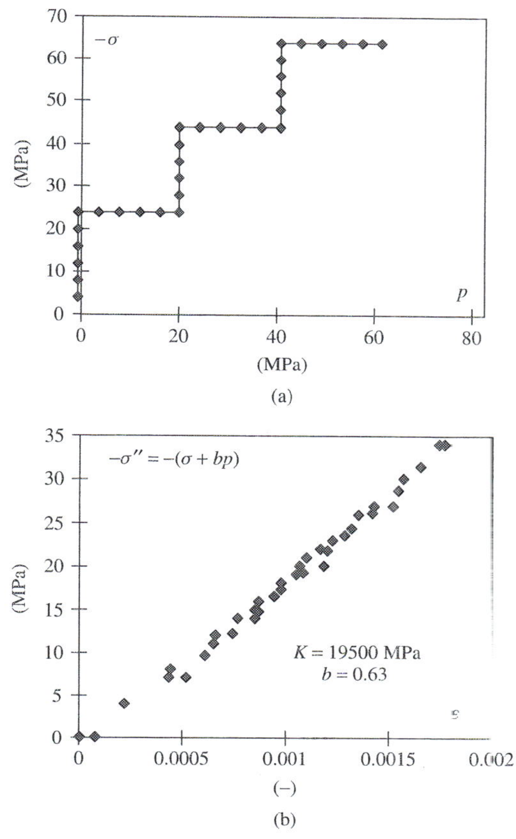

Figure 4.1: Determination of Biot's coefficient b and skeleton bulk modulus K for a limestone sample. (a) Loading path, (b) $-\sigma'' = -(\sigma + bp)$ plotted against $\varepsilon = -\epsilon$. Use of Biot's relation, that is the first relation in (4.35), gives the value $K_s = 52700$ MPa for the solid matrix bulk modulus (from Bouteca, Sarda (1995) © Sweets & Zeitlinger, see footnote).

Table 4.1: Order of magnitude of thermoporoelastic properties for different materials (after Heukamp, Ulm (2002), Bouteca, Sarda (1995) and Cowin (1998), see footnote)

Material	ϕ (%)	K (MPa × 10^3)	b (–)	N (MPa × 10^3)
Cement paste	40–63	15–2	0.07–0.37	1170–20
Mortar	27–40	15–3	0.04–0.35	2340–40
Bone	5	12	0.14	160
Granites	1–2	25–35	0.22–0.44	280–370
Marble	2	40	0.20	280
Sandstones	2–26	4.6–13	0.69–0.85	~17
Limestones	4–29	5–39	0.34–0.88	100–400

volumetric strain ϵ:

$$S_s - S_s^0 = 3\alpha(\sigma - \sigma_0) + 3(\alpha b - \alpha_\phi)(p - p_0) + \frac{C_\sigma}{T_0}(T - T_0) \tag{4.28}$$

In (4.28) $C_\sigma = C + 9T_0\alpha^2 K$ is the volumetric heat capacity at constant stress and is eventually the heat capacity measured in practice, although in most cases the difference $C_\sigma - C = 9T_0\alpha^2 K$ turns out to be negligible.

Let us finally recall that the thermodynamic stability of the skeleton requires the fulfilment of conditions $K > 0$, $\mu > 0$ or equivalently $E > 0$ and $-1 < \nu < \frac{1}{2}$, and $C > 0$, ensuring that unavoidable fluctuations in strain and temperature within the material will not spontaneously amplify (see §5.4.1). Coupling coefficients b and α are a priori not subjected to any particular condition. Typical values of poroelastic properties are given in Table 4.1.[3]

4.2.3 Relations Between Skeleton and Matrix Properties

So far the skeleton properties have been introduced without reference to the matrix properties. Nevertheless compatibility relations do exist between them, since the former necessarily result from the latter and from the geometry of the porous space. We will start first by addressing the compatibility relations between the elastic properties, not considering the thermal properties for a while.

Let us first recall the strain partition:

$$\epsilon = (1 - \phi_0)\,\epsilon_s + \phi - \phi_0 \tag{4.29}$$

[3]For data related to cement-based materials see Heukamp F., Ulm F.-J. (2002), 'Chemomechanics of Calcium Leaching of Cement-Based Materials at Different Scales: the Role of CH-Dissolution and C-S-H-Degradation on Strength and Durability Performance of Materials and Structures', MIT-CEE Report, **R002-03** (*PhD Thesis*, Massachusetts Institute of Technology). The range of values relates from an intact material to a material whose calcium has been completely leached out (see Fig. 7.1). For data related to rocks see Bouteca M., Sarda J.-P. (1995), 'Experimental measurements of thermoporoelastic coefficients', in *Mechanics of Porous Media*, ed. Charlez, P., Balkema, Rotterdam. For rocks see also Carmichäel R.S. (1982), *CRC Handbook of Physical Properties of Rocks*, volume II, CRC Press, Boca Raton, FL. For data related to bone see Cowin S.C. (1998), 'Bone fluid poroelasticity', *Poromechanics, A tribute to M.A. Biot*, Proceedings of the Biot Conference on Poromechanics, ed. J.-F. Thymus *et al.*, Balkema, Rotterdam.

and the stress partition (2.34), which, under the assumption of infinitesimal transformations, gives:

$$\sigma - \sigma^0 = (1 - \phi_0)\left(\sigma_s - \sigma_s^0\right) - \phi_0(p - p_0) \tag{4.30}$$

where σ_s stands for the mean stress related to the matrix at the mesoscale. We assume now that the matrix is homogeneous and linearly elastic, that is:

$$\sigma_s - \sigma_s^0 = K_s \epsilon_s \tag{4.31}$$

where K_s is the matrix bulk modulus. Under isothermal conditions (4.21a) reduces to:

$$\sigma - \sigma^0 = K\epsilon - b(p - p_0) \tag{4.32}$$

$$\phi - \phi_0 = b\epsilon + \frac{p - p_0}{N} \tag{4.33}$$

A combination of (4.29)–(4.33) gives:

$$(\sigma - \sigma^0)\left[\frac{1-b}{K} - \frac{1}{K_s}\right] = (p - p_0)\left[\frac{1}{N} + \frac{\phi_0}{K_s} - \frac{b\,(1-b)}{K}\right] \tag{4.34}$$

The loading variables $\sigma - \sigma^0$ and $p - p_0$ are independent state variables. Consequently both factors affecting $\sigma - \sigma^0$ and $p - p_0$ in (4.34) must be zero, yielding the compatibility relations:[4]

$$b = 1 - \frac{K}{K_s}; \quad \frac{1}{N} = \frac{b - \phi_0}{K_s} \tag{4.35}$$

In turn, since the bulk modulus K and Biot's coefficient b can be measured at the macroscopic level, the first of relations (4.35) provides the means to assess the value of matrix bulk modulus K_s (see Fig. 4.1). Alternatively, we can also perform an experiment where the mean stress is compressive and has a strength equal to the pore pressure intensity, that is an experiment where $-(\sigma - \sigma_0) = p - p_0$. For such an isothermal experiment, using (4.29)–(4.33) and the first of relations (4.35), we derive:

$$-(\sigma - \sigma^0) = p - p_0 : \ \epsilon = \epsilon_s; \quad \sigma - \sigma_0 = K_s \epsilon \tag{4.36}$$

Matrix modulus K_s can be determined as the coefficient linearly linking the volumetric dilation ϵ and the applied stress $-(\sigma - \sigma^0) = p - p_0$. It is worthwhile to note that a departure of K_s from the bulk modulus of the solid grains forming the matrix will infer the existence of an occluded porosity.

[4]These relations were originally derived in Biot M.A., Willis D.G. (1957), 'The elastic coefficients of theory of consolidation', *Journal of Applied Mechanics*, **24**, 594–601.

In addition to compatibility relations (4.35), K and μ admit upper bounds as a function of bulk and shear moduli K_s and μ_s of the matrix and initial porosity ϕ_0:

$$0 \leq K \leq (1-\phi_0)K_s \frac{4\mu_s}{4\mu_s + 3\phi_0 K_s} \tag{4.37a}$$

$$0 \leq \mu_s \leq (1-\phi_0)\mu_s \frac{9K_s + 8\mu_s}{9K_s + 8\mu_s + 6\phi_0(K_s + 2\mu_s)} \tag{4.37b}$$

There can be no question here of entering the details of the theories based on variational energy approaches carried out at the matrix scale leading to the above bounds.[5] However, the upper bound for K can be qualitatively explained. To this end consider a hollow sphere of external radius R_e and internal radius R_i. The constitutive material is linearly elastic with K_s and μ_s as the bulk and shear moduli. Starting from a free-stress state, this spherical envelope is submitted to a traction of strength σ on the external spherical envelop. Noting that $\phi_0 = R_i^3/R_e^3$, the void volume ratio with respect to the whole sphere, the radial displacement ξ of the external envelope is given by the standard elastic solution:

$$\xi = \frac{\sigma R_e}{1-\phi_0}\left(\frac{1}{3K_s} + \frac{\phi_0}{4\mu_s}\right) \tag{4.38}$$

Since the apparent volumetric dilation ϵ of the external envelop is:

$$\epsilon = \frac{(R_e+\xi)^3 - R_e^3}{R_e^3} \simeq \frac{3\xi}{R_e} \tag{4.39}$$

we write:

$$\sigma = (1-\phi_0)K_s \frac{4\mu_s}{4\mu_s + 3\phi_0 K_s}\epsilon \tag{4.40}$$

This relation finally applies to the overall behaviour of the porous material made up by the progressive filling of the space by hollow spheres whose radii are required to satisfy $R_i^3/R_e^3 = \phi_0$. According to (4.37a) and (4.40), the bulk modulus of such a material, as the most ideally isotropic porous material, eventually is an upper bound for all the possible values of the bulk modulus.

In order to state the compatibility relations concerning the thermal properties, analogously to (4.28), we write:

$$\mathfrak{S}_s - \mathfrak{S}_s^0 = 3\alpha_s(\sigma_s - \sigma_s^0) + \frac{C_{\sigma_s}}{T_0}(T - T_0) \tag{4.41}$$

where $\mathfrak{S}_s$ is the solid matrix entropy while C_{σ_s} is the solid matrix volumetric heat capacity at constant stress. Since entropy is an extensive quantity, we also write:

$$S_s - S_s^0 = (1-\phi_0)(\mathfrak{S}_s - \mathfrak{S}_s^0) \tag{4.42}$$

[5] See Hashin Z. (1983), 'Analysis of composite materials—A survey', *Journal of Applied Mechanics*, **50**, 481–504.

Combining (4.30), (4.41) and (4.42) we derive:

$$S_s - S_s^0 = 3\alpha_s(\sigma - \sigma^0) + 3\alpha_s\phi_0(p - p_0) + \frac{(1-\phi_0)C_{\sigma_s}}{T_0}(T - T_0) \tag{4.43}$$

A comparison between (4.28) and (4.43) provides the relations:

$$\alpha = \alpha_s; \quad \alpha_\phi = \alpha_s(b - \phi_0); \quad C_\sigma = (1-\phi_0)C_{\sigma_s} \tag{4.44}$$

It is instructive to note that, when deriving compatibility relations (4.44), no use has been made of skeleton state equations (4.21a) and (4.21c), nor the matrix constitutive equation:

$$\sigma_s - \sigma_s^0 = K_s\epsilon_s - 3\alpha_s K_s(T - T_0) \tag{4.45}$$

However, Maxwell's symmetry relations ensure that the first two relations in (4.44) can be used in (4.21a) and (4.21c).

4.2.4 Anisotropic Poroelasticity

In many common applications the porous material is anisotropic in the initial reference configuration. The origin of anisotropy may be twofold. In the first place the anisotropy may be due to the anisotropy of both the matrix and the porous geometry. In the second place it may also be due to an anisotropic pre-loading, as for instance gravity, which has induced a prestressed anisotropic reference state with regard to the original stress-free state. Two main kinds of linear anisotropy are presented below, namely orthotropy and transverse isotropy. We restrict ourselves to linear anisotropic poroelasticity. The extension to anisotropic thermoporoelasticity can be achieved by letting the temperature T play an analogous role to that of the pore pressure p.

Orthotropic poroelastic material

An orthotropic material admits three planes of material symmetry orthogonal to each other, as sketched in Fig. 4.2a. Let us adopt the orthonormal basis formed from the unit vectors $\mathbf{e}_1$, $\mathbf{e}_2$ and $\mathbf{e}_3$ parallel to the three intersection lines between a set of three material

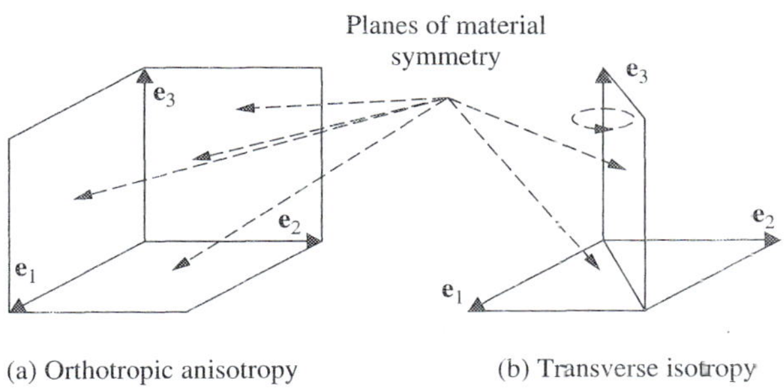

Figure 4.2: Different types of anisotropic materials.

symmetry planes. Since the constitutive equations of an orthotropic material must be irrespective of any change in orientation of vectors $\mathbf{e}_i$, in linear poroelasticity we write:

$$\sigma_{11} - \sigma_{11}^0 = c_{11}\varepsilon_{11} + c_{12}\varepsilon_{22} + c_{13}\varepsilon_{33} - b_1(p - p_0) \tag{4.46a}$$

$$\sigma_{22} - \sigma_{22}^0 = c_{12}\varepsilon_{11} + c_{22}\varepsilon_{22} + c_{23}\varepsilon_{33} - b_2(p - p_0) \tag{4.46b}$$

$$\sigma_{33} - \sigma_{33}^0 = c_{13}\varepsilon_{11} + c_{23}\varepsilon_{22} + c_{33}\varepsilon_{33} - b_3(p - p_0) \tag{4.46c}$$

$$\sigma_{12} - \sigma_{12}^0 = c_{66}\varepsilon_{12} \tag{4.46d}$$

$$\sigma_{13} - \sigma_{13}^0 = c_{55}\varepsilon_{13} \tag{4.46e}$$

$$\sigma_{23} - \sigma_{23}^0 = c_{44}\varepsilon_{23} \tag{4.46f}$$

$$\phi - \phi_0 = b_1\varepsilon_{11} + b_2\varepsilon_{22} + b_3\varepsilon_{33} + \frac{p - p_0}{N} \tag{4.46g}$$

where the components C_{ijkl} of the stiffness tensor and the moduli c_{ab} relate to each other according to the index correspondence:

$$ii = 11 \rightarrow a = 1; \quad ii = 22 \rightarrow a = 2; \quad ii = 33 \rightarrow a = 3$$

$$ij = 23 \text{ or } 32 \rightarrow a = 2; \quad ij = 13 \text{ or } 31 \rightarrow 5; \quad ij = 12 \text{ or } 21 \rightarrow 6$$

By analogy to the isotropic case we can invert (4.46a)–(4.46f) in the form:

$$\varepsilon_{11} = \frac{1}{E_1}\left(\sigma_{11}'' - \sigma_{11}''^0\right) - \frac{\nu_{21}}{E_2}\left(\sigma_{22}'' - \sigma_{22}''^0\right) - \frac{\nu_{31}}{E_3}\left(\sigma_{33}'' - \sigma_{33}''^0\right) \tag{4.47a}$$

$$\varepsilon_{22} = \frac{1}{E_2}\left(\sigma_{22}'' - \sigma_{22}''^0\right) - \frac{\nu_{12}}{E_1}\left(\sigma_{11}'' - \sigma_{11}''^0\right) - \frac{\nu_{32}}{E_3}\left(\sigma_{33}'' - \sigma_{33}''^0\right) \tag{4.47b}$$

$$\varepsilon_{33} = \frac{1}{E_3}\left(\sigma_{33}'' - \sigma_{33}''^0\right) - \frac{\nu_{13}}{E_1}\left(\sigma_{11}'' - \sigma_{11}''^0\right) - \frac{\nu_{23}}{E_2}\left(\sigma_{22}'' - \sigma_{22}''^0\right) \tag{4.47c}$$

$$\varepsilon_{12} = \frac{1}{2\mu_{12}}\left(\sigma_{12}'' - \sigma_{12}''^0\right) \tag{4.47d}$$

$$\varepsilon_{13} = \frac{1}{2\mu_{13}}\left(\sigma_{13}'' - \sigma_{13}''^0\right) \tag{4.47e}$$

$$\varepsilon_{23} = \frac{1}{2\mu_{23}}\left(\sigma_{23}'' - \sigma_{23}''^0\right) \tag{4.47f}$$

where σ_{ij}'' is Biot's effective stress, namely:

$$\sigma_{ij}'' = \sigma_{ij} + b_i\, p\, \delta_{ij} \quad \text{(no summation on repeated index)} \tag{4.48}$$

and where E_i and ν_{ij} are the skeleton Young moduli and the Poisson coefficients satisfying:

$$\nu_{ij}E_j = E_i\nu_{ji} \tag{4.49}$$

Transversely isotropic material

A transversely isotropic (or axisymmetric orthotropic) material admits a rotational material symmetry around an axis and the material planes of symmetry formed by the set of planes either orthogonal to the axis or including it. Let us adopt the orthonormal basis formed from the unit vectors $\mathbf{e}_1$, $\mathbf{e}_2$ and $\mathbf{e}_3$, so that $\mathbf{e}_1$ and $\mathbf{e}_2$ lie in the plane orthogonal to the axis of symmetry whereas $\mathbf{e}_3$ belongs to the latter, as sketched in Fig. 4.2b. Since the constitutive equations of a transversely isotropic material must be irrespective of any change in orientation of vectors $\mathbf{e}_i$ or any rotation around the axis of symmetry, in linear poroelasticity we write:

$$\sigma_{11} - \sigma_{11}^0 = c_{11}\varepsilon_{11} + c_{12}\varepsilon_{22} + c_{13}\varepsilon_{33} - b(p - p_0) \tag{4.50a}$$

$$\sigma_{22} - \sigma_{22}^0 = c_{12}\varepsilon_{11} + c_{11}\varepsilon_{22} + c_{13}\varepsilon_{33} - b(p - p_0) \tag{4.50b}$$

$$\sigma_{33} - \sigma_{33}^0 = c_{13}\varepsilon_{11} + c_{13}\varepsilon_{22} + c_{33}\varepsilon_{33} - b_3(p - p_0) \tag{4.50c}$$

$$\sigma_{12} - \sigma_{12}^0 = (c_{11} - c_{22})\varepsilon_{12} \tag{4.50d}$$

$$\sigma_{13} - \sigma_{13}^0 = c_{44}\varepsilon_{13} \tag{4.50e}$$

$$\sigma_{23} - \sigma_{23}^0 = c_{44}\varepsilon_{23} \tag{4.50f}$$

$$\phi - \phi_0 = b(\varepsilon_{11} + \varepsilon_{22}) + b_3\varepsilon_{33} + \frac{p - p_0}{N} \tag{4.50g}$$

By analogy to the isotropic case we can invert (4.50a)–(4.50f) in the form:

$$\varepsilon_{11} = \frac{1}{E}\left(\sigma_{11}'' - \sigma_{11}''^0\right) - \frac{\nu}{E}\left(\sigma_{22}'' - \sigma_{22}''^0\right) - \frac{\nu_3}{E}\left(\sigma_{33}'' - \sigma_{33}''^0\right) \tag{4.51a}$$

$$\varepsilon_{22} = \frac{1}{E}\left(\sigma_{22}'' - \sigma_{22}''^0\right) - \frac{\nu}{E}\left(\sigma_{11}'' - \sigma_{11}''^0\right) - \frac{\nu_3}{E_3}\left(\sigma_{33}'' - \sigma_{33}''^0\right) \tag{4.51b}$$

$$\varepsilon_{33} = \frac{1}{E_3}\left(\sigma_{33}'' - \sigma_{33}''^0\right) - \frac{\nu}{E}\left(\sigma_{11}'' - \sigma_{11}''^0\right) - \frac{\nu}{E}\left(\sigma_{22}'' - \sigma_{22}''^0\right) \tag{4.51c}$$

$$\varepsilon_{12} = \frac{1+\nu}{2E}\left(\sigma_{12}'' - \sigma_{12}''^0\right) \tag{4.51d}$$

$$\varepsilon_{13} = \frac{1+\nu}{2E}\left(\sigma_{13}'' - \sigma_{13}''^0\right) \tag{4.51e}$$

$$\varepsilon_{23} = \frac{1+\nu_3}{2E_3}\left(\sigma_{23}'' - \sigma_{23}''^0\right) \tag{4.51f}$$

where σ_{ij}'' is Biot's effective stress, namely:

$$\sigma_{ij}'' = \sigma_{ij} + b_i\, p\, \delta_{ij} \quad (b_1 = b_2 = b, \text{ no summation on repeated index}) \tag{4.52}$$

4.3 Thermoporoelastic Porous Material

4.3.1 Constitutive Equations of the Saturating Fluid

The skeleton constitutive equations are irrespective of the nature and of the constitutive equations of the saturating fluid, as far as only contact forces are considered between the skeleton and the fluid so that the latter exerts a pore pressure p on the internal walls forming the porous network. Conversely the constitutive equations of the fluid are irrespective of the skeleton ones. Their explicit expressions can be obtained by proceeding as for the skeleton. Indeed, differentiating fluid state equations (3.10), we write:

$$\frac{d\rho_f}{\rho_f} = \frac{dp}{K_f} - 3\alpha_f dT; \quad ds_f = -3\alpha_f \frac{dp}{\rho_f} + C_p \frac{dT}{T} \tag{4.53}$$

K_f is the fluid tangent bulk modulus, $3\alpha_f$ the fluid tangent coefficient of volumetric thermal dilation, while C_p is the fluid tangent volumetric specific heat capacity at constant pressure. The limit case of an incompressible fluid is obtained by letting $K_f \to \infty$ and $\alpha_f = 0$ in (4.53), yielding:

$$\rho_f = \rho_f^0; \quad s_f = s_f^0 + C_p \ln \frac{T}{T_0} \tag{4.54}$$

where index 0 relates to reference values.

If the saturating fluid is an ideal gas, we write:

$$p = \frac{RT}{\mathcal{M}_f} \rho_f \tag{4.55}$$

where $\mathcal{M}_f$ is the molar mass of the gas considered. Differentiation of (4.55) and a comparison of the resulting equation with the first equation in (4.53) provide the identification:

$$K_f = p; \quad 3\alpha_f = \frac{1}{T} \tag{4.56}$$

Assuming in addition that C_p is a constant, we can integrate the second equation of (4.53) to get:

$$s_f - s_f^0 = -\frac{R}{\mathcal{M}_f} \ln \frac{p}{p_0} + C_p \ln \frac{T}{T_0} \tag{4.57}$$

4.3.2 Constitutive Equations of the Porous Material

Constitutive equations (3.69) and (3.70) of the porous material viewed as an open thermodynamic system involve the current fluid mass content m_f. We can get rid of porosity ϕ to the benefit of m_f by using both the condition (1.64) of complete saturation and the fluid state equations when the latter are explicitly known over the whole range of variations of state variables p and T (e.g. (4.55) and (4.57)b in the case of an ideal gas).

Otherwise, since condition (1.64) for complete saturation remains constantly fulfilled, we differentiate the latter to get:

$$\frac{dm_f}{\rho_f} = d\phi + \phi\frac{d\rho_f}{\rho_f} \tag{4.58}$$

Combining (4.53) and (4.58) yields:

$$d\phi = \frac{dm_f}{\rho_f} - \phi\frac{dp}{K_f} + 3\phi\,\alpha_f dT \tag{4.59a}$$

$$d(m_f s_f) = s_f dm_f - 3\phi\,\alpha_f dp + m_f C_p\frac{dT}{T} \tag{4.59b}$$

The latter can be substituted into (4.13) to eliminate the porosity variation $d\phi$ to the benefit of the variation in fluid mass content dm_f, yielding:

$$d\sigma_{ij} = C_{ijkl}d\varepsilon_{kl} - b_{ij}dp - C_{ijkl}\alpha_{kl}dT \tag{4.60a}$$

$$\frac{dm_f}{\rho_f} = b_{ij}d\varepsilon_{ij} + \frac{1}{M}dp - 3\alpha_m dT \tag{4.60b}$$

$$dS = s_f dm_f + C_{ijkl}\alpha_{kl}d\varepsilon_{ij} - 3\alpha_m dp + C_d\frac{dT}{T} \tag{4.60c}$$

where $S = S_s + m_f s_f$ and where we note:

$$\frac{1}{M} = \frac{1}{N} + \frac{\phi}{K_f}; \quad \alpha_m = \alpha_\phi + \phi\alpha_f; \quad C_d = C + m_f C_p \tag{4.61}$$

Accordingly, for a linear isotropic poroelastic skeleton, we write:

$$d\sigma = K\,d\epsilon - b\,dp - 3\alpha K\,dT; \qquad ds_{ij} = 2\mu\,de_{ij} \tag{4.62a}$$

$$\frac{dm_f}{\rho_f} = b\,d\epsilon + \frac{dp}{M} - 3\alpha_m\,dT \tag{4.62b}$$

$$dS = s_f dm_f + 3\alpha K\,d\epsilon - 3\alpha_m dp + C_d\frac{dT}{T} \tag{4.62c}$$

With respect to the tangent properties of the porous material, the skeleton properties K, α and μ appear to be the 'drained' properties, that is the properties which can be measured in tests where the fluid pressure is held constant ($dp = 0$). In addition the drained tangent heat capacity C_d takes into account the fluid mass content through the term $m_f C_p$. In the case of a linear skeleton, the tangent drained properties K, α, μ and b are constant so that (4.62a) can be integrated, while incremental equations (4.62b) and (4.62c) can still be used when considering non-linear fluids. For this reason the incremental formulation is preserved in the forthcoming developments.

Alternatively, (4.60) combine to produce:

$$d\sigma_{ij} = C^u_{ijkl} d\varepsilon_{kl} - b_{ij} M \frac{dm_f}{\rho_f} - C^u_{ijkl}\alpha^u_{kl} dT \tag{4.63a}$$

$$dS = s_f dm_f + C^u_{ijkl}\alpha^u_{kl} d\varepsilon_{ij} - 3\alpha_m M \frac{dm_f}{\rho_f} + C_u \frac{dT}{T} \tag{4.63b}$$

where we note:

$$C^u_{ijkl} = C_{ijkl} + M b_{ij} b_{kl}$$

$$C^u_{ijkl}\alpha^u_{kl} = C_{ijkl}\alpha_{kl} + 3\alpha_m b_{ij} M; \quad C_u = C_d - 9T\alpha_m^2 M \tag{4.64}$$

Index u refers to 'undrained' tangent properties of the porous material, which are to be measured in tests where fluid mass changes are prevented ($dm_f = 0$). For a linear isotropic poroelastic skeleton, we write:

$$d\sigma = K_u d\epsilon - bM \frac{dm_f}{\rho_f} - 3\alpha_u K_u dT; \quad ds_{ij} = 2\mu de_{ij} \tag{4.65a}$$

$$dS = s_f dm_f + 3\alpha_u K_u d\epsilon - 3\alpha_m M \frac{dm_f}{\rho_f} + C_u \frac{dT}{T} \tag{4.65b}$$

where K_u and $3\alpha_u$ are the undrained bulk modulus and the undrained thermal volumetric dilation coefficient, respectively, namely:

$$K_u = K + b^2 M; \qquad 3\alpha_u K_u = 3\alpha K + 3\alpha_m M b \tag{4.66}$$

According to (4.61), $M > 0$, since $K_f > 0$ and $N > 0$ (owing to $b > \phi_0$ in (4.35)), resulting in $K_u > K$. Indeed the bulk porous material is stiffer for undrained conditions where, in contrast to drained conditions, the fluid participates in the overall response.

The constitutive equations (4.65) can be inverted in an analogous form to (4.25) to give:

$$d\varepsilon_{ij} = \frac{1+\nu_u}{E_u} d\sigma_{ij} - \frac{\nu_u}{E_u}(d\sigma_{kk})\delta_{ij} + \frac{1}{3} B \frac{dm_f}{\rho_f}\delta_{ij} - \alpha dT \delta_{ij} \tag{4.67}$$

where E_u and ν_u stand for the undrained Young modulus and the undrained Poisson coefficient, respectively, while B is the Skempton coefficient whose expression is:

$$B = \frac{bM}{K_u} \tag{4.68}$$

For an undrained isothermal experiment we let $dm_f = dT = 0$ in (4.65a), resulting in $d\epsilon = d\sigma / K_u$ so that (4.62b) gives:

$$dp = -B d\sigma \mid_{dm_f = dT = 0} \tag{4.69}$$

which provides the actual definition of B and allows us to derive a useful relation between coefficients B, b, ν_u and ν. Indeed (4.67) yields the actual definition of the undrained Poisson ratio:

$$i \neq j: \quad \nu_u = -\left(\frac{d\varepsilon_{jj}}{d\varepsilon_{ii}}\right)|_{dm_f=dT=d\sigma_{ii}=0} \tag{4.70}$$

Substitution of (4.69) into (4.25) allows us to express the strain ratio appearing in the right hand member of (4.70) as a function of B, b and ν. Putting the resulting expression equal to ν_u, we finally derive the relation:

$$\frac{bB(1-2\nu)}{3} = \frac{\nu_u - \nu}{1+\nu_u} \tag{4.71}$$

4.4 Advanced Analysis

4.4.1 Non-linear Isotropic Poroelasticity

A material is isotropic when no material frame is preferred to formulate its constitutive equations. Consequently, energy functions depend only on the first three invariants of the stress and strain tensors. In practice it is generally sufficient to consider only the two first invariants. This is then equivalent to considering the first invariants σ and ϵ defined in (4.7), together with the second invariant of the stress and strain deviators, namely:

$$\tau = \sqrt{\tfrac{1}{2}s_{ij}s_{ji}}; \qquad \gamma = \sqrt{\tfrac{1}{2}e_{ij}e_{ji}}; \tag{4.72}$$

Accordingly, when restricting consideration to isothermal evolutions and isotropic materials, the arguments of potential H_s defined by (4.10) reduce to σ, τ and p. Conveniently we write:

$$H_s = H_s(\sigma, \tau, p) \tag{4.73}$$

In the next two subsections we explore the consequences of (4.73) in the non-linear case.

Secant and tangent poroelastic properties and the effective stress

We now assume that energy H_s can be split into a volumetric part and a deviator part according to:

$$H_s(\sigma, \tau, p) = \phi_0 p + H_s^{vol}(\sigma, p) + H_s^{dev}(\tau^2) \tag{4.74}$$

where the reference state is taken free of pore pressure and stress. Use of (4.74) in (4.12) yields the non-linear poroelastic constitutive equations in the form:

$$\epsilon = \frac{1}{\mathcal{K}(\sigma, p)}\sigma + \frac{\beta(\sigma, p)}{\mathcal{K}(\sigma, p)}p \tag{4.75a}$$

$$\phi - \phi_0 = \frac{1}{\mathcal{P}(\sigma, p)}p + \frac{\beta(\sigma, p)}{\mathcal{K}(\sigma, p)}\sigma \tag{4.75b}$$

$$e_{ij} = \frac{1}{2\mu(\tau)}s_{ij} \tag{4.75c}$$

where, in contrast to the tangent properties (see §4.1.2), $\mathcal{K}(\sigma, p)$, $\mu(\tau)$ and $\mathcal{P}(\sigma, p)$ are secant moduli while $\beta(\sigma, p)$ is the secant Biot coefficient.[6]

Consider now a porous material whose matrix is both homogeneous and linearly elastic so that the volumetric strain ϵ_s and the mean stress σ_s related to the matrix are linearly linked according to:

$$\epsilon_s = \frac{\sigma_s}{K_s} \tag{4.76}$$

The volumetric loading applying on the skeleton is formed from the overall current volumetric mean stress σ and the current pore pressure p. It can be eventually achieved through two successive loading phases according to the decomposition $(\sigma, p) = (\sigma + p, 0) + (-p, p)$:

1. In phase I the loading $(\sigma + p, 0)$ is applied on the skeleton. According to (4.75a) and (4.75b), the loading phase I generates the volumetric strain ϵ_I and the porosity variation $(\phi - \phi_0)_I$ such as:

$$\epsilon_I = \frac{1}{\mathcal{K}(\sigma + p, 0)}(\sigma + p); \quad (\phi - \phi_0)_I = \frac{\beta(\sigma + p, 0)}{\mathcal{K}(\sigma + p, 0)}(\sigma + p) \tag{4.77}$$

2. In phase II the loading $(-p, p)$ is superimposed on the previous loading in order to carry out the current loading (σ, p). The loading phase II results in subjecting the matrix border to the pressure p. Since the matrix is homogeneous, this pressure generates within the latter the uniform volumetric strain ϵ_s whose expression is given by letting $\sigma_s = -p$ in (4.76) and which can be uniformly extended to the porous space. Accordingly, the macroscopic volumetric strain ϵ_{II} and the porosity variation $(\phi - \phi_0)_{II}$ caused by the second loading step are given by:

$$\epsilon_{II} = -\frac{p}{K_s}; \quad (\phi - \phi_0)_{II} = -\phi_0 \frac{p}{K_s} \tag{4.78}$$

Since the matrix is linearly elastic the bulk modulus K_s does not depend on the stress state. Hence, prior application of the non-linear phase I does not affect the volumetric strain and the porosity variation (4.78) generated by the loading phase II. Consequently, provided that the non-linear response $(\epsilon, \phi - \phi_0)$ to the loading (σ, p), is unique, the current volumetric strain and porosity variation are those resulting from the successive loading phases I and II whatever the actual order in which they are performed. Adding volumetric strains and porosity variations (4.77) and (4.78), we get:

$$\epsilon = \frac{1}{\mathcal{K}(\sigma + p, 0)}\sigma + \left(\frac{1}{\mathcal{K}(\sigma + p, 0)} - \frac{1}{K_s}\right)p \tag{4.79a}$$

$$\phi - \phi_0 = \left(\frac{\beta(\sigma + p, 0)}{\mathcal{K}(\sigma + p, 0)} - \frac{\phi_0}{K_s}\right)p + \frac{\beta(\sigma + p, 0)}{\mathcal{K}(\sigma + p, 0)}\sigma \tag{4.79b}$$

[6]Consistent triplets $\left[\mathcal{K}(\sigma, p), \beta(\sigma, p), \mathcal{P}(\sigma, p)\right]$ of poroelastic secant properties are provided by letting $H_s^{vol}(\sigma, p) = h_s^{vol}\left(\frac{1}{2}\sigma^2, \sigma p, \frac{1}{2}p^2\right)$. The secant moduli are then interrelated to h_s^{vol} through the following relations: $1/\mathcal{K}(\sigma, p) = \partial h_s^{vol}/\partial(\frac{1}{2}\sigma^2)$; $\beta(\sigma, p)/\mathcal{K}(\sigma, p) = \partial h_s^{vol}/\partial(\sigma p)$; $1/\mathcal{P}(\sigma, p) = \partial h_s^{vol}/\partial(\frac{1}{2}p^2)$; $G(\tau) = \partial h_s^{dev}/\partial(\tau^2)$. However, the non-uniqueness of function h_s^{vol} results in the non-uniqueness of consistent triplets of secant properties.

A comparison between (4.75) and (4.79) allows the identification:

$$\mathcal{K}(\sigma, p) = \mathcal{K}(\sigma + p, 0); \quad \frac{1}{\mathcal{P}(\sigma, p)} = \frac{\beta(\sigma + p, 0)}{\mathcal{K}(\sigma + p, 0)} - \frac{\phi_0}{K_s} \tag{4.80a}$$

$$\beta(\sigma, p) = \beta(\sigma + p, 0) = 1 - \frac{\mathcal{K}(\sigma + p, 0)}{K_s} \tag{4.80b}$$

It is instructive to note that any identification could not have been achieved in the case of a non-linear elastic matrix where matrix modulus K_s would have depended on mean stress σ_s in (4.76). The dependence of K_s on p on the right hand side of (4.80b) would have then not allowed for function β to depend only on $\sigma + p$, although it would have been required by the left hand side of (4.80b). Indeed, in the case of a non-linear elastic matrix the loading phase I would definitively affect the volumetric strain and the porosity variation resulting from the loading phase II.

Reorganization of (4.79) provides:

$$\sigma = K_{SEC}(\sigma')\epsilon - b_{SEC}(\sigma')p \tag{4.81a}$$

$$\phi - \phi_0 = \frac{p}{N_{SEC}(\sigma')} + b_{SEC}(\sigma')\epsilon \tag{4.81b}$$

where $K_{SEC}(\sigma' = \sigma + p) = \mathcal{K}(\sigma + p, 0)$ and $b_{SEC}(\sigma' = \sigma + p) = B(\sigma + p, 0)$ are the secant bulk modulus and secant Biot's coefficient with:

$$b_{SEC}(\sigma') = 1 - \frac{K_{SEC}(\sigma')}{K_s}; \quad \frac{1}{N_{SEC}(\sigma')} = \frac{b_{SEC}(\sigma') - \phi_0}{K_s} \tag{4.82}$$

Using (4.81) and (4.82) we can alternatively write:

$$d\sigma = K(\sigma')\,d\epsilon - b(\sigma')\,dp \tag{4.83a}$$

$$d\phi = \frac{dp}{N(\sigma')} + b(\sigma')d\epsilon \tag{4.83b}$$

where $K(\sigma')$ and $b(\sigma')$ are the tangent bulk modulus and tangent Biot's coefficient with:

$$K(\sigma') = \frac{K_{SEC}(\sigma')}{1 - \frac{\sigma'}{K_{SEC}}\frac{dK_{SEC}}{d\sigma'}}; \quad b(\sigma') = 1 - \frac{K(\sigma')}{K_s}; \quad \frac{1}{N(\sigma')} = \frac{b(\sigma') - \phi_0}{K_s} \tag{4.84}$$

Based solely on the assumption of a linearly elastic homogeneous matrix, (4.81)–(4.84) constitute a non-linear extension of linear isotropic poroelasticity.[7] It is noteworthy that the secant and the tangent properties depend on Terzaghi's effective stress $\sigma' = \sigma + p$, in spite of the matrix compressibility resulting in a value of Biot's coefficient different from unity. Such a non-linear extension of poroelasticity can adequately capture the non-linear behaviour of porous materials such as rocks whose porous network is often formed of

[7] The derivation of the non-linear results developed in this section is adapted from Dormieux L., Molinari A., Kondo D. (2002), 'Micromechanical approach to the behavior of poroelastic materials', *Journal of the Mechanics and Physics of Solids*, **50**, 2203–2231.

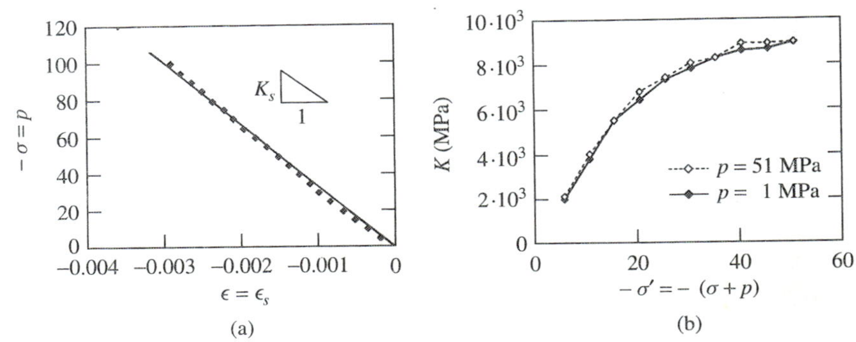

Figure 4.3: Experimental confirmation of non-linear constitutive equations (4.81)–(4.84) for a sandstone specimen: (a) confirmation of the linearity of the elastic behaviour of the matrix; (b) confirmation of the dependence of the drained tangent bulk modulus on Terzaghi's effective stress $\sigma' = \sigma + p$ (data from Bemer *et al.* (2001) reprinted by permission of Institut Français du Petrole, see footnote).

connected pores and cracks. The source of non-linearities is then due to the progressive opening of cracks under the pressurization of the saturating fluid. As illustrated in Fig. 4.3, the non-linear constitutive equations (4.81)–(4.84) are experimentally confirmed for a sandstone specimen.[8] An experiment where the loading varies according to phase II, that is $(\sigma, p) = (-p, p)$, first allows us to check the linear elastic behaviour of the matrix of the sandstone specimen by recording a value for the matrix modulus $K_s = -dp/d\epsilon$ insensitive to p (see Fig. 4.3a). The tangent moduli $K(\sigma')$ is then plotted against Terzaghi's effective stress $\sigma' = \sigma + p$ from experiments where the overall volumetric strain ϵ is measured as a function of σ, whereas the pore pressure p is held constant and takes different values. The two functions $K(\sigma')$ obtained for two distinct values 1 MPa and 51 MPa of the pore pressure p match in the limit of experimental accuracy as indicated in Fig. 4.3b, confirming the validity of the non-linear constitutive equations (4.81)–(4.84).

A non-linear isotropic model for clays

The approach to non-linear poroelastic behaviour of the previous section is not appropriate for poorly cohesive materials like soils. The non-linear behaviour of clay-like soils is generally well accounted for in the elastic range through the model originally developed to capture the elastic behaviour of the clay of the Cam river (see §8.4.4 for the plastic side of the model). The so-called Cam–Clay model is discussed below in order to ensure its thermodynamic consistency.[9]

[8]The experiments reported in Fig. 4.3 are due to Bemer E., Boutéca M., Vincké O., Hoteit N., Ozanam O. (2001), 'Poromechanics: from linear poroelasticity to non-linear poroelasticity and poroviscoelasticity', *Oil & Gaz Science and Technology, Rev. IFP*, **56**, (6), 531–544, where a theory of second-order poroelasticity is also presented in order to approach the non-linear constitutive equations.

[9]The approach to the model follows that of Bourgeois E. (1997), 'Mécanique des milieux poreux en transformation finie: position des problèmes et méthodes de résolution', *PhD Thesis*, Ecole Nationale des Ponts et Chaussées.

The solid grains forming the matrix of soils generally undergo negligible volume changes so that the constitutive equations can be expressed by means of Terzaghi's effective stress. Moreover, soils cannot sustain significant tensile stress, nor undergo significant dilation volumetric strain, so that we conveniently let:

$$p' = -\sigma' = -(\sigma + p)\,; \quad \varepsilon = -\epsilon \tag{4.85}$$

in order to deal with positive values, that is $p' \geq 0$ and $\varepsilon \geq 0$.

In the triaxial test the sample is submitted to an axial pressure $-\sigma_I$ in direction I and to a uniform pressure $-\sigma_{II} = -\sigma_{III}$ in the other directions. Since the material is isotropic the principal directions of the stress and strain tensors coincide, so that components s_{ij} and e_{ij} of their respective deviators are given by:

$$(s_{ij}) = \frac{1}{3}\begin{pmatrix} 2(\sigma_I - \sigma_{III}) & 0 & 0 \\ 0 & -(\sigma_I - \sigma_{III}) & 0 \\ 0 & 0 & -(\sigma_I - \sigma_{III}) \end{pmatrix} \tag{4.86a}$$

$$(e_{ij}) = \frac{1}{3}\begin{pmatrix} 2(\varepsilon_I - \varepsilon_{III}) & 0 & 0 \\ 0 & -(\varepsilon_I - \varepsilon_{III}) & 0 \\ 0 & 0 & -(\varepsilon_I - \varepsilon_{III}) \end{pmatrix} \tag{4.86b}$$

Substitution of (4.85) and (4.86) into (4.11), while letting $d\phi = d\epsilon$ (matrix incompressibility) and $dT = 0$, gives for the triaxial test:

$$\varepsilon\, dp' + \eta\, dq - dH_s = 0 \tag{4.87}$$

where η and q rely on the strain and the stress through:

$$\eta = -\frac{2}{3}(\varepsilon_I - \varepsilon_{III}); \quad q = -(\sigma_I - \sigma_{III}) \tag{4.88}$$

From (4.87) we derive:

$$H_s = H_s(p', q); \quad \varepsilon = \frac{\partial H_s}{\partial p'}; \quad \eta = \frac{\partial H_s}{\partial q} \tag{4.89}$$

so that:

$$d\varepsilon = \frac{dp'}{K(p', q)} + \frac{dq}{H(p', q)}; \quad d\eta = \frac{dp'}{H(p', q)} + \frac{dq}{3\mu(p', q)} \tag{4.90}$$

where the tangent properties $1/K$, $1/H$ and $1/3\mu$ satisfy:

$$\frac{\partial}{\partial q}\left(\frac{1}{K}\right) = \frac{\partial}{\partial p'}\left(\frac{1}{H}\right); \quad \frac{\partial}{\partial p'}\left(\frac{1}{3\mu}\right) = \frac{\partial}{\partial q}\left(\frac{1}{H}\right) \tag{4.91}$$

In experiments performed on clays, where the confining effective mean pressure p' is progressively increased while the deviator q is maintained at zero, the void ratio e (see (1.18)) is found to decrease linearly with respect to the logarithm of the mean effective pressure

p', while no significant variation of the strain deviator η is recorded (for experimental evidence see Fig. 6.7a). Accordingly we write:

$$q = 0: \quad de = -\kappa \frac{dp'}{p'}; \quad \eta = 0 \tag{4.92}$$

Combining (1.34), (4.85), (4.90) and (4.92), we derive:

$$K(p', q = 0) = \frac{p'}{k}; \quad H(p', q = 0) \to \infty \tag{4.93}$$

where we note $k = \kappa/(1 + e_0)$. In addition, for small values of q/p' the ratio $3\mu/K$ is experimentally found to be a material constant property:

$$q/p' \ll 1: \quad 3\mu(p', q) = \frac{p'}{g} \tag{4.94}$$

In the Cam–Clay elastic model $1/H$ terms in (4.90) are usually omitted, in contradiction with (4.91) and (4.94). In order to derive a model which is consistent with both thermodynamics and experimental observations (4.93) and (4.94), we first express tangent properties K, μ and H in the form:

$$K = \frac{p'}{k + l(\theta)}; \quad 3\mu = \frac{p'}{g}; \quad H = \frac{p'}{h(\theta)} \tag{4.95}$$

where θ denotes the stress inclination:

$$\frac{q}{p'} = \theta \tag{4.96}$$

Substitution of (4.95) into (4.91) and (4.93) provides:

$$-\frac{d}{d\theta}(\theta h(\theta)) = \frac{dl(\theta)}{d\theta}; \quad -g = \frac{dh(\theta)}{d\theta}; \quad l(0) = h(0) = 0 \tag{4.97}$$

whose solution substituted into (4.90) gives:

$$d\varepsilon = \left(k + g\frac{q^2}{p'^2}\right)\frac{dp'}{p'} - g\frac{q}{p'}\frac{dq}{p'} \tag{4.98a}$$

$$d\eta = -g\frac{q}{p'}\frac{dp'}{p'} + g\frac{dq}{p'} \tag{4.98b}$$

Integration of (4.98) yields:

$$\varepsilon = k \ln \frac{p'}{p'_0} - g\frac{q^2}{2p'^2}; \quad \eta = g\frac{q}{p'} \tag{4.99}$$

$$H_s(p', q) = kp'\left(\ln \frac{p'}{p'_0} - 1\right) + g\frac{q^2}{2p'} \tag{4.100}$$

where p'_0 is a reference value for the effective confining pressure p'. In order to invert relations (4.99), free energy $\Psi_s(\varepsilon, \eta)$ can be computed from H_s according to:

$$\Psi_s = \Psi_s(\varepsilon, \eta): \quad \Psi_s = p'\varepsilon + q\eta - H_s \tag{4.101}$$

yielding:

$$\Psi_s(\varepsilon, \eta) = kp'_0 \exp\frac{1}{k}\left(\varepsilon + \frac{\eta^2}{2g}\right); \quad p' = \frac{\partial \Psi_s}{\partial \varepsilon}; \quad q = \frac{\partial \Psi_s}{\partial \eta} \tag{4.102}$$

so that:

$$p' = p'_0 \exp\frac{1}{k}\left(\varepsilon + \frac{\eta^2}{2g}\right); \quad q = p'_0\frac{\eta}{g} \exp\frac{1}{k}\left(\varepsilon + \frac{\eta^2}{2g}\right) \tag{4.103}$$

Under triaxial stress conditions (4.86a), stress and strain invariants τ and γ defined in (4.72) can be expressed as a function of q and η according to:

$$\tau^2 = \frac{1}{6}q^2; \quad \gamma^2 = \frac{3}{8}\eta^2 \tag{4.104}$$

By expressing q and η as a function of τ and γ in the whole sequence above, potentials H_s and Ψ_s can be used to derive the constitutive equations for any state of stress.

4.4.2 Brittle Fracture of Fluid-infiltrated Materials

Brittle materials cannot store elastic or free energy beyond some critical threshold Ψ_{cr}. When the threshold is reached, brittle fracture occurs through the abrupt irreversible release of the whole stored energy. For a brittle porous material the limit in energy storage applies to the sole skeleton so that the failure criterion is:

$$\Psi_s = \Psi_{cr} \tag{4.105}$$

The elastic energy of a dry isotropic porous material ($p = 0$), subjected to the mean stress σ and undergoing the volumetric dilation ϵ, can be written:

$$\Psi_s = \frac{1}{2}\sigma\epsilon = \frac{1}{2}\frac{\sigma^2}{K} \tag{4.106}$$

Combining (4.105) and (4.106), an intrinsic dry or skeleton compressive strength ϖ_{cr} can be defined:

$$\varpi_{cr} = \sqrt{2K\Psi_{cr}} \tag{4.107}$$

Experiments performed on brittle fluid-infiltrated materials, such as shales or concretes,[10] have shown that the loading at failure depended on the loading rate. Let $-\sigma$ be the strength of the mean compressive stress applied to the geomaterial sample and increasing at constant rate $\overset{\circ}{\sigma}$:

$$-\sigma = \overset{\circ}{\sigma}t; \quad \overset{\circ}{\sigma} > 0 \tag{4.108}$$

In addition let ϖ_{fr} be the strength of the compressive stress $-\sigma = \varpi_{fr}$ provoking fracture. Its dependence on the loading rate $\overset{\circ}{\sigma}$ is formally written:

$$\varpi_{fr} = f\left(\overset{\circ}{\sigma}\right) \tag{4.109}$$

[10] See for instance Rossi P., Van Mier J.G.M., Toutlemonde F., Le Maou F., Boulay C. (1994), 'Effect of loading rate on the strength of concrete subjected to uniaxial tension', *Materials and Structures*, **27**, 260–264.

However, the dimensional consistency of the above relation requires us also to involve at least both a characteristic time and a characteristic strength both intrinsic to the material. The characteristic time is linked to some internal dissipative viscous process. For porous materials a good candidate is the squirt viscous flow mechanism. According to this mechanism the compression of the solid grains forming the matrix generates local gradients of the fluid pressure which in turn set the fluid in motion. The characteristic time τ_η scaling the process depends on the skeleton compressibility K and on the fluid viscosity η_f, as it concerns respectively the strength of the local gradients and the velocity of the fluid flow. Properties K and η_f eventually combine to form the characteristic time:

$$\tau_\eta \propto \frac{\eta_f}{K} \tag{4.110}$$

The intrinsic strength is the skeleton compressive strength ϖ_{cr} defined by (4.107) so that the characteristic time conveniently scaling the loading rate τ_σ is provided by:

$$\tau_\sigma = \frac{\varpi_{cr}}{\overset{\circ}{\sigma}} \tag{4.111}$$

Since η_f, K and ϖ_{cr} must enter the picture, dimensional analysis (see §3.3.1) shows that a dimensionally consistent form of (4.109) is:

$$\overline{\varpi}_{fr} = \frac{\varpi_{fr}}{\varpi_{cr}} = g\left(\frac{\tau_\eta}{\tau_\sigma}\right) \tag{4.112}$$

In a first approach the squirt flow can be captured by considering a porous material exhibiting a double porosity network. Considering isothermal evolutions and zero initial conditions for the stress and the fluid pressure, constitutive poroelastic equations (4.21a) and (4.21c) can be extended to materials exhibiting a double porosity network in the following form:

$$\sigma = K\epsilon - b_1 p_1 - b_2 p_2 \tag{4.113a}$$

$$\phi_1 - \phi_{01} = b_1\epsilon + \frac{p_1}{N_{11}} + \frac{p_2}{N_{12}} \tag{4.113b}$$

$$\phi_2 - \phi_{02} = b_2\epsilon + \frac{p_1}{N_{12}} + \frac{p_2}{N_{22}} \tag{4.113c}$$

where we anticipated $N_{12} = N_{21}$ owing to Maxwell's symmetry relations. Indeed, owing to the linear elasticity of the skeleton, its elastic or isothermal free energy can be written:

$$\begin{aligned}\Psi_s &= \frac{1}{2}\sigma\epsilon + \sum_{\alpha=1,2}\frac{1}{2}p_\alpha(\phi_\alpha - \phi_{0\alpha}) \\ &= \frac{1}{2}\frac{(\sigma + b_1 p_1 + b_2 p_2)^2}{K} + \frac{1}{2}\frac{p_1^2}{N_{11}} + \frac{p_2 p_1}{N_{12}} + \frac{1}{2}\frac{p_2^2}{N_{22}}\end{aligned} \tag{4.114}$$

According to the failure criterion (4.105) and (4.107) brittle fracture of the porous material occurs when the volumetric strain and the fluid pressures fulfil the relation:

$$\frac{(\sigma + b_1 p_1 + b_2 p_2)^2}{K} + \frac{p_1^2}{N_{11}} + \frac{2p_2 p_1}{N_{12}} + \frac{p_2^2}{N_{22}} = \frac{\varpi_{cr}^2}{K} \tag{4.115}$$

Moreover, remarking that $\phi = \phi_1 + \phi_2$ and taking into account the fluid pressures p_1 and p_2 in the appropriate form in the stress partition theorem (4.30), the same procedure we used in §4.2.3 allows us to extend the relations (4.35) in the form:

$$b_1 + b_2 = b = 1 - \frac{K}{K_s}; \quad \frac{1}{N_{12}} + \frac{1}{N_{\alpha\alpha}} = \frac{b_\alpha - \phi_{0\alpha}}{K_s} \tag{4.116}$$

In §6.4.3 it will be more precisely shown that:

$$b_\alpha = bS_\alpha \tag{4.117}$$

where S_α is the degree of saturation of the fluid associated with porous network α, that is:

$$S_\alpha = \frac{\phi_{0\alpha}}{\phi_0} \tag{4.118}$$

Finally, since the porous space is filled by the same fluid, we write:

$$\frac{m_1 - m_{01}}{\rho_f} = b_1\epsilon + \frac{p_1}{M_{11}} + \frac{p_2}{N_{12}} \tag{4.119a}$$

$$\frac{m_2 - m_{02}}{\rho_f} = b_2\epsilon + \frac{p_1}{N_{12}} + \frac{p_2}{M_{22}} \tag{4.119b}$$

where $m_\alpha = \rho_f \phi_\alpha$ is the fluid mass content related to network α, ρ_f being the reference fluid density. In addition we note:

$$\frac{1}{M_{\alpha a}} = \frac{1}{N_{\alpha a}} + \frac{\phi_{0\alpha}}{K_f} \tag{4.120}$$

The evolutions considered here are externally undrained so that the fluid motion within the porous material results only from an exchange of fluid mass between the two porous networks. Fluid mass conservation then implies:

$$m_1 + m_2 = m_{01} + m_{02} \tag{4.121}$$

and we write:

$$\frac{dm_1}{dt} = -\frac{dm_2}{dt} = -\overset{\circ}{r}_{1\to 2} = \rho_f \frac{d\zeta}{dt} \tag{4.122}$$

Combining (3.112) and (4.122) the rate $d\zeta/dt$ of fluid volume exchange between the two networks is finally governed by:

$$\frac{d\zeta}{dt} = -\frac{1}{\eta}(p_1 - p_2) \tag{4.123}$$

where η was found in §3.6.2 to be a viscosity coefficient proportional to the fluid viscosity η_f.

Prior to determining the normalized compressive stress strength at failure $\overline{\varpi}_{fr}$ as a function of the loading rate parameter τ_η/τ_σ, that is the function g in (4.112), we can determine its asymptotic values. For the sake of simplicity we assume from now on that the fluid flow is incompressible and that there is no coupling between the two porous networks, resulting in:[11]

$$\frac{1}{M_{\alpha a}} = \frac{1}{N_{\alpha a}}; \qquad \frac{1}{N_{12}} = 0 \tag{4.124}$$

- In the regime of slow loadings, that is $\tau_\eta/\tau_\sigma \gg 1$, the fluid exchange, which tends to decrease the pressure difference between the two porous networks, occurs infinitely rapidly when compared with the loading rate. As a consequence, in the limit of infinitely slow loading $\tau_\eta/\tau_\sigma \to 0$, the two pressures remain constantly equal, that is $p_1 = p_2$. Combining the latter equality with (4.113a), (4.116)–(4.120) and (4.124), we obtain:

$$\sigma = K_u\epsilon; \quad p_1 = p_2 = -bM_u\epsilon \tag{4.125}$$

  where:

$$K_u = K + b^2 M_u; \quad \frac{1}{M_u} = \frac{1}{M_{11}} + \frac{1}{M_{22}} \tag{4.126}$$

  Moduli K_u and M_u are eventually the undrained bulk modulus and the Biot modulus as if the porous networks were forming the same and unique porous network. Brittle fracture occurs when the criterion (4.115) is fulfilled, yielding:

$$\overline{\varpi}_{fr}^{slow} = \sqrt{\frac{K_u}{K}} \tag{4.127}$$

  where $\overline{\varpi}_{fr}^{slow}$ denotes the normalized compressive stress intensity provoking the fracture in the limit of infinitely slow loading.
- In the regime of fast loadings, that is $\tau_\eta/\tau_\sigma \gg 1$, the fluid exchange between the two porous networks occurs infinitely slowly when compared with the loading rate. As a consequence, in the limit of infinitely fast loading $\tau_\eta/\tau_\sigma \to \infty$, there is no fluid exchange between the two porous networks and the fluid mass content remains constant in each network, that is $m_\alpha = m_\alpha^0$. Combining the latter equality with (4.113a), (4.117), (4.119) and (4.124), we obtain:

$$\sigma = K_U\,\epsilon; \quad p_\alpha = -b_\alpha M_{\alpha a}\epsilon \tag{4.128}$$

  where:

$$K_U = K + b^2 M_U; \quad M_U = S_1^2 M_{11} + S_2^2 M_{22} \tag{4.129}$$

  Modulus $\mathcal{K}_U$ is eventually the bulk modulus under conditions that are undrained both externally and internally. Brittle fracture occurs when the criterion (4.115) is

[11] Consider separately two elements of porous material, each of them embedding one of the two porous networks. The constitutive equations can be written separately:

$$\phi_\alpha - \phi_{0\alpha} = b_\alpha \epsilon_\alpha + \frac{p_\alpha}{N_{\alpha a}}$$

The assumption $1/N_{12} = 0$ turns out to set the two porous elements in parallel so that the two elements undergo the same volumetric dilation $\epsilon_\alpha = \epsilon_1 = \epsilon_2$.

fulfilled, yielding:

$$\overline{\varpi}_{fr}^{fast} = \sqrt{\frac{K_U}{K}} \tag{4.130}$$

where $\overline{\varpi}_{fr}^{fast}$ denotes the normalized compressive stress strength provoking the fracture in the limit of infinitely fast loading.

Since $M_U > M_u$ it is instructive to note that $\overline{\varpi}_{fr}^{fast} \gg \overline{\varpi}_{fr}^{slow}$, as is experimentally observed. Indeed, owing to the different internal conditions of drainage, the faster the loading regime the greater the apparent stiffness of the porous material and, consequently, the greater the compressive stress intensity at failure supplying the same critical energy to the skeleton.[12]

The determination of $\overline{\varpi}_{fr}$ over the whole range of loading rates, that is for any value of τ_σ/τ_η, requires knowledge of the strain history in response to the stress history (4.108). From (4.113a), (4.119), (4.121)–(4.124), (4.126) and (4.129), eliminating the fluid pressures p_1 and p_2 to the benefit of ϵ and σ, we finally derive:

$$\epsilon + \tau_\eta \frac{d\epsilon}{dt} = \frac{\sigma}{K_u} + \frac{\tau_\eta}{K_U}\frac{d\sigma}{dt} \tag{4.131}$$

where the characteristic time τ_η is eventually identified as:

$$\tau_\eta = \frac{\eta}{M_{11} + M_{22}} \times \frac{K_U}{K_u} \tag{4.132}$$

Also eliminating the fluid pressures in the failure criterion (4.115), some added calculations carried out with (4.113a), (4.116)–(4.119), (4.121), (4.124), (4.126) and (4.129) provide the failure criterion in the form:

$$\frac{1}{K_U - K_u}\left(\frac{1}{2}K_u K_U \epsilon^2 - K_u \epsilon\sigma + \frac{1}{2}\sigma^2\right) = \frac{1}{2}\frac{\varpi_{cr}^2}{K} \tag{4.133}$$

Substitution of (4.108) into (4.131) leads to the differential equation:

$$\overline{\varepsilon} + \frac{\tau_\eta}{\tau_\sigma}\frac{d\overline{\varepsilon}}{d\overline{t}} = \frac{K}{K_u}\overline{t} + \frac{K}{K_U}\frac{\tau_\eta}{\tau_\sigma} \tag{4.134}$$

where we let:

$$\overline{\varepsilon} = -\frac{K\epsilon}{\varpi_{cr}}; \quad \overline{t} = \frac{t}{\tau_\sigma} \tag{4.135}$$

With zero initial conditions the solution of (4.134) is:

$$\overline{\varepsilon} = \frac{K}{K_u}\overline{t} - \frac{\tau_\eta}{\tau_\sigma}\left(\frac{K}{K_u} - \frac{K}{K_U}\right)\left[1 - \exp\left(-\frac{\tau_\sigma}{\tau_\eta}\overline{t}\right)\right] \tag{4.136}$$

[12] This phenomenon can also be encountered at the scale of the structure as for instance in the analysis of the borehole breakdown. See Garagash D., Detournay E. (1997), 'An analysis of the influence of the pressurization rate on the borehole breakdown pressure', *Journal of Solids and Structures*, **34**, (24), 3099–3118, which constituted the starting point of the analysis presented in this section.

Substitution of (4.108) into (4.133) and use of (4.135) give the failure criterion for the problem at hand in the form:

$$\bar{\varepsilon}^2 - 2\frac{K}{K_U}\bar{\varepsilon}\bar{t} + \frac{K^2}{K_u K_U}\bar{t}^2 = \frac{K}{K_u} - \frac{K}{K_U} \tag{4.137}$$

The dimensionless compressive stress intensity $\overline{\varpi}_{fr}$ provoking the fracture satisfies $\overline{\varpi}_{fr} = \bar{t}_{fr}$ where $\bar{t}_{fr}$ is the dimensionless time at which fracture occurs. The latter has to satisfy (4.136) and (4.137) simultaneously so that $\overline{\varpi}_{fr}$ is eventually given by the non-linear equation:

$$\frac{\overline{\varpi}^2_{fr} - \left(\overline{\varpi}^{slow}_{fr}\right)^2}{\left(\overline{\varpi}^{fast}_{fr}\right)^2 - \left(\overline{\varpi}^{slow}_{fr}\right)^2}$$
$$= 1 - \left(\frac{\overline{\varpi}_{fr}}{\overline{\varpi}^{slow}_{fr}}\right)^2 \left\{1 - \frac{1}{\overline{\varpi}_{fr}}\frac{\tau_\eta}{\tau_\sigma}\left[1 - \exp\left(-\frac{\tau_\sigma}{\tau_\eta}\overline{\varpi}_{fr}\right)\right]\right\}^2 \tag{4.138}$$

whose numerical solution is illustrated in Fig. 4.4.

4.4.3 From Poroelasticity to the Swelling of Colloidal Mixtures

Colloidal mixtures can be roughly defined as the mixtures resulting from the dispersion of interacting particles in a dispersive medium, the latter often reducing to an electrolytic solution. In most colloidal mixtures the particle surface carries an electric charge so that the particles interact with the electrolyte. Colloidal mixtures such as saturated clays are known to swell dramatically when they are immersed in pure water, while the overall pore pressure is kept constant. Such a spontaneous swelling can turn out to be detrimental. For

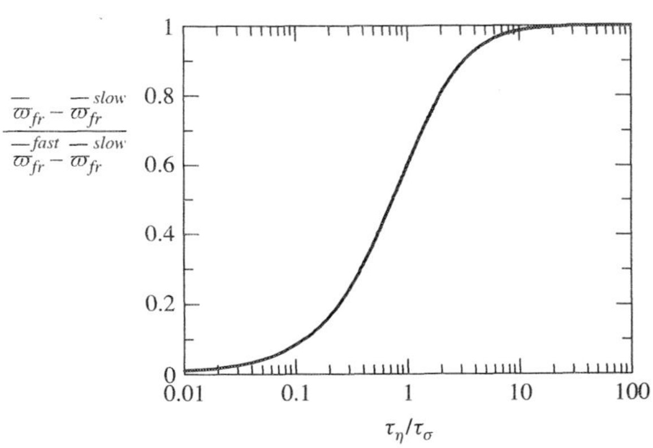

Figure 4.4: Influence of the loading rate on the stress at failure: normalized stress at failure $(\overline{\varpi}_{fr} - \overline{\varpi}^{slow}_{fr})/(\overline{\varpi}^{fast}_{fr} - \overline{\varpi}^{slow}_{fr})$ plotted against the ratio of the microdiffusion characteristic time τ_η and the loading characteristic time τ_σ, for $\overline{\varpi}^{slow}_{fr} = \sqrt{2}$ and $\overline{\varpi}^{fast}_{fr} = 2$.

instance, in petroleum or civil engineering, a hole or a tunnel bored in a clay layer can be seriously damaged when the clay comes into contact with the drilling fluid whose salt concentration is much lower than that of the interstitial solution. We show below how non-linear poroelasticity can be used to represent such a swelling process caused by the chemical activity induced by the electrostatic interaction between the charged particles of the colloid and the electrolyte.[13]

The swelling pressure

At the macroscopic scale the current elementary volume $d\Omega_t$ can be viewed as the superimposition of three particles: (i) the skeleton particle (s); (ii) the liquid water acting as a solvent (w); (iii) the salt acting as a solute. The salt is an electrolyte (e) formed from a cationic component (+) and an anionic component (−). The surface of the particles forming the internal solid walls of the porous network is electrically charged so that the electroneutrality holds only at the scale of the elementary volume $d\Omega_t$, the excess of charge of the electrolyte counterbalancing the charge of the particles. The inner solution is chemically active owing to the electrical interaction occurring between the electrolyte and the charged particle surface. Neglecting the activity of the liquid water in comparison with the electrolyte that of and restricting consideration to isothermal infinitesimal volumetric transformations, we apply (3.121) in the form:

$$\sigma\, d\epsilon + p_w\, d\phi + p_e\, d\left(\frac{\phi}{A}\right) - d\Psi_s = 0 \tag{4.139}$$

In (4.139), $A > 1$ stands for the electrolytic activity (see (3.117)) capturing the above-mentioned electrical long-range interaction whose energy is included in the skeleton free energy Ψ_s (see §3.6.3 for more details). Assuming in addition that the solid matrix does not undergo volumetric changes, as is actually the case for the platelets forming the clay particles, we substitute $d\phi = d\epsilon$ in (4.139), yielding:

$$(\sigma + p_w)\, d\epsilon + p_e\, d\left(\frac{\phi}{A}\right) - d\Psi_s = 0 \tag{4.140}$$

Alternatively, letting $G_s = \Psi_s - p_e(\phi/A)$, we write:

$$(\sigma + p_w)\, d\epsilon - \frac{\phi}{A} dp_e - dG_s = 0 \tag{4.141}$$

providing the state equations in the form:

$$G_s = G_s(\epsilon, p_e): \quad \sigma + p_w = \frac{\partial G_s}{\partial \epsilon}; \qquad \frac{\phi}{A} = -\frac{\partial G_s}{\partial p_e} \tag{4.142}$$

Differentiating the latter, we finally get:

$$d(\sigma + p_w) = K\, d\epsilon - b_e\, dp_e \tag{4.143a}$$

$$d\left(\frac{\phi}{A}\right) = b_e\, d\epsilon + \frac{dp_e}{N} \tag{4.143b}$$

[13]The approach is originally due to Dormieux L., Barboux P., Coussy O., Dangla P. (1995), 'A macroscopic modelling for the swelling phenomenon of a saturated clay', *European Journal of Mechanics A/Solids*, **14**, (6), 981–1004.

Since $d\phi = d\epsilon$, from (4.143b) we infer the instructive relations:

$$b_e = \frac{\partial\left(\phi/A\right)}{\partial\phi}\,|_{p_e};\quad \frac{1}{N} = \phi\frac{\partial\left(1/A\right)}{\partial p_e}\,|_{\phi} \tag{4.144}$$

According to (4.143a), in the absence of activity, that is $A = 1$, we recover the value $b_e = 1$ associated with an incompressible matrix. Hence the departure from one of tangent Biot's coefficient b_e captures the departure from ideality of the chemically active solution.

Equation (4.143a) can be rewritten in the form:

$$d\left(\sigma + p\right) = K\,d\epsilon - dp_{sw} \tag{4.145}$$

where $p = p_w + p_e$ is the pressure solution, while p_{sw} is the swelling pressure defined by:

$$p_{sw} = -\int_0^{p_e}\left(1 - b_e\right)dp_e \tag{4.146}$$

resulting in the independent and alternative expression for b_e:

$$b_e = \frac{\partial p_{sw}}{\partial p_e}\,|_{\phi} + 1 \tag{4.147}$$

In fact, the simultaneous validity of expressions (4.144) and (4.147) lies in Maxwell's symmetry relations attached to state equations (4.142). When the interstitial solution of a porous material such as a clay comes into contact with an external solution at the same overall pressure p but at a lower salt concentration, the electrolytic thermodynamic pressure varies negatively, that is $dp_e < 0$, so that the swelling pressure increases, that is $dp_{sw}\,|_{\phi} > 0$, provided that Biot's coefficient b_e is less than one as will soon be recognized, but as can already be expected from (4.144) with $A > 1$ and $\phi < 1$. If in addition the total mean stress σ is held constant, the porous material swells, with $d\epsilon = dp_{sw}/K$. On the contrary, the application of a compressive stress increment $d\sigma = -dp_{sw}$ is required for preventing any strain.

Chemical activity and tangent Biot's coefficient

In order to be actually operational, expression (4.144) of Biot's coefficient requires the explicit expression of the electrolytic activity A. The definition (3.117) of the latter can be specified here in the form:

$$\frac{1}{A} = \frac{\rho_e}{\rho_e^{eff}} = \frac{\rho_+ + \rho_-}{\rho_+^{eff} + \rho_-^{eff}} \tag{4.148}$$

where the subscript eff refers to the effective solution. Let us recall that the effective solution is defined such that the chemical potential of the actual electrolyte viewed as a whole, or the chemical potential of its cationic and anionic components considered separately, is equal to that of the effective solution (see §3.6.3). The question now arises of how the overall electrolytic activity A is related to the partial activities, A_+ and A_-, of respectively the cationic component and the anionic component, that is:

$$\frac{1}{A_+} = \frac{\rho_+}{\rho_+^{eff}}; \quad \frac{1}{A_-} = \frac{\rho_-}{\rho_-^{eff}} \tag{4.149}$$

The decomposition of A in terms of partial activities A_+ and A_- requires the analysis of the electrostatic interaction of the anionic and cationic components with the charge carried by the particle surface. The analysis is more conveniently carried out by introducing the molar concentrations c_+ and c_- of the ionic components + or −. The latter are linked to the mass density $\rho_{+ \text{ or } -}$ and to the molar mass $\mathcal{M}_{+ \text{ or } -}$ through the relations:

$$c_+ = \frac{\rho_+}{\mathcal{M}_+}; \quad c_- = \frac{\rho_-}{\mathcal{M}_-} \tag{4.150}$$

In view of (4.149) and (4.150) partial activities A_+ and A_- can be equivalently expressed in terms of molar concentrations according to:

$$\frac{1}{A_+} = \frac{c_+}{c_+^{eff}}; \quad \frac{1}{A_-} = \frac{c_-}{c_-^{eff}} \tag{4.151}$$

Let s be the constant overall particle surface per unit of macroscopic initial volume $d\Omega_0$;[14] let q be the electric charge carried per unit of surface of the internal walls of the porous network, and let $z_{+ \text{ or } -}$ be the valency of the ions forming the solute. The electroneutrality of the porous material as a whole is:

$$\phi F(z_+ c_+ - z_- c_-) + sq = 0 \tag{4.152}$$

where F is the Faraday constant (9.6486×10^4 C/mol), recalling that $F = \mathcal{N} e$ where $\mathcal{N} = 6.022 \times 10^{23}$ is the Avogadro number and $e = 1.6 \times 10^{-19}$C the elementary (protonic) charge. By contrast the electroneutrality of the effective solution requires:

$$z_+ c_+^{eff} - z_- c_-^{eff} = 0 \tag{4.153}$$

Combining (4.148)–(4.153) gives the decomposition:

$$\frac{1}{A} = \frac{1}{z_- \mathcal{M}_+ + z_+ \mathcal{M}_-} \left(\frac{z_- \mathcal{M}_+}{A_+} + \frac{z_+ \mathcal{M}_-}{A_-} \right) \tag{4.154}$$

while (4.151)–(4.153) provide the relation:

$$\frac{\phi}{A_+} - \frac{\phi}{A_-} = -\frac{sq}{F} \times \frac{z_+ + z_-}{z_+ z_-} \times \frac{1}{c_+^{eff} + c_-^{eff}} \tag{4.155}$$

The thermodynamic pressure p_e of the electrolyte is the pressure of the electrolyte in the effective solution, namely:

$$p_e = RT\left(c_+^{eff} + c_-^{eff}\right) \tag{4.156}$$

[14]So that $s/\rho_s(1 - \phi_0)$ is the specific surface, that is the surface per mass unit of the dry material.

where R is the ideal gas constant (8.314 J/(mol K)). As a consequence of (4.144), (4.154)–(4.156), the tangent Biot's coefficient b_e related to the electrolyte can be similarly obtained from the total or the partial cationic or anionic activities according to:

$$b_e = \frac{\partial\,(\phi/A)}{\partial\phi}\,|_{p_e} = \frac{\partial(\phi/A_+)}{\partial\phi}\,|_{p_e} = \frac{\partial(\phi/A_-)}{\partial\phi}\,|_{p_e} \tag{4.157}$$

Tangent Biot's coefficient and the electrical double layer theory

The explicit determination of b_e requires us to express partial activities A_+ and A_- and, consequently, the cationic and anionic concentrations c_+ and c_- as functions of ϕ, c_+^{eff} and c_-^{eff}. This can be achieved by using the standard electrical double layer theory.[15] To this end we first recall that c_+ and c_- are averaged macroscopic concentrations. Adopting the notation of §2.5.1, we write:

$$c_+ = \left\langle f_{\omega_f}(\mathbf{z})\, c_+(\mathbf{z})\right\rangle ; \quad c_- = \left\langle f_{\omega_f}(\mathbf{z})\, c_-(\mathbf{z})\right\rangle \tag{4.158}$$

where $c_+(\mathbf{z})$ and $c_-(\mathbf{z})$ are the ionic microscopic concentrations at points located at $\mathbf{z}$ in the actual solution saturating the porous space ω_f. From now on, for the sake of simplicity we restrict ourselves to a 1:1 electrolyte, that is $z_+ = z_- = 1$, typically a sodium chloride salt Na_+Cl_-. The molar electrochemical potential, $\widetilde{\mu}_{+\text{ or }-} = \mu_{+\text{ or }-}/\mathcal{M}_{+\text{ or }-}$, of the ionic components can be expressed in the form:

$$\widetilde{\mu}_+ = RT\ln c_+(\mathbf{z}) + FU(\mathbf{z}) ; \quad \widetilde{\mu}_- = RT\ln c_-(\mathbf{z}) - FU(\mathbf{z}) \tag{4.159}$$

where the first term accounts for the molecular entropy agitation while the second term accounts for the electrostatic energy, $U(\mathbf{z})$ being the local electric potential. In contrast to the actual solution, the effective solution is locally electroneutral and there is no electrostatic energy contribution to the molar effective electrochemical potential $\widetilde{\mu}_{+\text{ or }-}^{eff}$, the latter reducing to the standard molar Gibbs potential $\widetilde{g}^{eff}$. Since $c^{eff} = c_+^{eff} = c_-^{eff}$ because of the hypothesis $z_+ = z_- = 1$ and the electroneutrality condition (4.153) of the effective solution, we write:

$$\widetilde{\mu}_{+\text{ or }-}^{eff} = \widetilde{g}_+^{eff} = \widetilde{g}_-^{eff} = \widetilde{g}^{eff} = RT\ln c^{eff} \tag{4.160}$$

From the actual definition of the effective solution, we have $\widetilde{\mu}_{+\text{ or }-} = \widetilde{\mu}_{+\text{ or }-}^{eff}$ so that a combination of (4.159) and (4.160) gives:

$$c_+(\mathbf{z}) = c^{eff}\exp\left(-\frac{FU(\mathbf{z})}{RT}\right) ; \quad c_-(\mathbf{z}) = c^{eff}\exp\left(+\frac{FU(\mathbf{z})}{RT}\right) \tag{4.161}$$

Electroneutrality condition (4.152) can now be expressed in the form:

$$2\phi F c^{eff}\left\langle f_{\omega_f}(\mathbf{z})\sinh\frac{FU(\mathbf{z})}{RT}\right\rangle = sq \tag{4.162}$$

[15] See in particular Israelachvili J. (1991), *Intermolecular and Surface Forces*, Second Edition, Academic Press, London.

The determination of the macroscopic concentrations c_+ and c_- as a function of ϕ and c^{eff} eventually turns out to be equivalent to the determination of the electric potential field $U(\mathbf{z})$. To this end we introduce the local electric field $\mathbf{E}(\mathbf{z})$ which is derived from $U(\mathbf{z})$ according to:

$$\mathbf{E} = -\nabla_{\mathbf{z}} U \tag{4.163}$$

and which has to satisfy the Poisson electrostatic equation:

$$\nabla_{\mathbf{z}} \cdot (\varepsilon\varepsilon_0 \mathbf{E}) = q \tag{4.164}$$

The electric potential field $U(\mathbf{z})$ must therefore satisfy the field equation:

$$\nabla^2_{\mathbf{z}} U = -\frac{q}{\varepsilon\varepsilon_0} \tag{4.165}$$

where $\varepsilon\varepsilon_0$ stands for the permittivity of the medium, ε_0 being the permittivity of free space ($\varepsilon_0 = 8.854 \times 10^{-12}$ C^2 J/m) and ε, the assumed constant relative permittivity ($\varepsilon = 80$ at 293 K in an aqueous solution), while q is the local excess of charge, that is:

$$q(\mathbf{z}) = F(c_+(\mathbf{z}) - c_-(\mathbf{z})) \tag{4.166}$$

Taking into account the electroneutrality condition (4.152), integration of (4.165) over the porous volume ω_f and the use of the divergence theorem produce the useful relation:

$$\phi \int_{\partial\omega_f} (\nabla_{\mathbf{z}} \cdot U) \cdot \mathbf{n}_f da_\omega = \frac{sq\omega_f}{\varepsilon\varepsilon_0} \tag{4.167}$$

Equations (4.161), (4.165) and (4.166) finally produce the celebrated Poisson–Boltzmann equation governing the electric potential $U(\mathbf{z})$:

$$\nabla^2_{\mathbf{z}} \left(\frac{FU(\mathbf{z})}{RT} \right) = \frac{1}{\ell^2} \sinh \left(\frac{FU(\mathbf{z})}{RT} \right) \tag{4.168}$$

where ℓ is the Debye length whose expression is:

$$\ell = \sqrt{\frac{\varepsilon\varepsilon_0 RT}{2F^2 c^{eff}}} \tag{4.169}$$

The Debye length scales the 'thickness' of the electrical double layer, that is the charged layer close to the charged internal walls of the porous network and within which the electric potential $U(\mathbf{z})$ significantly varies (see Table 4.2 below for typical values of ℓ).

Once the morphology of the porous space is, specified the non-linear Poisson–Boltzmann equation (4.168) has to be solved. To this end let d be the length scaling the current porous volume defined according to:

$$s \times \frac{d}{2} = \phi \tag{4.170}$$

Table 4.2: Typical values of the parameters involved in the electric double layer theory, in an aqueous solution at $T = 293$ K, for strength $|q|$ of the electric charge ranging from 0.01 to 0.2 C/m^2

c^{eff} (mol/l^{-1})	ℓ (nm)	$2F\ell c^{eff}$ (C/m^{-2})	$\overline{q}$ (–)
10^{-4}	30.4	5.9×10^{-4}	17 – 340
10^{-3}	9.7	1.9×10^{-3}	5.4 – 107.6
10^{-2}	3.0	5.9×10^{-3}	1.70 – 34
10^{-1}	0.96	1.9×10^{-2}	0.54 – 10.8

In the case of colloidal mixtures whose solid particles are formed from platelets as for clays, d eventually scales the averaged current distance between the particles. Indeed, if N stands for the number of particles per unit of initial volume $d\Omega_0$, ϕ/N represents the averaged porous volume attached to a single particle. The latter can be alternatively assessed as $s/N \times d/2$, since s/N represents the averaged surface of a single particle. Let us then examine the 1D geometry where ω_f reduces to the gap d lying between two infinite planes (see Fig. 4.5), so that (4.168) is specified in the 1D form:

$$\frac{d^2}{d\overline{z}^2}\left(\frac{FU(z)}{RT}\right) = \sinh\left(\frac{FU(z)}{RT}\right) \tag{4.171}$$

where:

$$\overline{z} = \frac{z}{\ell}: \quad -\frac{d}{2\ell} < \overline{z} < \frac{d}{2\ell} \tag{4.172}$$

Since the unit of surface of the two planes carries the same charge q, the problem at hand is symmetric with respect to the midplane $\overline{z} = 0$ where the electric field must be zero, resulting in:

$$\frac{d}{d\overline{z}}\left(\frac{FU}{RT}\right)|_{\overline{z}=0} = 0 \tag{4.173}$$

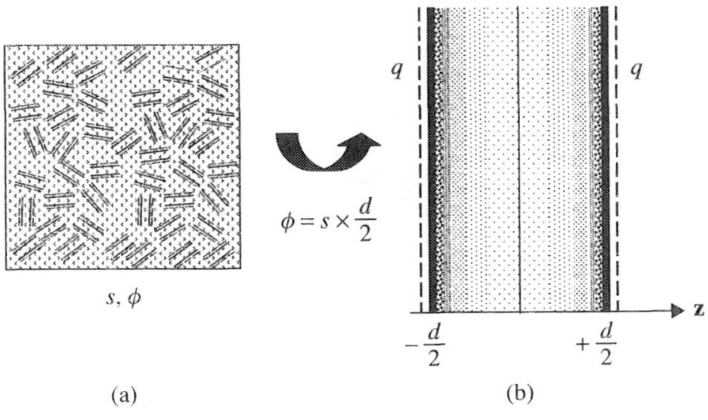

Figure 4.5: Colloidal mixture with porosity ϕ, whose solid particles are formed from platelets carrying the electric charge q per unit of surface and having s as overall particle surface per unit of macroscopic volume (a), idealized by two infinitely charged plates delimiting a gap of width d filled by the same solution (b).

whereas relation (4.167) is specialized in the form:

$$\frac{d}{d\overline{z}}\left(\frac{FU}{RT}\right)|_{\overline{z}=\pm d/2\ell}= \pm\text{sgn }(q)\,\overline{q} \tag{4.174}$$

where sgn (q) stands for the sign of the electric charge q, while $\overline{q}$ is the normalized strength of the latter:

$$\overline{q} = \frac{|q|}{2F\,\ell c^{eff}} \tag{4.175}$$

The operational approach to expression (4.157) of tangent Biot's coefficient b_e is required to solve the Poisson–Boltzmann equation (4.171). For this purpose, we first introduce the notation:

$$u_{d/2} = \frac{FU\left(\frac{d}{2\ell}\right)}{RT}; \quad u_0 = \frac{FU\,(0)}{RT} \tag{4.176}$$

Taking into account symmetry condition (4.173), together with (4.174) for the sign of $dU/d\overline{z}$, an integration of (4.171) premultiplied by $d/d\overline{z}(FU/RT)$ gives:

$$\overline{z} = \text{sgn }(q)\int_{u_0}^{\frac{FU}{RT}} \frac{du}{\sqrt{2}\,(\cosh u - \cosh u_0)} \tag{4.177}$$

Expressing boundary condition (4.174) with the help of (4.177), we require u_0 and $u_{d/2}$ to satisfy:

$$\sqrt{\cosh^2\frac{u_{d/2}}{2} - \cosh^2\frac{u_0}{2}} = \frac{1}{2}\overline{q} \tag{4.178}$$

The latter relation suggests the change of variable:

$$v = \sqrt{\cosh^2\frac{u}{2} - \cosh^2\frac{u_0}{2}} \tag{4.179}$$

Letting $\overline{z} = d/2\ell$ in (4.177), Eqs. (4.177)–(4.180) yield the relation leading to the determination of u_0:

$$\frac{d}{2\ell} = \frac{\phi}{s\ell} = \int_0^{\frac{1}{2}\overline{q}} \frac{dv}{\sqrt{v^2 + \cosh^2\frac{u_0}{2}} \times \sqrt{v^2 + \sinh^2\frac{u_0}{2}}} \tag{4.180}$$

The inverses $1/A_+$ and $1/A_-$ of the ionic activities are now specified in the form:

$$\frac{1}{A_+} = \frac{\int_{-\frac{d}{2\ell}}^{+\frac{d}{2\ell}} c_+\,(\overline{z})\,d\overline{z}}{\frac{d}{\ell}c^{eff}}; \quad \frac{1}{A_-} = \frac{\int_{-\frac{d}{2\ell}}^{+\frac{d}{2\ell}} c_-\,(\overline{z})\,d\overline{z}}{\frac{d}{\ell}c^{eff}} \tag{4.181}$$

By switching the integration variable from $\bar{z}$ to $u = FU/RT$, a combination of (4.161), (4.170) and (4.177)–(4.181) provides the following expressions:

$$\frac{1}{A_+} = 1 - \text{sgn}\ (q)\ \frac{s\ell}{\phi} \times \bar{q} + 2\frac{s\ell}{\phi} \times \int_0^{\frac{1}{2}\bar{q}} \sqrt{\frac{v^2 + \sinh^2 \frac{u_0}{2}}{v^2 + \cosh^2 \frac{u_0}{2}}} dv \tag{4.182a}$$

$$\frac{1}{A_-} = 1 + \text{sgn}\ (q)\ \frac{s\ell}{\phi} \times \bar{q} + 2\frac{s\ell}{\phi} \times \int_0^{\frac{1}{2}\bar{q}} \sqrt{\frac{v^2 + \sinh^2 \frac{u_0}{2}}{v^2 + \cosh^2 \frac{u_0}{2}}} dv \tag{4.182b}$$

From the definition (4.175) of $\bar{q}$, the upper limit of integration in the integrals involved in (4.180) and (4.182) does not depend on ϕ. Use of (4.157), where we let $p_e = RTc^{eff}$, together with (4.180) and (4.182) in order to obtain the derivatives of u_0 with respect to ϕ, finally allows us to express Biot's coefficient b_e in the semi-explicit form:

$$b_e = 1 - \frac{\displaystyle\int_0^{\frac{1}{2}\bar{q}} \left[v^2 + \sinh^2 \frac{u_0}{2}\right]^{-\frac{1}{2}} \left[v^2 + \cosh^2 \frac{u_0}{2}\right]^{-\frac{3}{2}} dv}{\displaystyle\int_0^{\frac{1}{2}\bar{q}} \left[v^2 + \frac{1}{2}\cosh u_0\right] \left[v^2 + \sinh^2 \frac{u_0}{2}\right]^{-\frac{3}{2}} \left[v^2 + \cosh^2 \frac{u_0}{2}\right]^{-\frac{3}{2}} dv} \tag{4.183}$$

The linearized approach to b_e consists in assuming $F\,|U|\,/RT \ll 1$ (in practice requiring $|U|$ to be less than 0.01 V), so that the differential system (4.171)–(4.174) approximately integrates in the form:

$$\frac{FU}{RT} \simeq u_0 \cosh \bar{z}; \quad u_0 = \text{sgn}\ (q)\ \frac{\bar{q}}{\sinh \frac{d}{2\ell}} \tag{4.184}$$

which allows us to derive explicit expressions for $c_+(\bar{z})$ and $c_-(\bar{z})$, and, consequently, for $1/A_+$ and $1/A_-$. In the limit $F|U|/RT \ll 1$ it can be finally shown that the last term in expressions (4.182) of $1/A_+$ or $1/A_-$ is approximated in the form:

$$2\frac{s\ell}{\phi} \times \int_0^{\frac{1}{2}\bar{q}} \sqrt{\frac{v^2 + \sinh^2 \frac{u_0}{2}}{v^2 + \cosh^2 \frac{u_0}{2}}} dv \simeq \left(1 + \frac{s\ell}{2\phi} \sinh \frac{2\phi}{s\ell}\right) \times \left(\frac{\bar{q}}{2 \sinh \frac{\phi}{s\ell}}\right)^2 \tag{4.185}$$

Again using (4.157) with $p_e = RTc^{eff}$, (4.182) and the linearized approximation give:

$$\frac{\bar{q}}{\tanh \frac{\phi}{s\ell}} \ll 1: \quad b_e \simeq 1 - \frac{\phi}{2s\ell} \frac{\cosh \frac{\phi}{s\ell}}{\sinh^3 \frac{\phi}{s\ell}} \times \bar{q}^2 \tag{4.186}$$

From Maxwell's symmetry relations the derivation of the semi-analytic expression (4.183) or the expression (4.186) for b_e can be independently and alternatively carried out from (4.147) provided that the expression of the swelling pressure p_{sw} is known. With the aim of determining the latter, let us first notice that, since the electric field is zero at the midplane $z = 0$, the upper half-layer $0 < z < d/2$ does not exert long-range electrostatic forces upon the lower half-layer $0 > z > -d/2$, and vice versa. As a consequence the interaction force between the two half-layers reduces to contact forces at $z = 0$. Accordingly the swelling pressure p_{sw} is the excess of pressure exerted at the mid-plane $z = 0$ by the cationic and the anionic components with respect to the thermodynamic pressure $p_e = 2RTc^{eff}$, resulting in:

$$p_{sw} = RTc_+(0) + RTc_-(0) - 2RTc^{eff} \tag{4.187}$$

yielding:

$$p_{sw} = \left[\cosh u_0 - 1\right] \times p_e \tag{4.188}$$

Using $p_e = 2RTc^{eff}$, (4.169), (4.175), (4.180) or (4.184) in order to provide the derivatives of u_0 with respect to p_e, expressions (4.183) or (4.186) can be independently retrieved from (4.147) and (4.188), although such a procedure turns out to be quite lengthy for the derivation of expression (4.183).

The values of u_0 and b_e, as functions of both $d/2\ell = \phi/s\ell$ and $\overline{q}$, can be successively computed from (4.180) and (4.183). Typical values for the set of parameters involved in the electrical double layer theory are given in Table 4.2. In addition, for small values of $\overline{q}$, (4.184) provides a convenient starting guess for u_0 when carrying out the numerical iterative procedure required to solve (4.180) with respect to u_0 for given values of $\phi/s\ell$ and $\overline{q}$. In Fig. 4.6 we plotted tangent Biot's coefficient b_e against the normalized electric charge $\overline{q}$ for various values of $\phi/s\ell$, the dashed lines representing the linearized approximation (4.186). As the effective concentration c^{eff} decreases, the molecular agitation and, consequently, the electrolyte thermodynamic pressure $p_e = 2RTc^{eff}$ decrease too. In the meantime, the electrostatic interactions become stronger in comparison with the effects of molecular agitation, and the normalized electric charge $\overline{q}$ increases, resulting in an increasing ratio of the swelling pressure p_{sw} versus the electrolyte thermodynamic pressure p_e (see (4.146)). As a result, tangent Biot's coefficient b_e decreases from one.

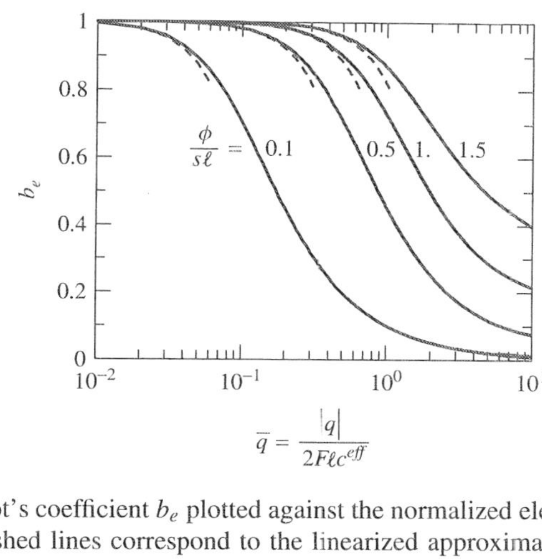

Figure 4.6: Tangent Biot's coefficient b_e plotted against the normalized electric charge $\overline{q}$ for various values of $\phi/s\ell$. The dashed lines correspond to the linearized approximation (4.186).

At the same values of the normalized electric charge $\overline{q}$, the smaller the value of $\phi/s\ell$, the larger the relative range of the electrostatic interactions and the smaller the value of b_e.

4.4.4 From Poroelasticity to Chemoelasticity and Ageing Materials

Poroelastic material subjected to a chemical reaction

Let us now address the case of a porous material within which a chemical reaction occurs, with no mass exchange with the outside (closed system). When considering infinitesimal elastic transformations, with no other internal variables than the extent of the reaction, state equations (3.150) associated with such a material reduce to:

$$\psi = \psi(\varepsilon_{ij}, T; \xi); \quad \sigma_{ij} = \frac{\partial \psi}{\partial \varepsilon_{ij}}; \quad S = -\frac{\partial \psi}{\partial T}; \quad \mathcal{A} = -\frac{\partial \psi}{\partial \xi} \tag{4.189}$$

while the kinetics of the reaction is governed by (see (3.151)):

$$\mathcal{A} = \eta \frac{d\xi}{dt} \exp\left(\frac{E_a}{RT}\right) \tag{4.190}$$

Alternatively, we can use energy $\mathcal{G}$ defined by:

$$\mathcal{G} = \sigma_{ij}\varepsilon_{ij} - \psi \tag{4.191}$$

and rewrite the state equations in the form:

$$\mathcal{G} = \mathcal{G}(\sigma_{ij}, T; \xi); \quad \varepsilon_{ij} = \frac{\partial \mathcal{G}}{\partial \sigma_{ij}}; \quad S = \frac{\partial \mathcal{G}}{\partial T}; \quad \mathcal{A} = \frac{\partial \mathcal{G}}{\partial \xi} \tag{4.192}$$

Differentiating (4.192) we can write:

$$d\varepsilon_{ij} = S_{ijkl}\, d\sigma_{kl} + \alpha_{ij}\, dT + \beta_{ij}\, d\xi \tag{4.193a}$$

$$dS = \alpha_{ij}\, d\sigma_{ij} + C\,\frac{dT}{T} - L\,\frac{d\xi}{T} \tag{4.193b}$$

$$d\mathcal{A} = \beta_{ij}\, d\sigma_{ij} - L\,\frac{dT}{T} - a\, d\xi \tag{4.193c}$$

In addition to the standard thermoelastic properties, $S_{ijkl} = C^{-1}_{ijkl}$, α_{ij} and C, three new properties, β_{ij}, L and a, are included. In (4.193a), β_{ij} is the ij^{th} component of the tensor of chemical dilation coefficients with symmetry $\beta_{ij} = \beta_{ji}$. As the chemical extent increases by quantity $d\xi$, the chemical reaction produces the strain $\beta_{ij}\, d\xi$. The associated volumetric strain $\beta_{ii}\, d\xi$ can be either a chemical dilation, that is $\beta_{ii} > 0$, or a chemical shrinkage, that is $\beta_{ii} < 0$. The latter case is illustrated in Fig. 4.7 by the shrinkage undergone by a concrete sample during its setting. In (4.193b), L is the latent heat associated with the reaction so that the reaction in progress produces the heat $L d\xi$ and can be either exothermic, that is $L > 0$, or endothermic, that is $L < 0$. Finally, in (4.193c), a accounts for the decrease in the chemical affinity per unit of reaction extent in an isostress and isothermal experiment.

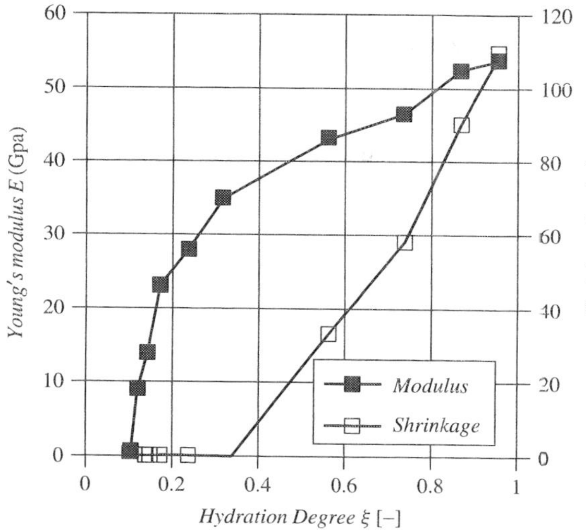

Figure 4.7: Young's modulus related to a concrete sample at early ages during the setting process and plotted against the hydration degree. The hydration degree of a concrete sample ξ is the ratio of the water mass currently bound to the hydrates to the maximal one which can be expected as time goes to infinity. Below some percolation threshold with regard to the hydration degree, here close to 0.1, Young's modulus is still zero because the hydrated cement grains have not yet come into contact to form the solid skeleton. During the setting process concrete undergoes a chemical shrinkage due both to the capillary forces arising during the process and to Le Châtelier contraction resulting from the difference in volume between the products (here the hydrates) and the reactants (here the liquid water and the anhydrous cement particles). The data are from Ulm, Coussy (1998), see footnote 17.

Poroelastic material subjected to a dissolution process

Returning to open systems, consider now a poroelastic material subjected to a dissolution process whose thermodynamics has been examined in §3.6.3. Restricting consideration to isothermal evolutions and isotropic materials, and accounting for the existence of the irreversible porosity ϕ_{ch} produced by the dissolution process, state equations (4.21) can be extended in the form:

$$\sigma - \sigma_0 = K_u \epsilon - bN(\phi - \phi_0 - \phi_{ch}) \tag{4.194a}$$

$$s_{ij} - s_{ij}^0 = 2\mu\, e_{ij} \tag{4.194b}$$

$$p - p_0 = N(-b\epsilon + \phi - \phi_0 - \phi_{ch}) \tag{4.194c}$$

where $K_u = K + b^2 N$. Skeleton poroelastic properties K, b, μ, N now depend on ϕ_{ch} (see for instance, in Table 4.1, the values of poroelastic properties related to intact and asymptotically leached cement-based materials) so that relations (4.35) can be extended in the form:

$$b = 1 - \frac{K(\phi_{ch})}{K_s}; \quad \frac{1}{N} = \frac{b(\phi_{ch}) - \phi_0 - \phi_{ch}}{K_s} \tag{4.195}$$

In the approximation of infinitesimal transformations, state equations (3.156) are:

$$\Psi_s = \Psi_s(\varepsilon_{ij}, \phi - \phi_{ch}) : \quad \sigma_{ij} = \frac{\partial \Psi_s}{\partial \varepsilon_{ij}}; \quad p = \frac{\partial \Psi_s}{\partial \phi} \tag{4.196}$$

Using (4.194), state equations (4.196) can be integrated in the form:

$$\begin{aligned}\Psi_s = W_s(\varepsilon_{ij}, \phi - \phi_{ch}, \phi_{ch}) &= \sigma_0 \epsilon + s_{ij}^0\, e_{ij} + p_0 \phi - \rho_s^0 \mu_s^0 \phi_{ch} \\ &+ \frac{1}{2} K_u\, (\phi_{ch}) \epsilon^2 - (bN)\ (\phi_{ch}) \epsilon\, (\phi - \phi_0 - \phi_{ch}) \\ &+ \frac{1}{2} N(\phi_{ch})(\phi - \phi_0 - \phi_{ch})^2 + \mu\, (\phi_{ch}) e_{ij} e_{ji}\end{aligned} \tag{4.197}$$

From (4.197) we derive:

$$-\frac{\partial \Psi_s}{\partial \phi_{ch}} = \rho_s^0 \left(\mu_s^0 + \Delta \mu_s\right) \tag{4.198}$$

where $\rho_s^0 \Delta \mu_s$ is expressed in the form:

$$\begin{aligned}\rho_s^0 \Delta \mu_s = p - p_0 &- \frac{1}{2} \frac{\partial K_u}{\partial \phi_{ch}} \epsilon^2 - \frac{\partial\, (bN)}{\partial \phi_{ch}} \epsilon\, (\phi - \phi_0 - \phi_{ch}) \\ &- \frac{1}{2} \frac{\partial N}{\partial \phi_{ch}} (\phi - \phi_0 - \phi_{ch})^2 - \frac{\partial \mu}{\partial \phi_{ch}}\, e_{ij} e_{ji}\end{aligned} \tag{4.199}$$

Comparing (4.198)–(4.199) and (3.167)–(3.168), $\Delta \mu_s$ identifies the increase of the initial chemical potential μ_s^0 of the solid matrix due to the elastic energy stored during the solid deformation. In the dissolution process the material releases this elastic energy and softens so that the derivatives of poroelastic properties K_u, bN, μ with respect to chemical porosity ϕ_{ch} are negative. In fact $\Delta \mu_s$ represents an averaged value within the solid matrix. Such a use of an averaged value is relevant with regard to the analysis of the dissolution process as long as there is no significant stress concentration, neither around possible sharply shaped faults nor along the bonds between grains as in the pressure-dissolution context examined in the next section. Indeed, in the approximation of infinitesimal transformations $\Delta \mu_s$ turns out to be quite negligible with respect to the initial chemical potential μ_s^0 of the solid matrix.

Pressure-dissolution process

According to (4.193c), the driving force of the reaction, that is the chemical affinity $\mathcal{A}$, a priori depends on the applied stress σ_{ij}. Roughly speaking, $\mathcal{A}$ stands for the difference between the overall chemical potential of the reactants and that of the products (see §3.6.3). In the pressure-dissolution process for instance, according to the analysis of §3.6.3 and as recalled in the previous section, $\mathcal{A}$ represents the difference between the chemical potential of the solid matrix constituent subjected to the dissolution process, and the chemical potential of the same constituent in solute form within the interstitial solution. When subjected to the macroscopic external loading σ_{ij}, stress locally concentrates along the bonds between the solid grains forming the matrix. Accordingly, for closed systems subjected to a pressure-dissolution process, the term $\beta_{ij}(\sigma_{kl}, T, \xi) d\sigma_{ij}$ in (4.193c) accounts for the effect of the stress concentration on the chemical potential of the matrix component which dissolves along these bonds. In the pressure-dissolution process the kinetics of dissolution is eventually controlled by the diffusion of the solute towards regions of the porous space where the solid in contact is poorly stressed. Equation (4.190), where $d\xi/dt$ can eventually be identified as the rate $d\phi_{ch}/dt$ of solid volume dissolving per unit of initial volume, can be interpreted this way. Assuming for the sake of simplicity that tangent chemoelastic properties are constant (linear approach) and disregarding thermal effects, we combine (4.190), (4.193a) and (4.193c) in the form:

$$\sigma_{ij} = C_{ijkl}(\varepsilon_{kl} - \varepsilon_{kl}^v); \quad \sigma_{ij} - C_{ijkl}\varepsilon_{kl}^v = \eta_{ijkl}\frac{d\varepsilon_{kl}^v}{dt} \tag{4.200}$$

where we assumed that the reference state is stress-free and associated with a zero affinity. In addition we note:

$$\begin{aligned} \varepsilon_{ij}^v &= \beta_{ij}\xi; & C_{ijkl} &= S_{ijkl}^{-1} \\ C_{ijkl}^{-1} &= a\beta_{ij}\beta_{kl}; & \eta_{ijkl}^{-1} &= \eta\beta_{ij}\beta_{kl} \end{aligned} \tag{4.201}$$

It is worthwhile to note that constitutive equations (4.200) are those of a material exhibiting a linear viscoelastic behaviour, ε_{kl}^v being the viscous strain (see Chapter 9). Based on experimental evidence, a Nabarro–Herring-type creep[16] is often assumed for the volumetric strain rate, namely:

$$\sigma_{ii} = k\frac{d\varepsilon_{ii}}{dt}; \quad k \propto K_s\frac{d^2}{D} \tag{4.202}$$

where d is the grain diameter and D the diffusion coefficient of the solute in the interstitial solution. Creep law (4.202) substitutes for (4.190). It finally turns out to neglect the elastic contribution to the strain, that is to let $\varepsilon_{ii} \simeq \varepsilon_{ii}^v$, and to express affinity $\mathcal{A}$ as a linear function of σ_{ii}.

Chemoelastic ageing materials

The pressure-dissolution process concerns high levels of stress. In most cases the contribution of the applied stress and that of the temperature to the chemical potentials can be neglected in comparison with the energy of the chemical bonds. Accordingly we integrate (4.193c) in the form:

$$\mathcal{A} = \mathcal{A}_0 - f(\xi) \tag{4.203}$$

where $\mathcal{A}_0$ stands for the initial affinity provoking the reaction. Substituting (4.203) into (4.190) we derive:

$$\mathfrak{M} = \int_0^\xi \frac{\eta\, dx}{\mathcal{A}_0 - f(x)} = \int_0^t \exp\left(-\frac{E_a}{RT}\right) d\tau \tag{4.204}$$

[16]Nabarro F.N. (1948), 'Report of a Conference on Strength of Solids', *Physical Society*, London, 75.

Since the mechanical tangent properties depend on ξ, the material will exhibit apparent ageing due to the internal chemical reaction. Figure 4.7 illustrates such an apparent ageing as it concerns the Young modulus at early ages of a concrete sample during the setting. Owing to (4.204), instead of reaction extent ξ, the defined maturity $\mathfrak{M}$ can conveniently play the role of the ageing variable. Indeed, when using the right hand member of (4.204) for its measurement, the form of $\mathfrak{M}$ can be achieved by following the temperature history. Two material samples having the same maturity $\mathfrak{M}$ will exhibit the same mechanical behaviour, irrespective of their specific temperature history.[17]

[17]The chemomechanics approach developed in this section is particularly well suited to the approach of the ageing of concrete during the hydration process producing its setting. See Ulm F.-J., Coussy O. (1998), 'Couplings in early-age concrete: from material modelling to structural design', *Journal of Solids and Structures*, **35**, (31–32), 4295–4311.

Chapter 5

Problems of Poroelasticity

This chapter is devoted to quasistatic problems of poroelasticity when the dynamic terms are not considered. The attention mainly focuses on the linearized problems of poroelasticity under the assumption of small perturbations. By choosing the displacement and the fluid pressure as principal unknowns, the general set of field equations to be solved irrespective of any specific problem can be firstly derived. The fluid diffusion process is then analyzed in order to highlight the short- and long-term approximations. Finally, standard and instructive problems of poroelasticity are solved.[1]

5.1 Linearized Poroelasticity Problems

5.1.1 The Hypothesis of Small Perturbations

The hypothesis of small perturbations consists in the whole set of hypotheses guaranteeing the linearity of the problem, as far as the constitutive equations of the constituents are not concerned. The small perturbation hypothesis mainly includes:

- The hypothesis of infinitesimal transformations:

$$\|\nabla \boldsymbol{\xi}\| \ll 1 \tag{5.1}$$

  Under this hypothesis the Lagrangian and Eulerian approaches coincide up to a first-order approximation. In particular the linearized strain tensor $\boldsymbol{\varepsilon}$ can be substituted for the Lagrange strain tensor $\boldsymbol{\Delta}$ (see (1.26)).
- The hypothesis of small displacements $\boldsymbol{\xi}$ for the skeleton particles, which is:

$$\|\boldsymbol{\xi}/L\| \ll 1 \tag{5.2}$$

  where L is the length scaling the dimensions of the porous structure. The hypothesis of small displacements allows us to merge the initial configuration and the

[1]For an exhaustive presentation of solved problems in linear poroelasticity see Wang H.F. (2000), *Theory of Linear Poroelasticity*, Princeton Series in Geophysics, Princeton University Press.

Poromechanics O. Coussy
 ISBN 0-470-84920-7

current configuration as far as the space argument of the unknown fields is concerned. This is equivalent to writing any unknown field f as a function of the initial location $\mathbf{X}$ or the current location of the skeleton particle $\mathbf{x}$, that is $\mathbf{X} \simeq \mathbf{x}$, $f(\mathbf{X}) \simeq f(\mathbf{x})$. Similarly we note that $\Omega = \Omega_0 \equiv \Omega_t$ and also for the border, $\partial\Omega \equiv \partial\Omega_0 \equiv \partial\Omega_t$.

- The hypothesis of small variations of Lagrangian porosity, which is:

$$\left|\frac{\phi - \phi_0}{\phi_0}\right| \ll 1 \tag{5.3}$$

where, as in what follows, the index 0 refers to the initial value.

- The hypothesis of small variations of the fluid mass density ρ_f, which is:

$$\left|\frac{\rho_f - \rho_f^0}{\rho_f^0}\right| \ll 1 \tag{5.4}$$

This hypothesis allows us to replace ρ_f by ρ_f^0 whenever required.

When adopting the small perturbation hypothesis and letting the body forces $\mathbf{F}$ be the gravity forces $\mathbf{g}$, the momentum balance is:

$$\text{in } \Omega: \quad \nabla \cdot \boldsymbol{\sigma} + \rho\mathbf{g} = 0 \tag{5.5}$$

where $\rho = \rho_s^0(1 - \phi_0) + \rho_f^0\phi_0$. Indeed the overall initial mass density is $\rho^0 = \rho_s^0(1 - \phi_0) + \rho_f^0\phi_0$, while the overall current mass density per unit of initial volume $d\Omega_0$ is $\rho = \rho_s^0(1 - \phi_0) + \rho_f\phi$. Assumptions (5.3) and (5.4) allow us to set $\rho = \rho_0$.

In the quasistatic limit Darcy's law (3.39) requires:

$$\mathcal{V} = \frac{\mathbf{w}}{\rho_f^0} = k(-\nabla p + \rho_f^0\mathbf{g}) \tag{5.6}$$

Furthermore, in the limit of small perturbations the fluid continuity equation (1.67) becomes:

$$\frac{\partial m_f}{\partial t} = -\nabla \cdot \mathbf{w} \tag{5.7}$$

where, as in the following, we adopted the notation $\partial(\cdot)/\partial t = (d(\cdot)/dt)(\mathbf{X}, t)$. Substitution of Darcy's law into the fluid continuity equation leads to:

$$\frac{\partial}{\partial t}\left(\frac{m_f}{\rho_f^0}\right) = k\nabla^2 p \tag{5.8}$$

where the initial fluid mass density ρ_f^0 is assumed to be uniform.

Except in regard to special points of fluid injection or application of concentrated loads, the fluid pressure is a regular function of space and $\nabla^2 p$ remains finite everywhere. Equation (5.8) then shows that the derivative $\partial m_f/\partial t$ has to remain finite too, so that fluid mass content m_f cannot undergo discontinuities with respect to time. Indeed, the resistant viscous forces associated with the fluid prevent any instantaneous displacements of the latter. The result is that, irrespective of specific constitutive equations, the instantaneous response of the porous medium to any external loading is undrained. We write:

$$m_f(t = 0^+) = m_f^0 \tag{5.9}$$

where m_f^0 stands for the initial fluid mass content prior to the application of the loading.

5.1.2 Field Equations and Boundary Conditions

The complete linearization requires the constitutive equations of the constituents to be linear and we add two hypotheses:

- The hypothesis of linear poroelasticity for the skeleton. With zero initial stress and fluid pressure, we write:

$$\sigma_{ij} = \left(K - \frac{2}{3}\mu\right)\epsilon\,\delta_{ij} + 2\mu\,\varepsilon_{ij} - b\,p\,\delta_{ij} \tag{5.10}$$

- The hypothesis of a linear behaviour for the fluid, which means to consider the tangent fluid bulk modulus K_f in (4.53) as a constant.

It is instructive to note that the hypothesis of small perturbations and the hypothesis of a linear behaviour for the constituents are not actually independent. For instance, there is always some range of strain and pressure such as the constitutive equations of a poroelastic skeleton that can be linearized in the form (5.10). Similarly there is always some range of the fluid mass density variation, that is $(\rho_f - \rho_f^0)/\rho_f^0$, such that K_f may be considered as nearly constant in (4.53). However, the validity of such ranges depends upon the skeleton and the fluid forming the porous material. For instance, this range will not be the same for a liquid or for a gas. In addition the range of validity of the hypothesis of small variations of any quantity can eventually be proven a posteriori, only when the linear solution has been derived for the problem at hand.

Using (1.26) and (1.27) and substituting (5.10) into (5.5), we derive:

$$\left(K + \frac{4}{3}\mu\right)\nabla\,(\nabla\cdot\boldsymbol{\xi}) - \mu\,\nabla\times(\nabla\times\boldsymbol{\xi}) - b\nabla p + \rho\mathbf{g} = 0 \tag{5.11}$$

which is known as Navier's equation. Under the assumption of small isothermal perturbations and zero initial fluid pressure, we can integrate (4.62b) in the form:

$$v_f = b\epsilon + \frac{p}{M} \tag{5.12}$$

where $v_f\,d\Omega$ is the current change in fluid volume content, that is:

$$v_f = \frac{m_f - m_f^0}{\rho_f^0} \tag{5.13}$$

while M is the modulus defined by:

$$\frac{1}{M} = \frac{1}{N} + \frac{\phi_0}{K_f} \tag{5.14}$$

Substitution of (5.12) into (5.8) yields:

$$b\frac{\partial \epsilon}{\partial t} + \frac{1}{M}\frac{\partial p}{\partial t} = k\,\nabla^2 p \tag{5.15}$$

which governs the fluid diffusion through the porous medium. According to (5.9), (5.12) and (5.13), the instantaneous undrained response can be expressed in the form:

$$\epsilon(t = 0^+) = -\frac{1}{bM}\,p(t = 0^+) \tag{5.16}$$

Field equations (5.11) and (5.15) have to be completed by boundary conditions on the border $\partial\Omega$ of the porous structure:

- The mechanical conditions concern either the displacement field or the stress vector. We write:

$$\text{on } \partial_T\Omega:\ \sigma_{ij}n_j = T_i^d;\quad \text{on } \partial_\xi\Omega:\ \xi_k = \xi_k^d\ (i \neq k,\ \partial\Omega = \partial_T\Omega \cup \partial_\xi\Omega) \tag{5.17}$$

- The hydraulic conditions concern either the fluid pressure or the fluid flow. We write:

$$\text{on } \partial_p\Omega:\ p = p^d;\quad \text{on } \partial_w\Omega:\ \mathcal{V}\cdot\mathbf{n} = \mathcal{V}^d\ (\partial\Omega = \partial_p\Omega \cup \partial_w\Omega) \tag{5.18}$$

  where T_i^d and ξ_k^d, p^d and $\mathcal{V}^d$, stand for the components of, respectively, the stress vector and the displacement vector, the fluid pressure and the flow of fluid volume, all of them being imposed on the specified part of the border $\partial\Omega$ of the porous domain.

5.1.3 The Diffusion Equation

Recalling that $\nabla \cdot \xi = \epsilon$ and applying the divergence operator ∇ to (5.11), we derive:

$$\left(K + \frac{4}{3}\mu\right)\nabla^2\epsilon - b\,\nabla^2 p = 0 \tag{5.19}$$

which relates the volumetric dilation ϵ to the fluid pressure p and is known as the dilation equation. Furthermore, the application of the Laplace operator ∇^2 to (5.12) gives:

$$\nabla^2 v_f = b\,\nabla^2\epsilon + \frac{1}{M}\nabla^2 p \tag{5.20}$$

With $K_u = K + b^2 M$, (5.8), (5.19) and (5.20) combine to give the diffusion equation:

$$\frac{\partial v_f}{\partial t} = c_f\,\nabla^2 v_f \tag{5.21}$$

where c_f is the fluid diffusivity coefficient:

$$c_f = kM \frac{K + \frac{4}{3}\mu}{K_u + \frac{4}{3}\mu} \tag{5.22}$$

where $K + \frac{4}{3}\mu$ and $K_u + \frac{4}{3}\mu$ are respectively the drained and the undrained and oedometric moduli (see (4.23)).

It is of interest to note that, whatever the problem considered, the same *uncoupled* diffusion equation (5.21) governs the evolution of the fluid content. By contrast, the diffusion equation (5.15) governing the fluid pressure remains generally coupled, owing to the presence of term $\partial\epsilon/\partial t$. However, there is a special case of interest where the diffusion equation for the fluid pressure p becomes uncoupled too. This is the case where the displacement is irrotational.[2] Indeed, when letting $\nabla \times \boldsymbol{\xi} = 0$ in (5.11), the latter integrates in the form:

$$\epsilon = \frac{b}{K + \frac{4}{3}\mu}\, p + f(t) \tag{5.23}$$

where $f(t)$ is an integration function and where for the sake of simplicity we did not consider the gravity forces. When the extent of the domain is infinite, both ϵ and p must vanish at infinity and the integration function $f(t)$ turns out to be zero so that:

$$\epsilon = \frac{b}{K + \frac{4}{3}\mu}\, p \tag{5.24}$$

Substitution of (5.24) into (5.12) gives:

$$p = M \frac{K + \frac{4}{3}\mu}{K_u + \frac{4}{3}\mu} v_f; \qquad \epsilon = \frac{bM}{K_u + \frac{4}{3}\mu} v_f \tag{5.25}$$

Use of the first relation of (5.25) in (5.21) eventually provides the uncoupled pressure diffusion equation:

$$\frac{\partial p}{\partial t} = c_f \nabla^2 p \tag{5.26}$$

5.2 Solved Problems of Poroelasticity

5.2.1 Injection of a Fluid

Point injection of fluid mass

The injection of liquid and its progressive diffusion is a problem of considerable interest in biomechanics or environmental geotechnics. This motivates us to consider the instantaneous injection at time $t = 0$ of a fluid mass M_f or, equivalently, of a fluid volume

[2]See in particular Detournay E., Cheng A. H.-D. (1993), 'Fundamentals of poroelasticity', *Comprehensive Rock Engineering*, Vol. II, ed. Hudson J., Pergamon Press, Oxford and New York.

$V_f = M_f/\rho_f^0$ at the origin of coordinates of a porous medium of infinite extent. Owing to the spherical symmetry we adopt spherical coordinates (r, θ, φ) (see §A.3) and write $v_f = v_f(r, t)$ so that diffusion equation (5.21) is specialized in the form:

$$\frac{\partial(r v_f)}{\partial t} = c_f \frac{\partial^2 (r v_f)}{\partial r^2} \tag{5.27}$$

whose solution must satisfy the instantaneous injection condition:

$$\forall r : \quad \int_0^r v_f(r, t)\, 4\pi r^2\, dr \,|_{t\to 0} = V_f \tag{5.28}$$

and can be looked for in the form:

$$v_f = f(r, t, V_f, c_f) \tag{5.29}$$

Since v_f has no dimension, arguments (r, t, V_f, c_f) of f must combine to form independent dimensionless quantities, yielding:

$$v_f = v\left(\frac{V_f}{(c_f t)^{3/2}}, \frac{r}{\sqrt{c_f t}}\right) \tag{5.30}$$

According to (5.27) and (5.28) v_f depends linearly on V_f so that (5.30) is specialized in the form:

$$v_f = \frac{V_f}{(c_f t)^{3/2}}\, v(u)\,; \quad u = \frac{r}{\sqrt{c_f t}} \tag{5.31}$$

Substitution of (5.31) into (5.27) and some further calculations lead to the ordinary differential equation:

$$\frac{d}{du}\left(u^2 \frac{dv}{du} + \frac{1}{2} u^3 v\right) = 0 \tag{5.32}$$

Imposing $v \to 0$ as $u \to 0$, a first integration provides:

$$\frac{dv}{v} = -\frac{1}{2} u \tag{5.33}$$

Integrating the previous equation and taking into account the instantaneous injection condition (5.28)[3] give

$$v_f = \frac{V_f}{(4\pi c_f t)^{3/2}} \exp\left(-\frac{r^2}{4 c_f t}\right) \tag{5.34}$$

[3]Using also the relation $\int_0^\infty u^2 \exp\left(-\frac{u^2}{4}\right) du = 2\sqrt{\pi}$.

In the problem at hand, the displacement field is radial and thus irrotational. Use of (5.25) and (5.34) then yields:

$$p = M\frac{K+\frac{4}{3}\mu}{K_u+\frac{4}{3}\mu} \times \frac{V_f}{8(\pi c_f t)^{3/2}} \exp\left(-\frac{r^2}{4c_f t}\right) \tag{5.35}$$

Letting $\boldsymbol{\xi} = \xi\,(r,t)\,\mathbf{e}_r$, the strain components can be expressed in the form (see §A.3):

$$\varepsilon_{rr} = \frac{\partial \xi}{\partial r}; \quad \varepsilon_{\theta\theta} = \varepsilon_{\varphi\varphi} = \frac{\xi}{r}; \quad \varepsilon_{ij} = 0 \text{ if } i \neq j \tag{5.36}$$

Equations (5.25), (5.34) and (5.36) give:

$$\epsilon = \frac{1}{r^2}\frac{\partial(r^2\xi)}{\partial r} = \frac{bM}{K_u+\frac{4}{3}\mu} \times \frac{V_f}{(4\pi c_f t)^{3/2}} \exp\left(-\frac{r^2}{4c_f t}\right) \tag{5.37}$$

Owing to the symmetry, we require the displacement to vanish as r goes to zero (or equivalently, as time goes to infinity) and we integrate (5.37) in the form:

$$\xi = \frac{bM}{K_u+\frac{4}{3}\mu} \times \frac{V_f}{4\pi r^2}\left[\operatorname{erf}\left(\frac{r}{2\sqrt{c_f t}}\right) - \frac{r}{\sqrt{\pi c_f t}}\exp\left(-\frac{r^2}{4c_f t}\right)\right] \tag{5.38}$$

where erf (u) is the error function:

$$\operatorname{erf}(u) = \frac{2}{\sqrt{\pi}}\int_0^u \exp(-\lambda^2)d\lambda \tag{5.39}$$

Finally the set of equations (5.35)–(5.38), together with (5.10), lead to the stress solution:

$$\sigma_{rr} = -\frac{bM\mu}{K_u+\frac{4}{3}\mu} \times \frac{V_f}{\pi r^3}\left[\operatorname{erf}\left(\frac{r}{2\sqrt{c_f t}}\right) - \frac{r}{\sqrt{\pi c_f t}}\exp\left(-\frac{r^2}{4c_f t}\right)\right] \tag{5.40}$$

$$\sigma_{\theta\theta} = \sigma_{\varphi\varphi} = \frac{bM\mu}{K_u+\frac{4}{3}\mu} \times \frac{V_f}{2\pi r^3}$$
$$\times\left[\operatorname{erf}\left(\frac{r}{2\sqrt{c_f t}}\right) - \frac{r}{\sqrt{\pi c_f t}}\left(1+\frac{r^2}{2c_f t}\right)\exp\left(-\frac{r^2}{4c_f t}\right)\right] \tag{5.41}$$

In Fig. 5.1 we represent the dimensionless functions $\overline{\xi}(r/\sqrt{c_f t})$, $\overline{p}(r/\sqrt{c_f t})$, $\overline{\sigma}_{rr}(r/\sqrt{c_f t})$ and $\overline{\sigma}_{\theta\theta}(r/\sqrt{c_f t})$ such as defined by:

$$\xi = \frac{bM}{K_u+\frac{4}{3}\mu} \times \frac{V_f}{4\pi r^2}\,\overline{\xi}\left(\frac{r}{\sqrt{c_f t}}\right) \tag{5.42a}$$

$$p = M\frac{K+\frac{4}{3}\mu}{K_u+\frac{4}{3}\mu} \times \frac{V_f}{8(\pi c_f t)^{3/2}}\,\overline{p}\left(\frac{r}{\sqrt{c_f t}}\right) \tag{5.42b}$$

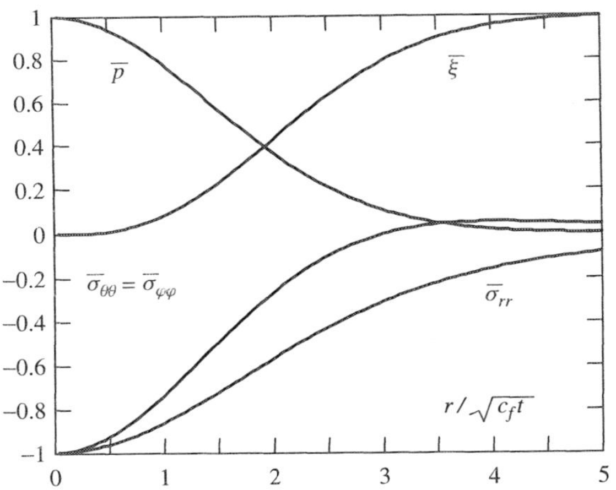

Figure 5.1: Normalized radial displacement $\overline{\xi}$, fluid pressure $\overline{p}$ and stresses $\overline{\sigma}_{rr}$ and $\overline{\sigma}_{\theta\theta} = \overline{\sigma}_{\varphi\varphi}$ plotted against $r/\sqrt{c_f t}$ for the point injection problem.

$$\sigma_{rr} = \frac{bM\mu}{K_u + \frac{4}{3}\mu} \times \frac{V_f}{6(\pi c_f t)^{3/2}} \overline{\sigma}_{rr}\left(\frac{r}{\sqrt{c_f t}}\right) \tag{5.42c}$$

$$\sigma_{\theta\theta} = \sigma_{\varphi\varphi} = \frac{bM\mu}{K_u + \frac{4}{3}\mu} \times \frac{V_f}{6(\pi c_f t)^{3/2}} \overline{\sigma}_{\theta\theta}\left(\frac{r}{\sqrt{c_f t}}\right) \tag{5.42d}$$

Line injection of fluid mass

In petroleum engineering liquid water is often injected into a reservoir from a primary well in order to recover the oil from a secondary well. This motivates us to consider the injection from a cylindrical well having negligible dimensions in comparison with those of the reservoir, so that the latter can be assimilated in a porous continuum of infinite extent.

The injection of the fluid is uniformly performed in all directions orthogonal to the well axis forming the Oz axis of coordinates. Owing to the cylindrical symmetry we adopt cylindrical coordinates (r, θ, z), each quantity spatially depending on r only. There is no fluid mass supply other than the rate provided by the injection from the well. In addition, we require the fluid flow to reduce to zero infinitely far from the well. This is expressed by the radiation condition $rw_r \rightarrow 0$ as $r \rightarrow \infty$, where w_r stands for the radial component of the relative flow vector of the fluid mass. Under these conditions, applying the fluid mass balance in integral form to a cylinder of infinite radius, we can write:

$$\int_0^\infty \dot{v}_f 2\pi r \, dr = q \tag{5.43}$$

where $\dot{v}_f$ stands for the time derivative of v_f, while q represents the constant flow rate of fluid injected per unit of vertical well length and per unit of initial fluid mass density ρ_f^0.

The time derivative $\dot{v}_f$ of v_f is governed by the time derivative of diffusion equation (5.21), here written in the form (see §A.2):

$$\frac{\partial \dot{v}_f}{\partial t} = c_f \frac{1}{r}\frac{\partial}{\partial r}\left(r\frac{\partial \dot{v}_f}{\partial r}\right) \qquad (r \neq 0) \tag{5.44}$$

We look for the solution by determining the conditions for which the differential system formed by (5.43) and (5.44) remains invariant in the linear transformation:

$$\dot{v}_f = V\dot{v}'_f;\; q = Qq';\; r = Rr';\; t = Tt';\; c_f = Cc'_f \tag{5.45}$$

Substitution of (5.45) into (5.43) and (5.44) leads to:

$$\frac{VR^2}{Q}\int_0^\infty \dot{v}'_f\, 2\pi r'\, dr' = q';\quad \frac{R^2}{CT}\frac{\partial\left(r'\dot{v}'_f\right)}{\partial t'} = c'_f\frac{\partial^2\left(r'\dot{v}'_f\right)}{\partial r'^2} \tag{5.46}$$

Inspecting (5.46), we conclude that system (5.43) and (5.44) will remain invariant through the transformation (5.45) provided that:

$$\frac{VR^2}{Q} = 1;\quad \frac{R^2}{CT} = 1 \tag{5.47}$$

or, equivalently:

$$\upsilon = \frac{\dot{v}_f r^2}{q} = \frac{\dot{v}'_f r'^2}{q'};\quad \varsigma = \frac{r^2}{c_f t} = \frac{r'^2}{c'_f t'} \tag{5.48}$$

so that the solution can be looked for in the form of a relation linking the two invariants υ and ς, that is:

$$\dot{v}_f = \frac{q}{c_f t}\upsilon(\varsigma) \tag{5.49}$$

In contrast to a sole dimensional analysis such as the one performed in the previous section, the search for the invariants has been carried out by the direct analysis of the governing equations. Therefore the procedure not only ensures the dimensional consistency of the solution, but also explores the linear or non-linear properties of the problem at hand. The method is quite general. For instance, if q is replaced by qt^n in (5.43), t must be replaced by t^{1-n} in (5.49).

Substitution of (5.49) into (5.44) leads to the ordinary differential equation:

$$\frac{d}{d\varsigma}\left[\varsigma\left(\upsilon + 4\frac{d\upsilon}{d\varsigma}\right)\right] = 0 \tag{5.50}$$

Requiring the order of magnitude of υ to be less than ς^{-1} as $\varsigma \to 0$ (i.e. as $t \to \infty$), a repeated integration provides:

$$\dot{v}_f = \frac{q}{4\pi c_f t}\exp\left(-\frac{r^2}{4c_f t}\right) \tag{5.51}$$

where the integration constant has been chosen in order to fulfil the injection condition (5.43). Integration of (5.51) with respect to time, with the initial condition $v_f(r, t = 0) = 0$, finally gives:

$$v_f = \frac{q}{4\pi c_f t} \mathrm{E}_1\left(-\frac{r^2}{4c_f t}\right); \quad \mathrm{E}_1(z) = \int_z^{\infty} \frac{\exp(-x)}{x} dx \tag{5.52}$$

The integral function $\mathrm{E}_1(z)$ can be approximated according to:

$$\mathrm{E}_1(z) \simeq -\gamma - \ln z; \quad 0 < z \ll 1 \tag{5.53}$$

where $\gamma \simeq 0.5772$ is the Euler constant. Substitution of (5.53) into (5.52) produces the approximation:

$$v_f \simeq \frac{q}{2\pi c_f t} \ln \frac{R(t)}{r}; \quad 0 < \frac{r^2}{4c_f t} \ll 1 \tag{5.54}$$

where:

$$R(t) = 2\sqrt{\exp(-\gamma)\, c_f t} \tag{5.55}$$

According to approximation (5.54), $R(t)$ represents an assessment of the current action radius of the injection process, that is the radius beyond which no significant injection has yet occurred.

The previous solution determines the history of the increase v_f of the fluid volume content consecutive to an injection performed at constant rate q, which starts at time $t = 0$ and is carried out indefinitely. Owing to the linearity of the problem at hand, the sudden interruption of the injection can be represented by imposing a counter-injection of opposite intensity $-q$ at the time $\mathcal{T}$ when the injection stops, yielding:

$$t > \mathcal{T} \quad v_f = \frac{q}{4\pi c_f t}\left[\mathrm{E}_1\left(-\frac{r^2}{4c_f t}\right) - \mathrm{E}_1\left(-\frac{r^2}{4c_f (t - \mathcal{T})}\right)\right] \tag{5.56}$$

The solution to the instantaneous line injection of a finite amount of fluid mass M_f or, equivalently, of the fluid volume $V_f = M_f/\rho_f^0$, is obtained by letting in (5.56):

$$\mathcal{T} \times q \to V_f \text{ as } \frac{\mathcal{T}}{t} \to 0 \tag{5.57}$$

The procedure eventually gives:

$$v_f = \frac{V_f}{4\pi c_f t} \exp\left(-\frac{r^2}{4c_f t}\right) \tag{5.58}$$

which should be compared with the instantaneous point injection solution (5.34).

The radial displacement field is irrotational and we can proceed as in the previous section to achieve determination of the whole solution. For the instantaneous injection, (5.25) and (5.58) combine to give the pressure solution:

$$p = M\frac{K + \frac{4}{3}\mu}{K_u + \frac{4}{3}\mu} \times \frac{V_f}{4\pi c_f t} \exp\left(-\frac{r^2}{4c_f t}\right) \tag{5.59}$$

Letting $\boldsymbol{\xi} = \xi\,(r, t)\,\mathbf{e}_r$, the strain components can be expressed in the form (see §A.2):

$$\varepsilon_{rr} = \frac{\partial \xi}{\partial r}; \quad \varepsilon_{\theta\theta} = \frac{\xi}{r}; \quad \varepsilon_{ij} = 0 \text{ if } i \neq j \tag{5.60}$$

Equations (5.25), (5.58) and (5.60) yield:

$$\epsilon = \frac{1}{r}\frac{\partial\,(r\xi)}{\partial r} = \frac{bM}{K_u + \frac{4}{3}\mu} \times \frac{V_f}{4\pi c_f t} \exp\left(-\frac{r^2}{4c_f t}\right) \tag{5.61}$$

Owing to the symmetry we require the displacement to vanish as r goes to zero (or equivalently, as time goes to infinity) and we integrate (5.61) in the form:

$$\xi = \frac{bM}{K_u + \frac{4}{3}\mu} \times \frac{V_f}{2\pi r}\left[1 - \exp\left(-\frac{r^2}{4c_f t}\right)\right] \tag{5.62}$$

Finally the set of equations (5.59)–(5.62), together with (5.10), yield:

$$\sigma_{rr} = -\frac{bM\mu}{K_u + \frac{4}{3}\mu} \times \frac{V_f}{\pi r^2}\left[1 - \exp\left(-\frac{r^2}{4c_f t}\right)\right] \tag{5.63a}$$

$$\sigma_{\theta\theta} = \frac{bM\mu}{K_u + \frac{4}{3}\mu} \times \frac{V_f}{\pi r^2}\left[1 - \left(1 + \frac{r^2}{2c_f t}\right)\exp\left(-\frac{r^2}{4c_f t}\right)\right] \tag{5.63b}$$

$$\sigma_{zz} = -\frac{bM\mu}{K_u + \frac{4}{3}\mu} \times \frac{V_f}{2\pi c_f t} \exp\left(-\frac{r^2}{4c_f t}\right) \tag{5.63c}$$

In Fig. 5.2 we represent the dimensionless functions $\overline{\xi}(r/\sqrt{c_f t})$, $\overline{p}(r/\sqrt{c_f t})$, $\overline{\sigma}_{rr}$ $(r/\sqrt{c_f t})$ and $\overline{\sigma}_{\theta\theta}(r/\sqrt{c_f t})$ such as defined by:

$$\xi = \frac{bM}{K_u + \frac{4}{3}\mu} \times \frac{V_f}{2\pi r}\,\overline{\xi}\left(\frac{r}{\sqrt{c_f t}}\right) \tag{5.64a}$$

$$p = M\frac{K + \frac{4}{3}\mu}{K_u + \frac{4}{3}\mu} \times \frac{V_f}{4\pi c_f t}\,\overline{p}\left(\frac{r}{\sqrt{c_f t}}\right) \tag{5.64b}$$

$$\sigma_{rr} = \frac{bM\mu}{K_u + \frac{4}{3}\mu} \times \frac{V_f}{4\pi c_f t}\,\overline{\sigma}_{rr}\left(\frac{r}{\sqrt{c_f t}}\right) \tag{5.64c}$$

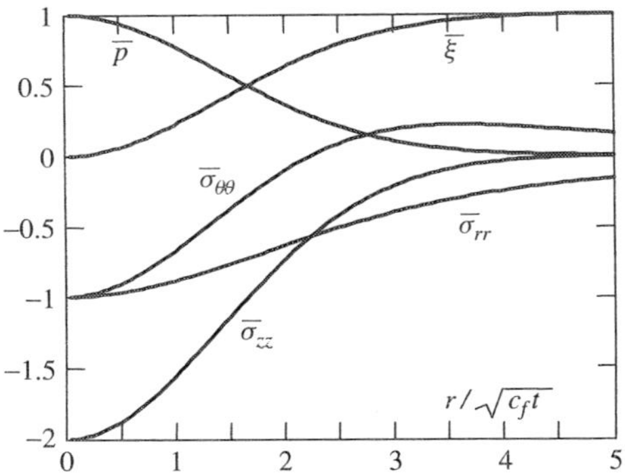

Figure 5.2: Normalized radial displacement $\overline{\xi}$, fluid pressure $\overline{p}$ and stresses $\overline{\sigma}_{rr}$, $\overline{\sigma}_{\theta\theta}$ and $\overline{\sigma}_{zz}$ plotted against $r/\sqrt{c_f t}$ for the line injection problem.

$$\sigma_{\theta\theta} = \frac{bM\mu}{K_u + \frac{4}{3}\mu} \times \frac{V_f}{4\pi c_f t} \overline{\sigma}_{\theta\theta}\left(\frac{r}{\sqrt{c_f t}}\right) \tag{5.64d}$$

$$\sigma_{zz} = -\frac{bM\mu}{K_u + \frac{4}{3}\mu} \times \frac{V_f}{4\pi c_f t} \overline{\sigma}_{zz}\left(\frac{r}{\sqrt{c_f t}}\right) \tag{5.64e}$$

The determination of the solution for the instantaneous injection proves to be of special interest since it allows the determination of the solution whatever the injection conditions. Indeed, let $s\,(r, t)$ be the solution obtained by letting $V_f = 1$ in the instantaneous injection solution. The solution $S\,(r, t)$, in response to any injection history $V_f\,(t)$, is obtained by superposition of the infinitesimal responses dS on the successive infinitesimal injections dV_f, giving:

$$S\,(r, t) = \int_0^t s\,(r, t-u)\; dV_f\,(u) \tag{5.65}$$

Now, if N injection lines are present, let (r_J, θ_J) be the polar coordinates of line J, and $V_f^{(J)}\,(t)$ its history of injection. The solution is then given by:

$$S\,(r, t) = \sum_{J=1}^{J=N} \int_0^t s_J\,(r, t-u)\; dV_f^{(J)}\,(u) \tag{5.66}$$

where $s_J\,(r, t) = s(\rho_J\,(r)\,, t)$, $\rho_J\,(r)$ being the distance from the observed point (r, θ) to the source (r_J, θ_J), that is:

$$\rho_J\,(r) = \sqrt{r^2 + r_J^2 - 2rr_J\cos(\theta - \theta_J)} \tag{5.67}$$

From a more mathematical point of view, the instantaneous localized injection, with $V_f = 1$, turns out to solve the generalized diffusion equation:

$$\frac{\partial v_f}{\partial t} = c_f\,\nabla^2 v_f + \frac{1}{2^n \pi r^n}\delta_{r=0}\,\delta_{t=0} \tag{5.68}$$

where δ stands for the Dirac δ-function (or delta distribution) and where $n = 2$ for the point injection (previous section) and $n = 1$ for the line injection (this section).

5.2.2 Consolidation of a Soil Layer

When a soil layer is subjected to an extra loading, the saturating fluid undergoes an overpressurization. Subsequently this overpressure progressively vanishes, owing to the diffusion process of the fluid towards the boundary of the soil layer which remains drained. In turn, the skeleton has progressively to sustain alone the extra loading and a delayed settlement of the soil layer occurs. The whole process is the celebrated consolidation problem, principally attached to the name of Terzaghi.[4] It is a key problem in soil mechanics since an unexpected consolidation can endanger the underlying construction. A similar problem in petroleum geophysics is the sudden release of an abnormal fluid overpressure following the breakthrough of a reservoir layer by a drilling tool.

Consider a soil layer resting on a rigid impervious base at depth $z = h$, while its upper surface at $z = 0$ remains drained; that is, at zero pressure if we take the hydrostatic pressure induced by the atmospheric pressure and the gravity forces as the reference pressure. The hydraulic boundary conditions are:

$$z = 0 : \quad p = 0; \qquad z = h : \quad \frac{\partial p}{\partial z} = 0 \tag{5.69}$$

Beyond the instantaneous application of a vertical constant load $\sigma_{zz} = -\varpi$ at the upper surface $z = 0$, the equilibrium in the vertical direction requires:

$$\sigma_{zz} = -\varpi \tag{5.70}$$

Furthermore the displacement in the soil layer is vertical, namely $\xi = \xi\,(z, t)\,\mathbf{e}_z$, so that the only non-zero strain component is ε_{zz}. We write:

$$\varepsilon_{zz} = \epsilon = \frac{\partial \xi}{\partial z} \tag{5.71}$$

Constitutive equations (5.10) for σ_{zz} and the two above equations provide:

$$\frac{\partial \xi}{\partial z} = \epsilon = \frac{bp - \varpi}{K + \frac{4}{3}\mu} \tag{5.72}$$

Combining (5.16) and (5.72) gives the instantaneous response of the soil layer:

$$p(z, t = 0^+) = \frac{bM\varpi}{K_u + \frac{4}{3}\mu}; \quad \frac{\partial \xi}{\partial z}(z, t = 0^+) = -\frac{\varpi}{K_u + \frac{4}{3}\mu} \tag{5.73}$$

Substitution of (5.72) into (5.15) gives:

$$\frac{\partial p}{\partial t} = c_f \frac{\partial^2 p}{\partial z^2} \tag{5.74}$$

[4]Terzaghi K. (1923), 'Die Berechnung der Durchlassigkeitsziffer des Tcnes aus dem Verlauf der hydrodynamischen Spannungsercheinungen', *Sitzungsberichte Akademie der Wissenschatten, Wien Mathematisch-Naturwisseuschaftliche Klasse, Abteilung IIa* **132**, 105–124.

where c_f is the diffusion coefficient (5.22). Indeed, for the problem at hand the displacement field is irrotational so that (5.23) applies. The integration function $f(t)$ in (5.23) eventually turns out to be zero so that (5.26) is recovered.

The fluid pressure p is appropriately scaled by the initial pressure (5.73) induced by the instantaneous loading. The depth z is scaled by the finite thickness h of the soil layer, the latter being the relevant length required to define an overall diffusion characteristic time τ. We write:

$$p = \frac{bM\varpi}{K_u + \frac{4}{3}\mu}\overline{p}; \quad z = h\overline{z}; \quad t = \tau\overline{t}; \quad \tau = \frac{h^2}{c_f} \tag{5.75}$$

Accordingly we rewrite (5.74) in the dimensionless form:

$$\frac{\partial \overline{p}}{\partial \overline{t}} = \frac{\partial^2 \overline{p}}{\partial \overline{z}^2} \tag{5.76}$$

with the boundary and initial conditions (5.69) and (5.73) now reading:

$$\overline{z} = 0 : \ \overline{p} = 0; \quad \overline{z} = 1 : \ \frac{\partial \overline{p}}{\partial \overline{z}} = 0; \quad \overline{t} = 0 : \ \overline{p} = 1 \tag{5.77}$$

Substituting (5.75) into (5.72) and integrating over the layer thickness, while taking into account the boundary condition $\xi(z = h, t) = 0$, we derive the settlement $s(t) = \xi(z = 0, t)$ in the form:

$$s(t) = s_\infty + (s_0 - s_\infty) \int_0^1 \overline{p}\left(\overline{z}, \overline{t}\right) dz \tag{5.78}$$

In the above equation s_0 stands for the instantaneous undrained settlement ($\overline{p} = 1$), while s_∞ stands for the drained settlement achieved as time goes to infinity so that the initial fluid overpressure has vanished ($\overline{p} = 0$). They are expressed in the form:

$$s_0 = \frac{\varpi h}{K_u + \frac{4}{3}\mu}; \quad s_\infty = \frac{\varpi h}{K + \frac{4}{3}\mu} \tag{5.79}$$

Early time solution. The early time solution can be determined by exploring times much smaller than the overall characteristic diffusion time τ in letting:

$$\vartheta \ll 1 : \quad \overline{t} = \vartheta t^*; \quad \overline{z} = \sqrt{\vartheta}\, z^* \tag{5.80}$$

so that (5.76) and (5.77) can be rewritten in the form:

$$\frac{\partial \overline{p}}{\partial t^*} = \frac{\partial^2 \overline{p}}{\partial z^{*2}} \tag{5.81}$$

$$z^* = 0 : \ \overline{p} = 0; \quad z^* = \frac{1}{\sqrt{\vartheta}} \gg 1 : \ \frac{\partial \overline{p}}{\partial z^*} = 0; \quad t^* = 0 : \ \overline{p} = 1 \tag{5.82}$$

For early times such as $\sqrt{\vartheta} \ll 1$, (5.82) shows that the impermeability condition of the rigid base is repelled at $z^* \rightarrow \infty$. At early times the diffusion process involves only the region close to the drained upper surface where it tends to restore a zero fluid pressure. Indeed the early time solution behaves the same as if the soil layer was semi-infinite. The search for invariants of (5.81) and (5.82), like the one carried out in the line injection problem, shows that the early time solution can be looked for in the form $\overline{p} = \overline{p}(z^*/\sqrt{t^*} = \overline{z}/\sqrt{\overline{t}})$, which, when substituted into (5.81) and (5.82), leads to an ordinary differential equation whose standard solution is:

$$\overline{t} \ll 1 : \quad \overline{p} = \operatorname{erf}\left(\frac{\overline{z}}{2\sqrt{\overline{t}}}\right) \tag{5.83}$$

where erf stands for the error function (5.39).

Any time solution. A general solution to (5.76) and (5.77) can be investigated in the form of the infinite series:

$$\overline{p}\left(\overline{z}, \overline{t}\right) = \sum_{n=0}^{n=\infty} \overline{p}_n \sin\left(\frac{(2n+1)\,\pi}{2}\overline{z}\right) \times \exp\left(-\frac{(2n+1)^2\,\pi^2}{4}\overline{t}\right) \tag{5.84}$$

where each term of the series satisfies the diffusion equation (5.76) and the boundary conditions (5.77) at $\overline{z} = 0$ and $\overline{z} = 1$. The initial condition requires $\overline{p}\,(\overline{z}, 0) = 1$, that is:

$$\sum_{n=0}^{n=\infty} \overline{p}_n \sin\left(\frac{(2n+1)\,\pi}{2}\overline{z}\right) = 1 \tag{5.85}$$

When multiplying (5.85) by $\sin\left(\frac{1}{2}(2m+1)\pi\overline{z}\right)$ and integrating from 0 to 1, the left hand side member of the resulting equation is non-zero only for $n = m$, giving:

$$\overline{p}_n = \frac{4}{\pi\,(2n+1)} \tag{5.86}$$

so that:

$$\overline{p}\left(\overline{z}, \overline{t}\right) = \sum_{n=0}^{n=\infty} \frac{4}{\pi\,(2n+1)} \sin\left(\frac{(2n+1)\,\pi}{2}\overline{z}\right) \times \exp\left(-\frac{(2n+1)^2\,\pi^2}{4}\overline{t}\right) \tag{5.87}$$

In Fig. 5.3 the normalized fluid pressure $\overline{p}$ is plotted against the location $\overline{z}$ for various normalized times $\overline{t}$.

Substitution of (5.87) into (5.78) gives:

$$s\,(t) = s_\infty + (s_0 - s_\infty) \sum_{n=0}^{n=\infty} \frac{8}{\pi^2\,(2n+1)^2} \times \exp\left(-\frac{(2n+1)^2\,\pi^2}{4}\frac{t}{\tau}\right) \tag{5.88}$$

providing the slope at the origin of the consolidation process in the form:

$$\frac{ds}{dt}(t = 0^+) = 2(s_\infty - s_0)\frac{c_f}{h^2} \tag{5.89}$$

which can be used when aiming at the experimental assessment of the diffusivity coefficient c_f.

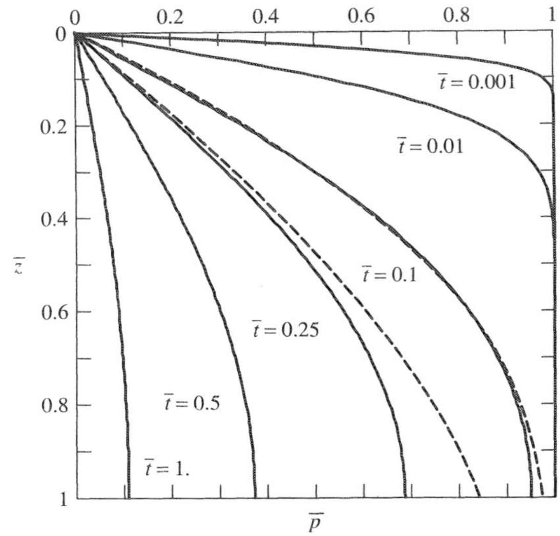

Figure 5.3: Normalized overpressure $\overline{p}$ during the consolidation process plotted against normalized depth $\overline{z} = z/h$ for various normalized times $\overline{t} = c_f t/h^2$. For $\overline{t} = 0.1, 0.25$ the dashed lines represent the early time solution that cannot be distinguished from the exact solution for $\overline{t} = 0.001, 0.01$.

At the layer scale the consolidation process appears as a creep process under constant load. Invoking the linearity of the whole set of equations governing the consolidation process, we derive the settlement due to any history $\varpi(t)$ of the overload in the form:

$$s(t) = \int_0^t s_{\varpi=1}(t-u)\, d\varpi(u) \tag{5.90}$$

where $s_{\varpi=1}(t)$ is the settlement history obtained by letting $\varpi = 1$ in (5.88).

At the material scale the consolidation process is due to the progressive dissipation through fluid diffusion of the overpressure induced by the instantaneous applied load. Under constant load ϖ, the energy required for the fluid to diffuse against the resistant viscous forces is supplied by a part of the elastic free energy instantaneously stored in the skeleton and the compressible fluid. Indeed, the energy $D_f = \int_0^\infty dt \int_\Omega \varphi_f \, d\Omega$, which is dissipated through the fluid conduction throughout the whole consolidation process, is specialized in the form (see §3.3.1 for the expression of φ_f):

$$D_f = \left(\frac{3bM\varpi}{3K_u + 4\mu}\right)^2 \frac{k\tau}{h} \times \mathcal{I}; \quad \mathcal{I} = \int_0^\infty d\overline{t} \int_0^1 \left(\frac{\partial \overline{p}}{\partial \overline{z}}\right)^2 d\overline{z} \tag{5.91}$$

Integrating by parts with respect to space, making use of (5.76) and (5.77) and reversing the integration order, we successively derive:

$$\mathcal{I} = \int_0^\infty d\overline{t} \left(\left[\frac{\partial \overline{p}}{\partial \overline{z}} \overline{p}\right]_0^1 - \int_0^1 \frac{\partial^2 \overline{p}}{\partial \overline{z}^2} \overline{p}\, d\overline{z} \right) = -\int_0^\infty d\overline{t} \int_0^1 \frac{\partial \overline{p}}{\partial \overline{t}} \overline{p}\, d\overline{z} = \frac{1}{2} \tag{5.92}$$

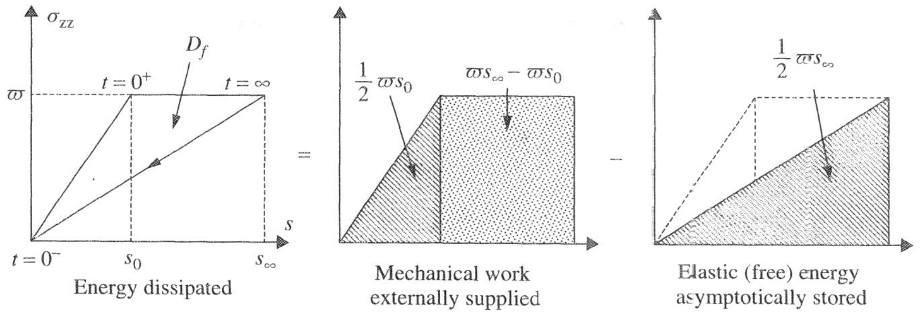

Figure 5.4: Analysis of the dissipated energy in the consolidation process. The path with an arrow represents a drained unloading process that would be performed infinitely slowly in order to keep the fluid at zero pressure and so that no dissipation occurs. The energy picture as sketched is analogous to the one relative to the creep process (see Fig. 9.1).

With the help of (5.79) the factor affecting $\mathcal{I}$ in expression (5.91) for D_f can be identified as $\varpi s_\infty - \varpi s_0$. Collecting the above results, we finally write:

$$D_f = \frac{1}{2}\varpi s_\infty - \frac{1}{2}\varpi s_0 \tag{5.93}$$

The interpretation of dissipation D_f can be more easily grasped through the decomposition $D_f = \frac{1}{2}\varpi s_0 + (\varpi s_\infty - \varpi s_0) - \frac{1}{2}\varpi s_\infty$. The first term $\frac{1}{2}\varpi s_0$ accounts for the elastic energy stored by both the skeleton and the compressible fluid during the instantaneous application of the load, that is between $t = 0^-$ and $t = 0^+$. The second term $\varpi s_\infty - \varpi s_0$ accounts for the external mechanical work supplied to the layer from $t = 0^+$ to $t \to \infty$. The expression for the dissipation is eventually obtained by subtracting the last term $\frac{1}{2}\varpi s_\infty$ which represents the (elastic) free energy finally stored by the sole skeleton at the end of the consolidation process. This is sketched in Fig. 5.4 where the path with an arrow represents a drained unloading process that would be performed infinitely slowly in order to keep the fluid at zero pressure and avoid any dissipation.

5.2.3 Drilling of a Borehole

Another problem of special interest in petroleum engineering is the drilling of a vertical borehole in a porous layer subjected to the weight of the upper layers. In a first approach the drilling can be considered as instantaneous, while the effect of the upper layers is accounted for through a hydrostatic initial stress field.

Owing to the cylindrical symmetry we adopt cylindrical coordinates (r, θ, z), each quantity spatially depending only on r. Since the domain is infinite and the displacement is radial and thus irrotational, diffusion equation (5.26) applies to the difference of fluid pressure $p - p_0$ induced by the drilling of the borehole with respect to the initial pressure p_0. We write:

$$\frac{\partial(p - p_0)}{\partial t} = c_f \frac{1}{r}\frac{\partial}{\partial r}\left(r\frac{\partial(p - p_0)}{\partial r}\right) \tag{5.94}$$

to which we add the initial and boundary conditions:

$$p\,(r, t=0) - p_0 = 0; \qquad p\,(r=a, t) - p_0 = p_1 - p_0 \tag{5.95}$$

where a is the borehole radius and p_1 is the fluid pressure on the wall of the borehole after drilling.

Let $f^*\,(r, s)$ be the Laplace transform with respect to time of the function $f\,(r, t)$ according to:

$$f^*\,(r, s) = \int_0^\infty f\,(r, t) \exp\,(-st)\, dt \tag{5.96}$$

Applying the Laplace transform to (5.94) and (5.95) yields:

$$\frac{1}{r}\frac{\partial}{\partial r}\left(r\frac{\partial (p-p_0)^*}{\partial r}\right) - \frac{s}{c_f}(p-p_0)^* = 0; \quad p^*\,(r=a, s) - p_0 = \frac{p_1 - p_0}{s} \tag{5.97}$$

whose solution, which is bounded at infinity, is:

$$(p - p_0)^* = (p_1 - p_0)\,\frac{\mathrm{K}_0(r\sqrt{s/c_f})}{s\,\mathrm{K}_0(a\sqrt{s/c_f})} \tag{5.98}$$

where K_n is the modified Bessel function of the second kind and order n, while $\sqrt{s}$ stands for the square root of s having a positive real part. Returning to the time domain, we derive:

$$p\,(r, t) = p_0 + (p_1 - p_0)\,\overline{p}\,(r, t) \tag{5.99}$$

where $\overline{p}\,(r, t)$ is the inverse Laplace transform of the following function $\overline{p}^*\,(s, t)$:

$$\overline{p}^*\,(s, t) = \frac{\mathrm{K}_0(r\sqrt{s/c_f})}{s\,\mathrm{K}_0(a\sqrt{s/c_f})} \tag{5.100}$$

Based on standard inversion techniques of the Laplace transform, further calculations allow us to express $\overline{p}\,(r, t)$ in the form:

$$\overline{p}\,(r, t) = 1 - \frac{2}{\pi}\int_0^\infty \frac{\mathrm{J}_0\,(\nu)\,\mathrm{Y}_0\left(\nu\frac{r}{a}\right) - \mathrm{J}_0\left(\nu\frac{r}{a}\right)\mathrm{Y}_0\,(\nu)}{\mathrm{J}_0^2\,(\nu) + \mathrm{Y}_0^2\,(\nu)} \exp\left(-c_f t\frac{\nu^2}{a^2}\right)\frac{d\nu}{\nu} \tag{5.101}$$

where J_n and Y_n stand for the Bessel functions of order n, respectively, of the first and second kind.

The only non-zero component is the radial one, namely $\boldsymbol{\xi} = \xi\,(r, t)\,\mathbf{e}_r$, so that the displacement is irrotational and (5.24) applies. Combining of the latter with (5.60) and (5.98) gives:

$$\epsilon^* = \frac{1}{r}\frac{d(r\xi^*)}{dr} = \frac{b(p_1 - p_0)}{K + \frac{4}{3}\mu} \times \frac{\mathrm{K}_0(r\sqrt{s/c_f})}{s\,\mathrm{K}_0(a\sqrt{s/c_f})} \tag{5.102}$$

Using the relation:

$$\frac{1}{x}\frac{d(x\mathrm{K}_1)}{dx} = -\mathrm{K}_0\,(x) \tag{5.103}$$

we integrate (5.102) in the form:

$$\xi^* = \frac{b(p_1 - p_0)}{K + \frac{4}{3}\mu} \times \frac{\sqrt{c_f}}{s\sqrt{s}\mathrm{K}_0(a\sqrt{s/c_f})}\left(\frac{a}{r}v\,(s) - \mathrm{K}_1(r\sqrt{s/c_f})\right) \tag{5.104}$$

where $v\,(s)$ is a yet unknown function.

Let ϖ be the strength of the hydrostatic stress due to the weight of the upper layers, so that the initial stress prior to drilling can be expressed in the form:

$$\sigma_{ij}\,(r, t = 0) = -\varpi\,\delta_{ij}; \quad \varpi > 0 \tag{5.105}$$

After the drilling, the fluid pressure p_1 applies to the wall newly created of the borehole. The boundary condition related to the stress and its Laplace transform are:

$$\sigma_{rr}\,(r = a, t) = -p_1; \quad \sigma_{rr}^*\,(r = a, s) = -\frac{p_1}{s} \tag{5.106}$$

Taking into account the non-zero initial conditions (5.105), the Laplace transform of (5.10) where we use (5.60) gives for the radial stress:

$$\sigma_{rr}^*\,(r, s) = -\frac{\varpi}{s} - 2\mu\,\frac{\xi^*}{r} \tag{5.107}$$

Equations (5.104), (5.106) and (5.107) allow us to determine the unknown function $v\,(s)$, resulting in:

$$\begin{aligned}\xi^* = {} & \frac{p_1 - \varpi}{2\mu s}\frac{a^2}{r} + \frac{b(p_1 - p_0)}{K + \frac{4}{3}\mu} \times \frac{a\sqrt{c_f}}{s\sqrt{s}\,\mathrm{K}_0(a\sqrt{s/c_f})} \\ & \times \left(\frac{1}{r}\mathrm{K}_1(a\sqrt{s/c_f}) - \frac{1}{a}\mathrm{K}_1(r\sqrt{s/c_f})\right)\end{aligned} \tag{5.108}$$

Returning to the time domain, we derive:

$$\xi\,(r, t) = \frac{p_1 - \varpi}{2\mu}\frac{a^2}{r} + a\,\frac{b(p_1 - p_0)}{K + \frac{4}{3}\mu}\,\overline{\xi}\,(r, t) \tag{5.109}$$

with $\overline{\xi}\,(r, t)$ the inverse Laplace transform of function $\overline{\xi}^*\,(s, t)$:

$$\overline{\xi}^*\,(s, t) = \frac{\sqrt{c_f}}{s\sqrt{s}\,\mathrm{K}_0(a\sqrt{s/c_f})}\left(\frac{1}{r}\mathrm{K}_1(a\sqrt{s/c_f}) - \frac{1}{a}\mathrm{K}_1(r\sqrt{s/c_f})\right) \tag{5.110}$$

Based on standard inversion techniques of Laplace transforms and on the asymptotic properties of Bessel functions,[5] further calculations allow us to express $\overline{\xi}(r, t)$ in the form:

$$\overline{\xi}(r,t) = \frac{1}{2}\left(\frac{r}{a} - \frac{a}{r}\right) + \frac{2}{\pi}\int_0^{\infty} \frac{\left[\mathrm{J}_1\left(\frac{r}{a}\right) - \frac{r}{a}\mathrm{J}_1(\nu)\right]\mathrm{Y}_0(\nu) - \left[\mathrm{Y}_1\left(\frac{r}{a}\right) - \frac{r}{a}\mathrm{Y}_1(\nu)\right]\mathrm{J}_0(\nu)}{\mathrm{J}_0^2(\nu) + \mathrm{Y}_0^2(\nu)} \times \exp\left(-c_f t\frac{\nu^2}{a^2}\right)\frac{d\nu}{\nu^2} \tag{5.111}$$

Finally, with the help of (5.10), (5.24) and (5.60), the stresses can be expressed in the form:

$$\sigma_{rr} = -\varpi - 2\mu\frac{\xi}{r} \tag{5.112a}$$

$$\sigma_{\theta\theta} = -\varpi - \frac{2b\mu}{K + \frac{4}{3}\mu}(p - p_0) + 2\mu\frac{\xi}{r} \tag{5.112b}$$

$$\sigma_{zz} = -\varpi - \frac{2b\mu}{K + \frac{4}{3}\mu}(p - p_0) \tag{5.112c}$$

The near-field/long-term solution is:

$$\frac{c_f t}{(r-a)^2} \gg 1: \quad \overline{p}(r,t) \simeq 1; \quad \overline{\xi}(r,t) \simeq \frac{1}{2}\left(\frac{r}{a} - \frac{a}{r}\right) \tag{5.113}$$

The early time solution can be obtained by using the equivalence $c_f t/a^2 \to 0 \iff a^2 s/c_f \to \infty$. An expansion of expressions (5.100) and (5.110) for $\overline{p}^*(s,t)$ and $\overline{\xi}^*(s,t)$ with respect to $a^2 s/c_f \to \infty$ can be obtained by using standard asymptotic expansions of Bessel functions. Returning to the time domain, we eventually obtain the early time solution in the form:

$$t \ll \tau = \frac{a^2}{c_f}:$$

$$\overline{p}(r,t) = \sqrt{\frac{a}{r}}\left[1 - \mathrm{erf}\left(\frac{r-a}{2\sqrt{c_f t}}\right)\right] - \frac{1}{8}\sqrt{\frac{a}{r}}\frac{r-a}{2r} \times \left\{\frac{2}{\sqrt{\pi}}\frac{\sqrt{c_f t}}{a}\exp\left(-\frac{(r-a)^2}{4c_f t}\right) - \frac{r-a}{a}\left[1 - \mathrm{erf}\left(\frac{r-a}{2\sqrt{c_f t}}\right)\right]\right\} \tag{5.114}$$

$$\overline{\xi}(r,t) = \frac{2}{\sqrt{\pi}}\frac{\sqrt{c_f t}}{r} + \sqrt{\frac{a}{r}} \times \left\{\frac{r-a}{a}\left[1 - \mathrm{erf}\left(\frac{r-a}{2\sqrt{c_f t}}\right)\right] - \frac{2}{\sqrt{\pi}}\frac{\sqrt{c_f t}}{a}\exp\left(-\frac{(r-a)^2}{4c_f t}\right)\right\} \tag{5.115}$$

[5] See for instance *Handbook of mathematical functions* (1964), ed. Abramowitz M. and Stegun I.A., Dover, New York.

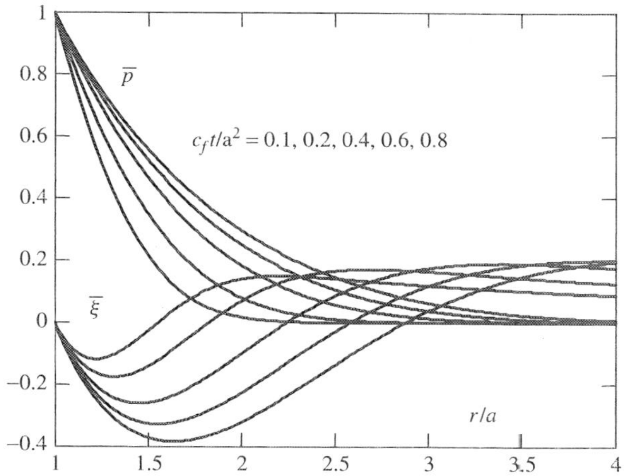

Figure 5.5: Normalized variations of fluid pressure $\overline{p}$ and radial displacement $\overline{\xi}$ as functions of the normalized radius r/a for increasing normalized times $c_f t/a^2$.

where erf stands for the error function (5.39). For $t = 0$ we found that $\overline{p}(r, t)$ and $\overline{\xi}(r, t)$ are zero. In fact, the instantaneous drilling of the borehole does not induce any instantaneous variation of the fluid pressure, nor volumetric change, since ϵ is proportional to the variation of the fluid pressure. Indeed the instantaneous response of the surrounding porous medium to the drilling of the borehole is undrained and corresponds to the standard elastic solution, that is:

$$\xi(r, t = 0^+) = \frac{p_1 - \varpi}{2\mu} \frac{a^2}{r} \tag{5.116}$$

As a consequence, according to (5.99) and (5.109), functions $\overline{p}(r, t)$ and $\overline{\xi}(r, t)$ capture the evolution of the fluid pressure and of the displacement beyond the instantaneous undrained response. In Fig. 5.5 we plot the early time variations of $\overline{p}(r, t)$ and $\overline{\xi}(r, t)$ against the dimensionless radius r/a, for increasing values of time normalized by the characteristic diffusion time a^2/c_f.

5.3 Thermoporoelasticity Problems

5.3.1 Field Equations

The previous linear analysis can be extended in order to include thermal effects. Under the assumption of small perturbations we first rewrite the thermal equation (3.36) in the form:

$$T\left(\frac{\partial S}{\partial t} + \nabla \cdot (s_f \mathbf{w})\right) = -\nabla \cdot \mathbf{q} + \Phi_M \tag{5.117}$$

We now assume small variations of temperature such as $|T - T_0| / T_0 \ll 1$, in order to replace T by T_0 whenever required. Dissipation Φ_M of mechanical energy occurs only through fluid flow so that $\Phi_M \simeq \varphi_f = \mathcal{V}^2/k$ (see (3.40)). Use of continuity equation (5.7), together with substitution of Fourier's law (3.57) into (5.117), gives:

$$T\left(\frac{\partial S}{\partial t} - s_f \frac{\partial m_f}{\partial t}\right) = \kappa \nabla^2 T - \rho_f T_0 \mathcal{V} \cdot \nabla s_f + \frac{\mathcal{V}^2}{k} \tag{5.118}$$

A complete linearization requires us to integrate the second of fluid state equations (4.53) in the form:

$$s_f - s_f^0 = -3\alpha_f \frac{p - p_0}{\rho_f^0} + \frac{C_p}{T_0}(T - T_0) \tag{5.119}$$

which, substituted into (5.118), yields:

$$T\left(\frac{\partial S}{\partial t} - s_f \frac{\partial m_f}{\partial t}\right) = \kappa \nabla^2 T - \rho_f C_p \mathcal{V} \cdot \nabla T + \frac{1}{k}(1 - 3\alpha_f T_0)\mathcal{V}^2 \tag{5.120}$$

The complete linearization can eventually be performed provided that the last two terms on the right hand side of (5.120) can be neglected with respect to the first term. This can be achieved under the twofold assumption:

$$Pe = \frac{\rho_f C_p L \mathcal{V}^d}{\kappa} \ll 1; \qquad Br = \frac{1 - 3\alpha_f T_0}{\kappa k \theta^d}(L\mathcal{V}^d)^2 \ll 1 \tag{5.121}$$

where L, $\mathcal{V}^d$ and θ^d stand for a characteristic length, a characteristic fluid flow and a characteristic difference of temperature associated with the problem at hand, respectively. Pe is the Péclet number and quantifies the order of magnitude of the heat convectively transported by the fluid in comparison with the heat supplied by diffusion through the porous medium. Br is the Brinkman number and quantifies the order of magnitude of the heat source due to the viscous dissipation in comparison with the heat supplied by conduction. When both numbers are much less than one, the heat is mainly supplied by diffusion and (5.120) reduces to:

$$T\left(\frac{\partial S}{\partial t} - s_f \frac{\partial m_f}{\partial t}\right) = \kappa \nabla^2 \theta \tag{5.122}$$

where $\theta = T - T_0$ represents the temperature variation, the initial temperature T_0 being assumed constant within the medium.

Linear (isotropic) thermoporoelastic constitutive equations are obtained by integrating (4.62a) and (4.62b) in the form:

$$\sigma_{ij} = \left(K - \frac{2}{3}\mu\right)\epsilon\delta_{ij} + 2\mu\varepsilon_{ij} - bp\delta_{ij} - 3\alpha K\theta\delta_{ij} \tag{5.123a}$$

$$v_f = b\epsilon + \frac{p}{M} - 3\alpha_m \theta \tag{5.123b}$$

Using (1.26) and (1.27) and substituting (5.123a) into (5.5), we derive:

$$\left(K + \frac{4}{3}\mu\right)\nabla(\nabla \cdot \boldsymbol{\xi}) - \mu \nabla \times (\nabla \times \boldsymbol{\xi}) - b\nabla p - 3\alpha K \nabla\theta + \rho \mathbf{g} = 0 \tag{5.124}$$

Use of definition (5.13) and substitution of (5.123b) into (5.8) furnish:

$$b\frac{\partial \epsilon}{\partial t} + \frac{1}{M}\frac{\partial p}{\partial t} - 3\alpha_m \frac{\partial \theta}{\partial t} = k\nabla^2 p \tag{5.125}$$

Finally substitution of constitutive equation (4.62c), that is:

$$dS = s_f\, dm_f + 3\alpha K\, d\epsilon - 3\alpha_m\, dp + C_d \frac{dT}{T} \tag{5.126}$$

into (5.122) gives:

$$3\alpha T_0 K \frac{\partial \epsilon}{\partial t} - 3\alpha_m T_0 \frac{\partial p}{\partial t} + C_d \frac{\partial \theta}{\partial t} = \kappa \nabla^2 \theta \tag{5.127}$$

The mechanical and hydraulic boundary conditions (5.17) and (5.18) have then to be completed by the thermal boundary conditions:

$$\text{on } \partial_\theta \Omega :\ \theta = \theta^d; \quad \text{on } \partial q \Omega :\ \mathbf{q} \cdot \mathbf{n} = q^d \quad (\partial \Omega = \partial_\theta \Omega \cup \partial_q \Omega) \tag{5.128}$$

5.3.2 Half-space Subjected to a Change in Temperature

Let us illustrate the thermo/hydro/mechanical couplings which can be encountered by considering the following instructive problem.[6] At time $t = 0$ a temperature variation θ^d is suddenly applied at the border $x = 0$ of a semi-infinite porous medium lying in the region $x \geq 0$, the border remaining drained and free of stress. The initial and boundary conditions read:

$$t = 0 :\ \sigma_{xx} = p = \theta = 0; \qquad x = 0 :\ \sigma_{xx} = p = 0,\ \theta = \theta^d \tag{5.129}$$

The mechanical equilibrium requires:

$$\sigma_{xx} = 0 \tag{5.130}$$

Moreover, the displacement is unidimensional, namely $\boldsymbol{\xi} = \xi\,(z, t)\,\mathbf{e}_x$, so that the only non-zero strain component is ε_{xx}. We write:

$$\varepsilon_{xx} = \epsilon = \frac{\partial \xi}{\partial x} \tag{5.131}$$

Use of (5.123a) and the two equations above yield:

$$\frac{\partial \xi}{\partial x} = \epsilon = \frac{bp + 3\alpha K \theta}{K + \frac{4}{3}\mu} \tag{5.132}$$

[6] For thermoporoelastic problems see in particular McTigue D.F. (1986), 'Thermoelastic response of fluid-saturated porous rock', *Journal of Geophysical Research*, **91**, 9533–9542.

Substituting (5.132) into the 1D forms of (5.125) and (5.127), we derive:[7]

$$\frac{1}{c_f}\frac{\partial p}{\partial t}+\frac{d}{k}\frac{\partial \theta}{\partial t}=\frac{\partial^2 p}{\partial x^2} \tag{5.133}$$

$$\frac{dT_0}{\kappa}\frac{\partial p}{\partial t}+\frac{1}{c_\theta}\frac{\partial \theta}{\partial t}=\frac{\partial^2 \theta}{\partial x^2} \tag{5.134}$$

where c_f is the diffusion coefficient given by (5.22) and where we note:

$$\frac{d}{k}=\frac{3\alpha K\theta}{K+\frac{4}{3}\mu}-3\alpha_m;\quad \frac{1}{c_\theta}=\frac{1}{\kappa}\left(C_d+\frac{9\alpha^2K^2T_0}{K+\frac{4}{3}\mu}\right)\simeq\frac{C_d}{\kappa} \tag{5.135}$$

Letting $p=p^d\,\overline{p}$ where:

$$p^d=\frac{c_f d}{k}\,\theta^d \tag{5.136}$$

and $\theta=\theta^d\,\overline{\theta}$, we can rewrite (5.133) and (5.134) in the form:

$$\frac{\partial \overline{p}}{\partial t}+\frac{\partial \overline{\theta}}{\partial t}=c_f\,\frac{\partial^2 \overline{p}}{\partial x^2} \tag{5.137}$$

$$c_f c_\theta\,\frac{d^2T_0}{k\kappa}\,\frac{\partial \overline{p}}{\partial t}+\frac{\partial \overline{\theta}}{\partial t}=c_\theta\,\frac{\partial^2 \overline{\theta}}{\partial x^2} \tag{5.138}$$

The first term on the left hand side of (5.138) accounts for the latent heat associated with the skeleton and the fluid volumetric expansions. It turns out be negligible under the condition $c_f c_\theta(d^2T_0/k\kappa)\ll 1$, the latter being generally fulfilled for the usual porous materials. Neglecting the first term in (5.138), the latter reduces to the standard uncoupled thermal diffusion equation whose solution meeting the initial and boundary conditions (5.129) is:

$$\overline{\theta}=1-\operatorname{erf}\left(\frac{x}{2\sqrt{c_\theta t}}\right) \tag{5.139}$$

where erf stands for the error function (5.39). Substituting (5.139) into (5.137) and solving the resulting equation with initial and boundary conditions (5.129), we finally derive:

$$\overline{p}=\left(1-\frac{c_f}{c_\theta}\right)^{-1}\left[\operatorname{erf}\left(\frac{x}{2\sqrt{c_f t}}\right)-\operatorname{erf}\left(\frac{x}{2\sqrt{c_\theta t}}\right)\right] \tag{5.140}$$

[7] Whatever the problem at hand, by extending the analysis of §5.1.3 to account for thermal effects, it can be shown more generally that:

$$\begin{pmatrix}\frac{1}{c} & \frac{dT_0}{\kappa}\\ \frac{d}{k} & \frac{1}{c_\theta}\end{pmatrix}\begin{pmatrix}\frac{\partial v_f}{\partial t}\\ \frac{\partial\left(s-s_f^0 m_f\right)}{\partial t}\end{pmatrix}=\begin{pmatrix}\nabla^2 v_f\\ \nabla^2(s-s_f^0 m_f)\end{pmatrix}$$

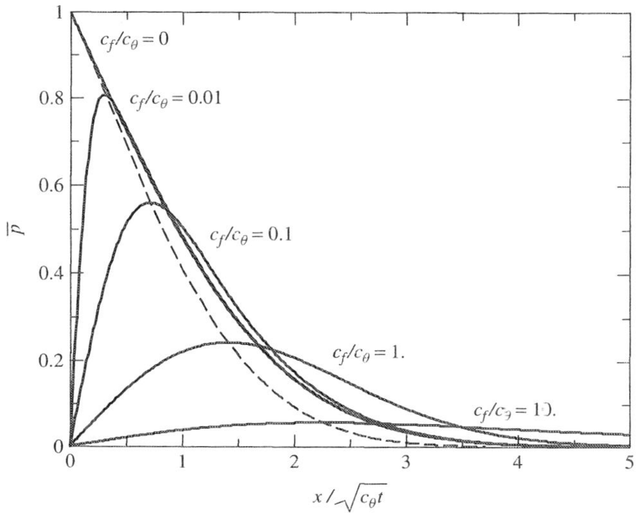

Figure 5.6: Normalized fluid pressure induced by a thermal loading applied on the border of a half-space. The profiles $\overline{p}$ are plotted against $x/\sqrt{c_\theta t}$ for different values of the ratio c_f/c_θ of the hydraulic and the thermal diffusivities. The dashed curve represents the peak of fluid pressure obtained when varying the ratio c_f/c_θ.

For $c_f/c_\theta \to 0$, no fluid diffusion occurs and, consequently, no time lag occurs between the change in temperature and the change in fluid pressure. The mechanical response is undrained and $\overline{p} = \overline{\theta}$ everywhere. In contrast, for $c_f/c_\theta \to \infty$, the fluid diffusion occurs infinitely rapidly in comparison with the thermal diffusion. The response is drained and $\overline{p} = 0$ everywhere. For intermediate values of c_f/c_θ a peak of fluid pressure propagates within the porous medium, whose location $x_{\max}(t)$ nullifies the derivative of $\overline{p}$ with respect to x, yielding:

$$x_{\max}(t) = 2\sqrt{\chi t} \quad \text{where } \chi = \frac{c_f c_\theta}{c_f - c_\theta} \ln\sqrt{c_f/c_\theta} \tag{5.141}$$

In Fig. 5.6 we present the normalized fluid pressure profiles $\overline{p}(x/\sqrt{c_\theta t})$ and the curve $\overline{p}_{\max} = \overline{p}(x_{\max}/\sqrt{c_\theta t})$ of the peak of fluid pressure obtained when varying the diffusivity ratio c_f/c_θ.

5.4 Advanced Analysis

5.4.1 Uniqueness of Solution

Theorem of virtual work with two fields

Let $\mathbf{V}^*$ be any virtual velocity field of the skeleton particles and let $\dot{\boldsymbol{\varepsilon}}^*$ be the associate virtual strain rate according to:

$$\dot{\boldsymbol{\varepsilon}}^* = \frac{1}{2}\left[\nabla\mathbf{V}^* + {}^t\nabla\mathbf{V}^*\right] \tag{5.142}$$

Multiplication of momentum equation (5.5) by $\mathbf{V}^*$ and integration of the resulting equation over volume Ω give:

$$\int_\Omega \boldsymbol{\sigma} : \dot{\boldsymbol{\varepsilon}}^* \, d\Omega = \int_{\partial\Omega} \mathbf{T} \cdot \mathbf{V}^* \, da \qquad (\boldsymbol{\sigma} \cdot \mathbf{n} = \mathbf{T}) \tag{5.143}$$

where, for the sake of simplicity, we did not consider body forces.

Similarly let $\mathcal{V}^* = \mathbf{w}^*/\rho_f^0$ be any virtual filtration vector and let $\dot{v}_f^* = \dot{m}_f^*/\rho_f^0$ be the virtual rate of change in fluid volume content associated with $\mathcal{V}^*$ through the continuity equation (5.7) according to:

$$\dot{v}_f^* = -\nabla \cdot \mathcal{V}^* \tag{5.144}$$

Multiplication of (5.144) by fluid pressure p and integration over volume Ω give:

$$\int_\Omega p \, \dot{v}_f^* \, d\Omega - \int_\Omega \nabla p \cdot \mathcal{V}^* \, d\Omega = \int_{\partial\Omega} -p\mathcal{V}^* \cdot \mathbf{n} \, da \tag{5.145}$$

Finally, adding (5.143) and (5.145) we derive:

$$\int_\Omega (\boldsymbol{\sigma} : \dot{\boldsymbol{\varepsilon}}^* + p \, \dot{v}_f^* - \nabla p \cdot \mathcal{V}^*) d\Omega = \int_{\partial\Omega} (\mathbf{T} \cdot \mathbf{V}^* - p\mathcal{V}^* \cdot \mathbf{n}) \, da \tag{5.146}$$

Equation (5.146) represents the theorem of virtual work with two fields, $\mathbf{V}^*$ and $\mathcal{V}^*$, stating that the balance of mechanical work rates extends to virtual velocity fields.

Conditions of material stability

Consider a material volume Ω, with no loading or fluid pressure on its border $\partial\Omega$. The conditions of material stability are the conditions ensuring that material volume Ω cannot spontaneously and inexorably amplify unavoidable internal fluctuations occurring within Ω. To determine these conditions, let us apply the theorem of virtual work (5.146) to actual fluctuations $\dot{\boldsymbol{\varepsilon}}^* = \dot{\boldsymbol{\varepsilon}}$ and $\dot{v}_f^* = \dot{v}_f$ which can occur under zero boundary conditions, that is $\mathbf{T} = p = 0$. We write:

$$\int_\Omega (\boldsymbol{\sigma} : \dot{\boldsymbol{\varepsilon}} + p \, \dot{v}_f) d\Omega = \int_\Omega \nabla p \cdot \mathcal{V} \, d\Omega \tag{5.147}$$

The right-hand member of (5.147) is negative, since it represents the integral over Ω of the opposite of the positive viscous dissipation φ_f associated with the actual fluid flow (see (3.37)). We get:

$$\int_\Omega (\boldsymbol{\sigma} : \dot{\boldsymbol{\varepsilon}} + p \, \dot{v}_f) d\Omega \leq 0 \tag{5.148}$$

The constitutive equations of linear poroelasticity (4.60), where tangent properties are held constant, allow us to write:

$$\sigma_{ij} = \frac{\partial W}{\partial \varepsilon_{ij}}; \qquad p = \frac{\partial W}{\partial v_f} \tag{5.149}$$

where $W(\varepsilon_{ij}, v_f)$ is the reduced potential defined by:

$$W(\varepsilon_{ij}, v_f) = \frac{1}{2}\varepsilon_{ij}C_{ijkl}\varepsilon_{kl} + \frac{1}{2}M(b_{ij}\varepsilon_{ij} - v_f)^2 \tag{5.150}$$

so that:

$$\sigma : \dot{\varepsilon} + p\,\dot{v}_f = \frac{dW}{dt} \tag{5.151}$$

The reduced potential $W(\varepsilon_{ij}, v_f)$ turns out to be the elastic energy per unit of initial volume of the mass-closed system formed from the skeleton and from the same fluid mass as the one initially contained in volume $d\Omega_0$, whatever the actual fluid particles currently contained in $d\Omega_t$ are. Combining (5.148) and (5.151) we derive:

$$\frac{d}{dt}\int_\Omega W\,(\varepsilon_{ij}, v_f)d\Omega \leq 0 \tag{5.152}$$

The sufficient conditions of material stability are therefore:

$$\frac{1}{2}\varepsilon_{ij}C_{ijkl}\varepsilon_{kl} > 0; \quad M > 0 \tag{5.153}$$

Indeed, under conditions (5.153), $W(\varepsilon_{ij}, v_f)$ has a zero lower bound, whereas it cannot increase according to (5.152). Consequently, if fluctuations in strain and fluid content, ε_{ij} and v_f, actually occur within volume Ω, the latter cannot, at least, spontaneously amplify. It is noteworthy that stability conditions (5.153) do not involve the coupling properties b_{ij}.[8] For an isotropic material the stability conditions (5.153) are specialized in the form:

$$K > 0,\ \mu > 0 \Longleftrightarrow E > 0,\ -1 < \nu < \frac{1}{2}; \quad M > 0 \tag{5.154}$$

As expected, they are automatically satisfied as soon as the solid matrix and the fluid properties fulfil the material stability conditions $K_s > 0$, $\mu_s > 0$ and $K_f > 0$.

Uniqueness of solution

Let $(\sigma\,(t)\,, p\,(t))$ and $(\sigma'\,(t)\,, p'\,(t))$ be two possible solutions for the stress and fluid pressure. Since these solutions are associated with the same boundary conditions and since the field equations they have to satisfy are linear, the difference $(\sigma'\,(t) - \sigma\,(t)\,, p'\,(t) - p\,(t))$ meets all the requirements for a solution associated with zero boundary conditions for the stress vector and fluid pressure. Similarly, if $(\mathbf{V}\,(t)\,, \mathcal{V}\,(t))$ and $(\mathbf{V}'\,(t)\,, \mathcal{V}'\,(t))$ denote the solution associated with $(\sigma\,(t)\,, p\,(t))$ and $(\sigma'\,(t)\,, p'\,(t))$, the difference $(\mathbf{V}'\,(t) - \mathbf{V}\,(t)\,, \mathcal{V}'\,(t) - \mathcal{V}\,(t))$ meets all the requirements for a solution associated with

[8] Analogously, the conditions of material stability with regard to fluctuations in temperature do not involve the thermal dilation coefficients. They only require the positiveness of drained volumetric heat capacity C_σ such as defined in (4.28).

zero boundary conditions. Applying the theorem of virtual work rate (5.146) to fields $(\boldsymbol{\sigma}' - \boldsymbol{\sigma}, p' - p)$ and $(\mathbf{V}' - \mathbf{V}, \mathcal{V}' - \mathcal{V})$ leads to:

$$\int_{\Omega} \left((\boldsymbol{\sigma}' - \boldsymbol{\sigma}) : (\dot{\boldsymbol{\varepsilon}}' - \dot{\boldsymbol{\varepsilon}}) + (p' - p)\left(\dot{v}_f' - \dot{v}_f\right) \right) d\Omega$$
$$= \int_{\partial\Omega} (\nabla p' - \nabla p) \cdot (\mathcal{V}' - \mathcal{V})\, da \tag{5.155}$$

Letting $\mathcal{L}_{\dot{\chi}} = -\nabla p$ and $\dot{\chi} = \mathcal{V}$ in (3.89) gives:

$$(\nabla p' - \nabla p) \cdot (\mathcal{V}' - \mathcal{V}) \leq 0 \tag{5.156a}$$

so that the right hand member of (5.155) is negative. Irrespective of specific skeleton constitutive equations, we derive:

$$\int_{\Omega} \left((\boldsymbol{\sigma}' - \boldsymbol{\sigma}) : (\dot{\boldsymbol{\varepsilon}}' - \dot{\boldsymbol{\varepsilon}}) + (p' - p)\left(\dot{v}_f' - \dot{v}_f\right) \right) d\Omega \leq 0 \tag{5.157}$$

Proceeding as in the previous section, from (5.157) we conclude:

$$\frac{d}{dt} \int_{\Omega} W\left(\varepsilon_{ij}' - \varepsilon_{ij}, v_f' - v_f\right) d\Omega \leq 0 \tag{5.158}$$

where W is still defined by (5.150). According to stability conditions (5.153), energy $W\left(\varepsilon_{ij}' - \varepsilon_{ij}, v_f' - v_f\right)$ is always positive as soon as $\boldsymbol{\varepsilon}' \neq \boldsymbol{\varepsilon}$ and $v_f' \neq v_f$. Indeed, W can be used to express the distance between two solutions. Inequality (5.158) then shows that the distance with regard to energy between two possible solutions can only decrease. Consequently, since both of them must meet the same initial conditions, their distance being zero at $t = 0$ remains so: the solution is unique.

5.4.2 The Beltrami–Michell Equations

The solution to poroelasticity problems can be looked for by adopting the stress components instead of the displacement components in the set of principal unknowns. The solution $\boldsymbol{\sigma}$ to choose among the stress fields satisfying the momentum equation (5.5) has to be associated with a strain field $\boldsymbol{\varepsilon}$ through the constitutive equation according to:

$$\varepsilon_{ij} = \frac{1 + \nu}{E}(\sigma_{ij} + bp\,\delta_{ij}) - \frac{\nu}{E}(\sigma + bp)\,\delta_{ij} \tag{5.159}$$

In (5.159) the six components ε_{ij} defining the symmetric strain tensor field $\boldsymbol{\varepsilon}$ are required to respect the continuity of matter by being derived from the three components of the displacement field $\boldsymbol{\xi}$ according to:

$$\varepsilon_{ij} = \frac{1}{2}\left(\frac{\partial \xi_i}{\partial x_j} + \frac{\partial \xi_j}{\partial x_i}\right) \tag{5.160}$$

Successive derivations of (5.160) give:

$$\frac{\partial^2 \varepsilon_{kl}}{\partial x_i \partial x_j} + \frac{\partial^2 \varepsilon_{ij}}{\partial x_k \partial x_l} = \frac{\partial^2 \varepsilon_{jl}}{\partial x_i \partial x_k} + \frac{\partial^2 \varepsilon_{ik}}{\partial x_j \partial x_l} \tag{5.161}$$

Conversely, successive integrations of (5.161) lead to the existence of a displacement field $\boldsymbol{\xi}$ fulfilling (5.160). More precisely, through partial integrations it can be shown that only three among the six relations (5.161) are actually independent provided that the other three are satisfied on the boundary of the considered domain. Conditions (5.161) are therefore the necessary and sufficient conditions ensuring that components ε_{ij} of any symmetric tensor field ε are actually those of a strain tensor field. Conditions (5.161) are called the kinematical compatibility relations.

Substitution of constitutive equations (5.159) into (5.161) and use of momentum equation (5.5) give:

$$\nabla^2\left[(1+\nu)\,\sigma_{ij} - 3\sigma\,\delta_{ij}\right] + 3\frac{\partial^2 \sigma}{\partial x_i \partial x_j} + b\,(1-2\nu)\left[\nabla^2 p + \frac{\partial^2 p}{\partial x_i \partial x_j}\right] = 0 \tag{5.162}$$

The above six equations are known as the Beltrami–Michell equations. If and only if they are satisfied by the stress field $\boldsymbol{\sigma}$, the latter being in addition the solution to momentum equation (5.5), the kinematical compatibility of the strain field $\boldsymbol{\varepsilon}$ associated with $\boldsymbol{\sigma}$ through (5.159) is automatically ensured. In many problems it can be convenient to consider the equation resulting from the tensorial contraction of (5.162) by letting $i = j$. We get:

$$\nabla^2 \sigma_{kk} + 2b\,\frac{1-2\nu}{1-\nu}\,\nabla^2 p = 0 \tag{5.163}$$

Combining now constitutive equations (5.10) and (5.12), while using relation $K_u = K + b^2 M$, we derive:

$$v_f = \frac{b}{K}\left(\frac{1}{3}\sigma_{kk} + \frac{p}{B}\right) \tag{5.164}$$

where B is the Skempton coefficient defined by (4.68). Substitution of (5.164) into (5.21) finally provides the diffusion equation in the form:

$$\frac{\partial}{\partial t}\left[\frac{1}{3}\sigma_{kk} + \frac{p}{B}\right] = c_f\,\nabla^2\left[\frac{1}{3}\sigma_{kk} + \frac{p}{B}\right] \tag{5.165}$$

Equations (5.5), (5.162) and (5.165) eventually constitute the set of field equations to be satisfied by the stress and the fluid pressure solutions with no further reference to the displacement field.

5.4.3 Mandel's Problem

In order to illustrate the stress approach to problems of linear poroelasticity developed in the previous section, let us consider Mandel's problem.[9] A slab, of extent $2a$ in the x direction and infinitely long in the z direction, is sandwiched between two rigid impermeable plates (see Fig. 5.7). The slab edges $x = \pm a$ are stress-free and drained. By the intermediary of the plates a compressive force, normal to the plates in the y direction and of intensity $2a\varpi$ per unit length in the z direction, is suddenly applied to the slab and is eventually held constant.

Owing to the problem symmetry the fields depend only on x and t. Assuming no friction at the slab–plate interface, there is no shear so that $\sigma_{xy} = 0$ everywhere. Moreover, since the slab edges are stress-free, equilibrium in the x direction eventually requires $\sigma_{xx} = 0$. Furthermore, the strain is zero in the infinite z direction. Using $\varepsilon_{zz} = 0$ and $\sigma_{xx} = 0$ in constitutive equations (4.25)–(4.26), we derive:

$$\sigma_{kk} = (1 + \nu)\,\sigma_{yy} - (1 - 2\nu)\,bp \tag{5.166}$$

The instantaneous response of the slab is undrained everywhere so that the initial stress is uniform. Use of (5.164) and (5.166), where we let $\sigma_{yy} = -\varpi$ and $v_f = 0$, gives the fluid overpressure immediately after the application of the loading. Using (4.71), the instantaneous undrained fluid pressure $p(\overline{x}, \overline{t} = 0^+)$ can be expressed in the form:

$$p(\overline{x}, \overline{t} = 0^+) = \frac{1}{3}B(1 + \nu_u)\varpi \tag{5.167}$$

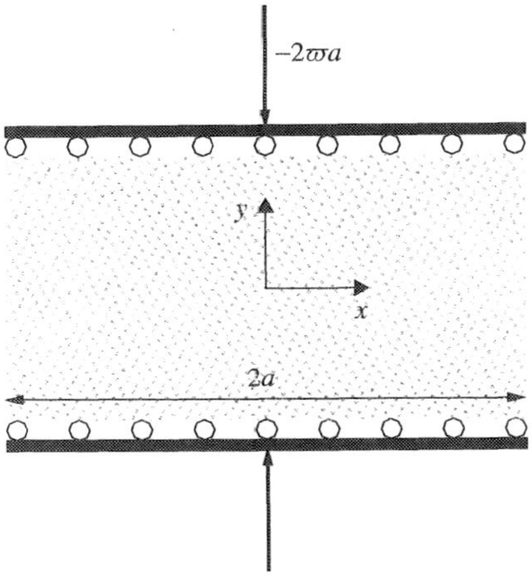

Figure 5.7: Mandel's problem.

[9]See Mandel J. (1953), 'Consolidation des sols (étude mathématique)', *Géotechnique* **3**, 287–299. See also Cheng A.H.-D., Detournay E. (1988), 'A direct boundary element method for plane strain poroelasticity', *International Journal of Numerical and Analytical Methods in Geomechanics*, **12**, 551–572, and Abousleiman Y., Cheng A H.-D., Cui L., Detournay E., Roegiers J.-C. (1996), 'Mandel's problem revisited', *Géotechnique*, **46**, 187–195.

The stress and the fluid overpressure are appropriately scaled by the loading strength and the instantaneous undrained response, respectively. We note:

$$\overline{\sigma}_{ij} = \frac{\sigma_{ij}}{\varpi}; \quad \overline{p} = \frac{p}{\frac{1}{3}B(1+\nu_u)\varpi} \tag{5.168}$$

Substitution of (5.166) into (5.163) and use of (4.71) give:

$$\frac{\partial^2}{\partial x^2}\left(\overline{\sigma}_{yy} + \frac{\nu_u - \nu}{1-\nu}\overline{p}\right) = 0 \tag{5.169}$$

The previous equation can be integrated in the form:

$$\overline{\sigma}_{yy} = \overline{\sigma}_{yy}\left(1, \overline{t}\right) - \frac{\nu_u - \nu}{1-\nu}\overline{p} \tag{5.170}$$

where the integration function $\overline{\sigma}_{yy}\left(1, \overline{t}\right)$ has been chosen to depend only on time, since a change from x to $-x$ does not affect the field value. In addition, the integration function is recognized to be $\overline{\sigma}_{yy}\left(1, \overline{t}\right)$ in order to match the boundary condition $\overline{p}\left(1, \overline{t}\right) = 0$ at the drained edges. Substitution of (5.170) into (5.166) allows us to express $\overline{\sigma}_{kk}$ as a function of $\overline{\sigma}_{yy}\left(1, \overline{t}\right)$ and $\overline{p}$. Substitution of the resulting expression into (5.165) while using (4.71) and (5.167) finally provides the diffusion equation in the form:

$$\frac{\partial}{\partial \overline{t}}\left[\overline{\sigma}_{yy}\left(1, \overline{t}\right) + \frac{1-\nu_u}{1-\nu}\overline{p}\right] = \frac{\partial^2}{\partial \overline{x}^2}\left[\overline{\sigma}_{yy}\left(1, \overline{t}\right) + \frac{1-\nu_u}{1-\nu}\overline{p}\right] \tag{5.171}$$

where we note:

$$\overline{x} = \frac{x}{a}; \quad \overline{t} = \frac{c_f t}{a^2} \tag{5.172}$$

A general solution to (5.171) can be investigated in the form of the infinite series:

$$\overline{\sigma}_{yy}\left(1, \overline{t}\right) + \frac{1-\nu_u}{1-\nu}\overline{p} = A_0 + \sum_{n=1}^{n=\infty} A_n \cos(\alpha_n \overline{x}) \exp\left(-\alpha_n^2 \overline{t}\right) \tag{5.173}$$

where each term of the series satisfies the diffusion equation (5.171) and the invariance of the solution with regard to a change from x to $-x$. The boundary condition $\overline{p}\left(1, \overline{t}\right) = 0$ at the drained edges requires:

$$\overline{\sigma}_{yy}\left(1, \overline{t}\right) = A_0 + \sum_{n=1}^{n=\infty} A_n \cos\alpha_n \exp\left(-\alpha_n^2 \overline{t}\right) \tag{5.174}$$

The set of equations (5.170), (5.173) and (5.174) allows us to express the normalized stress in the form:

$$\overline{\sigma}_{yy}\left(\overline{x}, \overline{t}\right) = A_0 + \frac{\nu_u - \nu}{1-\nu_u}\sum_{n=1}^{n=\infty} A_n \left[\frac{1-\nu}{\nu_u - \nu}\cos\alpha_n - \cos(\alpha_n \overline{x})\right] \exp\left(-\alpha_n^2 \overline{t}\right) \tag{5.175}$$

Equilibrium of the slab in the y direction requires:

$$\int_0^{+1} \overline{\sigma}_{yy}\left(\overline{x}, \overline{t} \geq 0\right) d\overline{x} = -1 \tag{5.176}$$

Substitution of (5.175) into (5.176) gives:

$$A_0 + \frac{\nu_u - \nu}{1 - \nu_u} \sum_{n=1}^{n=\infty} A_n \left[\frac{1 - \nu}{\nu_u - \nu} - \frac{\tan\alpha_n}{\alpha_n}\right] \cos\alpha_n \exp\left(-\alpha_n^2 \overline{t}\right) = -1 \tag{5.177}$$

This equation must hold irrespective of time values, yielding:

$$A_0 = -1; \quad \frac{\tan\alpha_n}{\alpha_n} = \frac{1 - \nu}{\nu_u - \nu} \tag{5.178}$$

With the help of (5.173) and (5.174), the instantaneous overpressure condition, that is $\overline{p}(\overline{x}, \overline{t} = 0^+) = 1$, can be expressed in the form:

$$\sum_{n=1}^{n=\infty} A_n \left[\cos(\alpha_n \overline{x}) - \cos\alpha_n\right] = \frac{1 - \nu_u}{1 - \nu} \tag{5.179}$$

When multiplying (5.179) by $\cos(\alpha_m \overline{x})$ and integrating, the left hand side of the resulting equation is non-zero only for $n = m$, giving:

$$A_n = 2 \frac{1 - \nu_u}{1 - \nu} \times \frac{\sin\alpha_n}{\alpha_n - \sin\alpha_n \cos\alpha_n} \tag{5.180}$$

Collecting the results, we eventually express the fluid pressure in the form:

$$\overline{p}\left(\overline{x}, \overline{t}\right) = 2 \sum_{n=1}^{n=\infty} \frac{\cos(\alpha_n \overline{x}) - \cos\alpha_n}{\alpha_n - \sin\alpha_n \cos\alpha_n} \sin\alpha_n \times \exp\left(-\alpha_n^2 \overline{t}\right) \tag{5.181}$$

In Fig. 5.8 we plot the dimensionless pressure profiles $\overline{p}$ for incompressible constituents, that is for $\nu_u = 0.5$, and for two different values of the drained Poisson coefficient ν. For small times the fluid pressure rises above the initial undrained value in the central region of the slab. This unexpected non-monotonic behaviour can be explained by inspecting the diffusion equation governing the fluid pressure. Using (5.171), (5.174) and (5.180), we obtain:

$$\frac{\partial \overline{p}}{\partial \overline{t}} - \frac{\partial^2 \overline{p}}{\partial \overline{x}^2} = 2 \sum_{n=1}^{n=\infty} \frac{\alpha_n^2 \sin\alpha_n \cos\alpha_n}{\alpha_n - \sin\alpha_n \cos\alpha_n} \exp\left(-\alpha_n^2 \overline{t}\right) \tag{5.182}$$

The right hand side of (5.182) arises from the stress application and the strain compatibility of the slab with regard to the rigid plates. This term acts as a uniform source term in the diffusion equation governing the fluid pressure. Close to the edges the drainage operates immediately and, in spite of the source term, the fluid pressure is monotonically decreasing with time. By contrast, at early times the central region is poorly drained and the source term is the most intense. As a consequence, the fluid pressure in the central region of the slab rises above the initial undrained value for small times. This pressure buildup is

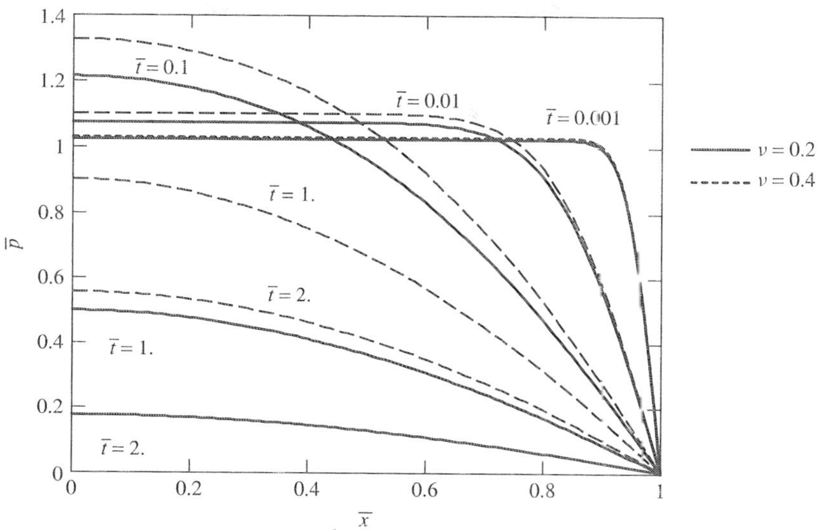

Figure 5.8: Pressure profiles across the slab in Mandel's problem for incompressible constituents and two different values of drained Poisson's coefficient ν. At early times the fluid pressure increases in the central region owing to the source term in the diffusion equation arising from the strain compatibility of the slab with regard to the rigid plates.

enhanced by higher values of the drained Poisson's coefficient ν, allowing less extension in the x direction at the drained edges of the slab. As time passes the middle zone is subjected to a gradient of fluid pressure increasing with time while the source term is exponentially decreasing with time, so that the fluid pressure finally decreases everywhere.

5.4.4 Non-linear Sedimentation

Under the effect of gravity the solid particles of a suspension precipitate and progressively accumulate on the impermeable base to constitute eventually the successive solid layers of a porous medium. Under the weight of the overlying layers the skeleton so formed consolidates. It results in a porosity decreasing with depth, whereas the trapped fluid, which is subjected to a vertical gradient of pressure, flows upwards. In the meantime the skeleton settles and the extent of the region of particles still precipitating decreases. Finally the precipitation zone disappears and a pure self-weighted consolidation process takes place. This roughly depicts the sedimentation process, as sketched in Fig. 5.9. The process is central to the sedimentation of successive layers accumulating on the seafloor. It is also often invoked to explain the non-uniformity in porosity of a concrete block resulting from the setting of the cement particles.We will give below a non-linear unidimensional approach to the sedimentation process. The approach is non-linear with regard to four aspects: (i) the displacements and the strain are finite;[10] (ii) the stress–strain relationship

[10]The approach to the consolidation in the finite transformation of the layers having already sedimented is adapted from Bourgeois E., Dormieux L. (1996), 'Consolidation of a non-linear poroelastic layer in finite deformations', *European Journal of Mechanics, A/Solids*, **15**, (4) 575–598.

is non-linear; (iii) the permeability depends on the current porosity; (iv) there is a moving discontinuity between the region formed of particles still precipitating and the layers of sediments currently consolidating.

Consider a vertical porous layer of initial thickness h_0, resting on a rigid impervious base taken as the origin of coordinates $z = 0$. Equilibrium in the vertical direction requires:

$$\frac{\partial \sigma_{zz}}{\partial z} - ((1-n)\,\rho_s + n\rho_f)g = 0 \tag{5.183}$$

while Darcy's law is written in the form:

$$\frac{w_z}{\rho_f} = n\left(V_z^f - V_z^s\right) = k_0\,\delta\,(n)\left(-\frac{\partial p}{\partial z} - \rho_f g\right) \tag{5.184}$$

where k_0 is a reference permeability while function $\delta\,(n)$ accounts for the dependence of the permeability on the Eulerian porosity (see (3.47)). We assume now that the solid particles forming the skeleton undergo negligible volume changes so that the results derived in §3.4.1 for the vertical effective stress σ'_{zz} apply. Adopting for the $\mathbf{e}_3$ direction as the vertical direction, that is $z = x_3$, we write (3.80) in the form:

$$\sigma'_{zz} = \sigma_{zz} + p = -\mathsf{E}\,\varpi\,(n) \tag{5.185}$$

where E is a reference oedometric modulus (see (4.23)). Getting rid of σ_{zz} and p from the three equations above, we derive:

$$\frac{w_z}{\rho_f} = k_0\,\delta\,(n)\left(\mathsf{E}\,\frac{\partial \varpi}{\partial z}\,(n) + (1-n)\,(\rho_s - \rho_f)g\right) \tag{5.186}$$

A skeleton particle, which was located at Z at time $t = 0$ (Lagrangian coordinate), is located at $z = z\,(Z, t)$ at current time t (Eulerian coordinate) (see Fig. 5.9). Letting $Z = X_3$, from (3.72) and (3.79) we get:

$$J = \frac{\partial z}{\partial Z} = 1 + \phi - \phi_0 = \frac{1-n_0}{1-n} \tag{5.187}$$

Since the skeleton surface normal to the fluid flow does not undergo any change, (1.65) reduces to $w_z = M_Z$. Combining (5.186) and (5.187) then gives:

$$\frac{M_Z}{\rho_f} = k_0\,\frac{1-n}{1-n_0}\,\delta\,(n)\left(\mathsf{E}\,\frac{\partial \varpi}{\partial Z}\,(n) + (1-n_0)(\rho_s - \rho_f)g\right) \tag{5.188}$$

As the fluid flow is assumed to be incompressible, (1.64) and (1.67) allow us to write the fluid mass conservation in the form:

$$\rho_f\,\frac{\partial \phi}{\partial t} = -\frac{\partial M_Z}{\partial Z} \tag{5.189}$$

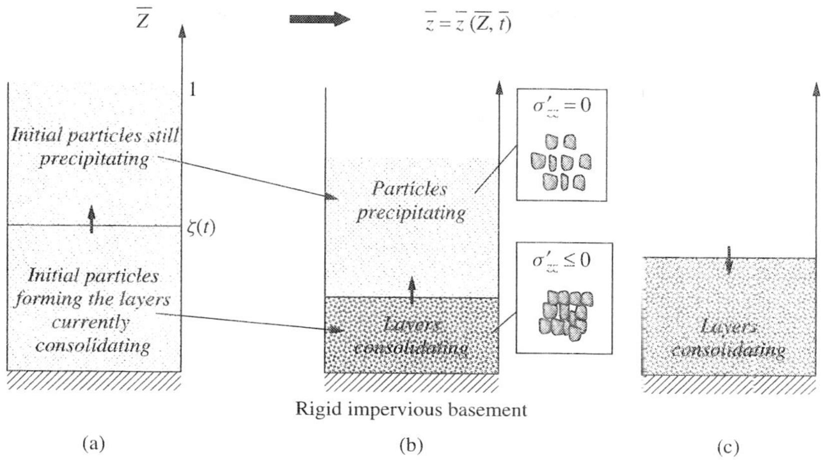

Figure 5.9: Sedimentation in dimensionless coordinates: (a) Lagrangian initial configuration; (b) Eulerian current configuration at a time when the precipitation and the consolidation processes are still both active; (c) Eulerian current configuration at a later time when the precipitation has stopped and only the consolidation remains active. The solid line delimits the current height of the column of particles already sedimented and consolidating.

In addition, since both constituents are assumed to be incompressible, the mass conservation implies that the volume of skeleton particles flowing downwards is opposite to the volume of fluid flowing upwards, that is:

$$V_z^s = -\frac{M_Z}{\rho_f} \tag{5.190}$$

Substitution of (5.188) into (5.189) and use of (5.187) eventually give:

$$\frac{\partial}{\partial \bar{t}}\left(\frac{1}{1-n}\right) + \frac{\partial}{\partial \overline{Z}}\left[(1-n)\,\delta(n)\left(\frac{\partial \varpi}{\partial \overline{Z}}(n) + \alpha\right)\right] = 0 \tag{5.191}$$

where we let:

$$\bar{t} = \frac{t}{\tau};\quad \overline{Z} = \frac{Z}{h_0};\quad \tau = \frac{h_0^2(1-n_0)^2}{k_0 \mathsf{E}};\quad \alpha = (1-n_0)(\rho_s - \rho_f)\frac{g h_0}{\mathsf{E}} \tag{5.192}$$

During the whole process the upper skeleton particles of the sedimenting column remain subjected to the fluid pressure exerted by the overlying fluid so that the effective stress σ'_{zz} is equal to zero at the top of the column of the skeleton particles. Assuming in addition a uniform initial porosity, the initial and boundary conditions to be added to (5.191) are:

$$n \mid_{\bar{t}=0} = n_0;\quad \varpi \mid_{\overline{Z}=1} = 0 \tag{5.193}$$

which have to be completed by the condition expressing the impermeability of the base:

$$\left(\frac{\partial \varpi}{\partial \overline{Z}}(n) + \alpha\right) \mid_{\overline{Z}=0} = 0 \tag{5.194}$$

For porosities larger than some threshold n_{cr}, the skeleton is formed of isolated particles or isolated sets of particles subjected to the fluid pressure p, so that the intrinsic partial stress related to both the matrix and the fluid reduces to the same hydrostatic pressure:

$$n \geq n_{cr}: \quad \boldsymbol{\sigma}^s = \boldsymbol{\sigma}^f = -p\mathbf{1} \tag{5.195}$$

Accordingly, (2.34) results in $\sigma_{zz} = (1-n)\,\sigma_{zz}^s + n\sigma_{zz}^f = -p$ so that (5.185) entails $\varpi\,(n \geq n_{cr}) = 0$. For porosities lower than the above-mentioned threshold n_{cr}, the skeleton is formed of jointed particles able to support stresses of their own. We finally write:

$$\varpi\,(n \geq n_{cr}) = 0; \quad \varpi\,(n < n_{cr}) > 0 \tag{5.196}$$

Constitutive equation (5.196) leads us to consider a first stage where an upper zone exists in which the particles still precipitate and the porosity remains constantly equal to the initial uniform value n_0 in order that (5.191) remains satisfied. Note that n_0 cannot be less than n_{cr}, since otherwise the boundary condition at $Z = 0$ could not be initially met. Indeed n_{cr} is expected to be the surface porosity at the end of the sedimentation process. Therefore we write:

$$1 \geq \overline{Z} > \zeta\left(\overline{t}\right): \quad n = n_0 \tag{5.197}$$

while (5.188) and (5.190) provide the related solid particle velocity in the form:

$$1 \geq \overline{Z} > \zeta\left(\overline{t}\right): \quad V_z^s = -\frac{M_Z}{\rho_f} = -k_0\delta(n_0)(1-n_0)(\rho_s - \rho_f)g \tag{5.198}$$

The remaining zone $\zeta\,(t) \geq Z \geq 0$ is governed by a consolidation process with:

$$n\left(\overline{Z} = \zeta\left(\overline{t}\right), \overline{t}\right) = n_{cr} \tag{5.199}$$

The surface $\overline{Z} = \zeta\,(t)$ is therefore a surface of discontinuity with regard to the porosity. Owing to the fluid incompressibility, the jump condition (1.78) here is specialized in the form:

$$\left[\!\left[\frac{\tau}{h_0}\frac{M_Z}{\rho_f} - \phi\frac{d\zeta}{d\overline{t}}\right]\!\right] = 0 \tag{5.200}$$

The surface of discontinuity is initially located on the base, that is $\zeta\,(0) = 0$. Combining (5.187)–(5.188) and (5.196)–(5.200) allows us to determine the condition governing the current location $\zeta\left(\overline{t}\right)$ of the moving surface of discontinuity in the initial $\overline{Z}$ configuration:

$$\left(\frac{1}{1-n_{cr}} - \frac{1}{1-n_0}\right)\frac{d\zeta}{d\overline{t}} + (1-n_{cr})\delta(n_{cr})\frac{\partial\varpi}{\partial\overline{Z}}\,|_{\overline{Z}=\zeta^-}$$
$$+\,\alpha\big[(1-n_{cr})\delta(n_{cr}) - (1-n_0)\delta(n_0)\big] = 0 \tag{5.201}$$

The consolidation process in the zone $0 \leq Z \leq \zeta\,(t)$ is eventually governed by the non-linear diffusion equation (5.191) with boundary conditions (5.194), (5.199) and (5.201).

Once the Lagrangian porosity field $n\left(\overline{Z}, \overline{t}\right)$ is known, the observable Eulerian current porosity field $n(\overline{z} = z/h_0, \overline{t})$ is determined by noting that the associated current dimensionless height $\overline{z}\left(\overline{Z}, t\right)$ (Eulerian coordinate) formed by the solid particles which were initially at height $\overline{Z}$, is provided by integrating (5.187) according to:

$$\overline{z}\left(\overline{Z}, \overline{t}\right) = \int_0^{\overline{Z}} \frac{1 - n_0}{1 - n\left(\mathcal{Z}, \overline{t}\right)} d\mathcal{Z} \tag{5.202}$$

A quantity of interest is the dimensionless current height of solid particles, which is the ratio $\overline{h} = h/h_0$ between the current height h and the initial height h_0. The dimensionless height history is derived from the previous equation in the form:

$$\overline{h}\left(\overline{t}\right) = 1 - \zeta\left(\overline{t}\right) + \overline{h}_{sed}\left(\overline{t}\right); \quad \overline{h}_{sed}\left(\overline{t}\right) = \int_0^{\zeta(\overline{t})} \frac{1 - n_0}{1 - n\left(\mathcal{Z}, \overline{t}\right)} d\mathcal{Z} \tag{5.203}$$

where $\overline{h}_{sed}\left(\overline{t}\right)$ is the dimensionless current sedimentation height. Noting that $dh/dt = V_z^s\,(Z = 1, t)$ and using (5.198), we also derive:

$$\zeta\left(\overline{t}\right) < 1: \quad \frac{d\overline{h}}{d\overline{t}} = -\alpha\delta(n_0)(1 - n_0)^2 \tag{5.204}$$

It is therefore worthwhile to remark that the rate of decrease of the overall height of solid particles is constant as long as a precipitation zone still exists. In addition, since at the end of the sedimentation process the time derivative must vanish in (5.191) and the fluid flow is zero, the asymptotic porosity field $n_\infty\left(\overline{Z}\right) = n\left(\overline{Z}, \overline{t} \to \infty\right)$ can be directly determined by solving:

$$\frac{\partial \varpi}{\partial \overline{Z}}(n_\infty\left(\overline{Z}\right)) + \alpha = 0; \quad n_\infty\left(\overline{Z} = 1\right) = n_{cr} \tag{5.205}$$

The asymptotic field $n_\infty\left(\mathcal{Z}\right)$ substituted for $n\left(\mathcal{Z}, \overline{t}\right)$ in (5.203) leads to the determination of the asymptotic height $h_\infty = h\left(\overline{t} \to \infty\right)$ irrespective of the previous height history $h\left(\overline{t} < \infty\right)$. Conversely, the experimental determination of the observable Eulerian asymptotic porosity field $n_\infty\left(\overline{z}\right)$ permits the experimental determination of the possibly unknown function $\varpi\left(n\right)$. Indeed, the combination of (5.187) and (5.205) furnishes:

$$\varpi\left(n\right) = -\alpha \int_{n_{cr}}^{n} \frac{1 - n_\infty}{1 - n_0} \frac{d\overline{z}}{dn_\infty} dn_\infty \tag{5.206}$$

For spherical particles the still unspecified functions $\delta\left(n\right)$ and $\varpi\left(n\right)$ can be written in the form:

$$\delta\left(n\right) = \frac{n^3}{1 - n^2}; \quad \varpi\left(n\right) = \langle n_{cr} - n\rangle^{3/2} \tag{5.207}$$

The first relation is Kozeny–Carman's formula (see (3.48)). The second relation, where $\langle n_{cr} - n\rangle = \frac{1}{2}(|n_{cr} - n| + n_{cr} - n)$ stands for the positive part of quantity $n_{cr} - n$, is

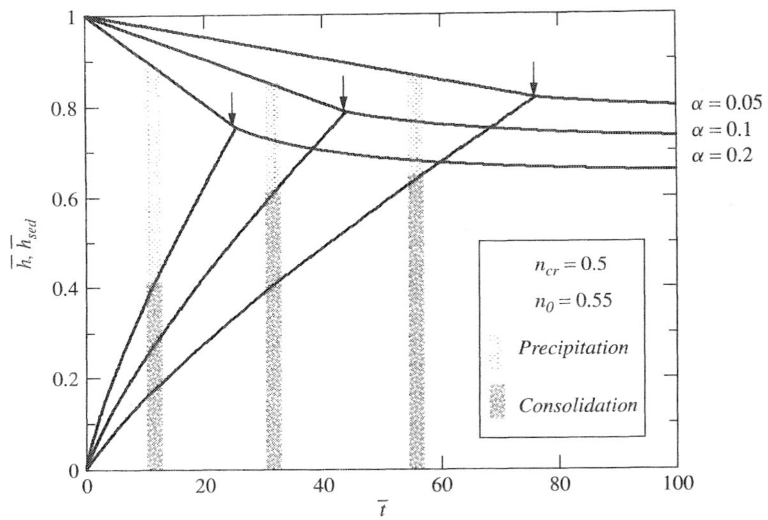

Figure 5.10: Dimensionless overall and sedimentation heights, $\overline{h}$ and $\overline{h}_{sed}$, plotted against dimensionless time $\overline{t}$ for different values of the ratio α of characteristic times of precipitation and consolidation, both referring to the initial height h_0. The initial porosity is $n_0 = 0.55$ and the critical porosity $n_{cr} = 0.5$ (see (5.207)). The arrow indicates the end of the precipitation process. The asymptotic dimensionless height $\overline{h}_\infty$ is numerically found to be equal to 0.78, 0.72 and 0.65 for $\alpha = 0.05$, 0.1 and 0.2 respectively.

based on Hertz's solution to the contact between two elastic spheres subjected to a vertical loading. Substituting (5.207) into (5.205) we derive:

$$n_{cr} - n_\infty \left(\overline{Z}\right) = \left[\alpha \left(1 - \overline{Z}\right)\right]^{2/3} \tag{5.208}$$

In Fig. 5.10, adopting (5.207) and numerically solving the resulting sedimentation equation (5.191), we plot the dimensionless overall and sedimentation heights $\overline{h}$ and $\overline{h}_{sed}$ against the dimensionless time $\overline{t}$ for various values of parameter α. Inspection of (5.191) with definitions (5.192) reveals that parameter α is eventually the ratio of the characteristic time associated with a pure precipitation process to the characteristic time associated with a pure consolidation process, both times referring to the same height h_0.

Chapter 6

Unsaturated Thermoporoelasticity

In many processes, such as the drying, the drainage or the imbibition of materials, the porous space becomes filled by several fluids so that the porous material is said to be unsaturated with regard to a reference fluid of principal concern, generally chosen in liquid form. In most cases two fluids coexist within the porous space, either two liquids, for instance oil and water in petroleum engineering, or a liquid and a gas, for instance liquid water and wet air in the drying of materials. The unsaturated context introduces new thermo/hydro/mechanical couplings mainly associated with the surface tension or the energy related to each fluid–fluid or fluid–solid interface. They cause macroscopic effects such as the drying shrinkage of materials that can occur at zero overall stress. The aim of this chapter is to explore these new couplings by extending the macroscopic energy approach to the unsaturated situation.[1]

6.1 Mass and Momentum Balance

6.1.1 Partial Porosities and Degree of Saturation

When the porous space is filled by several fluids referred to by index $\alpha = 1, 2, \ldots$ Eulerian and Lagrangian porosities n and ϕ can be split into partial porosities n_α and ϕ_α, so that the volume occupied by fluid α is $n_\alpha d\Omega_t = \phi_\alpha d\Omega_0$. We write:

$$n = \sum_\alpha n_\alpha; \quad \phi = \sum_\alpha \phi_\alpha \tag{6.1}$$

The degree of saturation S_α relative to fluid α is defined by:

$$S_\alpha = \frac{n_\alpha}{n} = \frac{\phi_\alpha}{\phi}; \quad \sum_\alpha S_\alpha = 1 \tag{6.2}$$

[1]The approach developed in this chapter is the result of a collaboration with P. Dangla. See in particular Coussy O., Dangla P. (2002), 'Approche énergétique du comportement des sols non saturés', *Mécanique des Sols Non Saturés*, ed. Coussy O. and Fleureau J.-M., Hermès, Paris. See also the references given in the general bibliography.

Poromechanics O. Coussy
© 2004 John Wiley & Sons, Ltd ISBN 0-470-84920-7

Let ρ_α and m_α be respectively the intrinsic mass density and the Lagrangian fluid mass content related to fluid α. We write:

$$\rho_\alpha n_\alpha = \rho_\alpha n S_\alpha \tag{6.3}$$

$$m_\alpha = \rho_\alpha \phi_\alpha = \rho_\alpha \phi S_\alpha \tag{6.4}$$

6.1.2 Mass and Momentum Balance

Extending the notation and the results of Chapter 1, the Eulerian continuity equation relative to fluid α is:

$$\frac{\partial(\rho_\alpha n S_\alpha)}{\partial t} + \nabla_x \cdot \left(\rho_\alpha n S_\alpha \mathbf{V}^\alpha\right) = 0 \tag{6.5}$$

while its Lagrangian counterpart is:

$$\frac{dm_\alpha}{dt} + \nabla_X \cdot \mathbf{M}^\alpha = 0 \tag{6.6}$$

Adopting in the following the hypothesis of infinitesimal transformations (see Chapter 1), the previous equation can be rewritten in the form:

$$\frac{dm_\alpha}{dt} + \nabla \cdot \mathbf{w}^\alpha = 0 \tag{6.7}$$

where $\mathbf{w}^\alpha = \rho_\alpha n_\alpha(\mathbf{V}^\alpha - \mathbf{V}^s)$ is the Eulerian relative flow vector of fluid mass related to fluid α.

Still denoting the total stress by $\boldsymbol{\sigma}$ and disregarding dynamic terms (quasistatic approximation), the overall momentum balance is still:

$$\nabla \cdot \boldsymbol{\sigma} + \rho \mathbf{f} = 0 \tag{6.8}$$

where ρ stands for the overall mass density of the porous material. With regard to the saturated situation, in addition to the partial stresses related to the solid and the fluids, the membrane stresses, which act along all the interfaces separating each component from the other, will contribute to the total stress $\boldsymbol{\sigma}$ (see §6.6 of this chapter for further details).

6.1.3 Mass and Momentum Balance with Phase Change

We now consider the case where the fluid components are formed from a liquid (index l), its vapour (index v) and the (dry) air (index a). Since the vapour and the air occupy the same porous space and form the gas phase (index g), we write:

$$\phi_v = \phi_a = \phi_g; \quad \phi_l + \phi_g = \phi; \quad S_g = \frac{\phi_g}{\phi}; \quad S_l = \frac{\phi_l}{\phi} \tag{6.9}$$

Owing to the possible change of the liquid into its vapour the mass balance equation relative to each component can be written:

$$\frac{dm_l}{dt} + \nabla \cdot \mathbf{w}^l = -\overset{\circ}{m}_{l\to v} \tag{6.10a}$$

$$\frac{dm_v}{dt} + \nabla \cdot \mathbf{w}^v = \overset{\circ}{m}_{l\to v} \tag{6.10b}$$

$$\frac{dm_a}{dt} + \nabla \cdot \mathbf{w}^a = 0 \tag{6.10c}$$

where $\overset{\circ}{m}_{l\to v}$ stands for the rate of liquid mass changing into vapour per unit of initial volume $d\Omega_0$. In addition, still excluding dynamic effects, the momentum balance equation is written in the same way as (6.8).[2]

6.2 Thermodynamics

6.2.1 Energy and Entropy Balance for the Porous Material

In the context of infinitesimal and quasistatic transformations, the energy balance equation (3.21) can be extended in the form:

$$\frac{dE}{dt} = \boldsymbol{\sigma} : \frac{d\boldsymbol{\varepsilon}}{dt} - \nabla \cdot \left(\sum_{\alpha} h_\alpha \mathbf{w}^\alpha + \mathbf{q} \right) + \mathbf{f} \cdot \mathbf{w} \tag{6.11}$$

where E is the overall density of internal energy per unit of volume, $\mathbf{q}$ is the overall heat flow vector, while h_α is the specific enthalpy related to fluid α.

The second law applied to the unsaturated situation turns out to extend the entropy balance (3.23) in the form:

$$\frac{dS}{dt} \geq -\nabla \cdot \left(\sum_{\alpha} s_\alpha \mathbf{w}^\alpha + \frac{\mathbf{q}}{T} \right) \tag{6.12}$$

where S is the overall density of entropy per unit of volume, while s_α is the specific entropy related to fluid α. The first law (6.11) and second law (6.12) combine to give the following form of the fundamental inequality:

$$\Phi = \Phi_{int} + \varphi_f + \varphi_{th} \geq 0 \tag{6.13}$$

[2]There is an analogy between the present case, where one of the fluid components is a liquid which can transform into its vapour, and the double porous network case where the fluid component of one network can transform into the fluid component of the other network. The derivation of balance equations can then be extended from the double porous network case to the phase change case. For the mass and momentum balance see for instance §1.5.3 and §2.5.2. In particular, (2.70) can be applied in order to include dynamic effects in the phase change case. The same remark holds for the laws of thermodynamics as developed in the next section.

where:

$$\Phi_{int} = \boldsymbol{\sigma} : \frac{d\boldsymbol{\varepsilon}}{dt} - \sum_{\alpha} g_\alpha (\nabla \cdot \mathbf{w}^\alpha) - S\frac{dT}{dt} - \frac{d\Psi}{dt} \tag{6.14a}$$

$$\varphi_{th} = -\frac{\mathbf{q}}{T} \cdot \nabla T \tag{6.14b}$$

$$\varphi_f = \sum_{\alpha} (-(\nabla g_\alpha)_T + \mathbf{f}) \cdot \mathbf{w}^\alpha \tag{6.14c}$$

where $(\nabla g_\alpha)_T$ stands for the gradient, taken at constant temperature, of the specific Gibbs potential g_α related to fluid α, while $\Psi = E - T$ is the free energy. Extending the analysis of §3.2.3, Φ_{int} is anticipated to be the dissipation related to the skeleton, while φ_f and φ_{th} are the dissipations related to fluid mass and heat transfer respectively. The entropy balance is finally expressed in the form of the thermal equation:

$$T\left(\frac{dS}{dt} + \nabla \cdot \left(\sum_{\alpha} s_\alpha \mathbf{w}^\alpha\right)\right) = -\nabla \cdot \mathbf{q} + \Phi_M \tag{6.15}$$

where $\Phi_M = \Phi_{int} + \varphi_f$ stands for the mechanical dissipation and acts as a spontaneous heat source term.

6.2.2 Skeleton State Equations. Averaged Fluid Pressure and Capillary Pressure

Use of mass balance equation (6.7) allows us to rewrite Φ_{int} given by (6.14a) in the form:

$$\Phi_{int} \equiv \Phi_s = \boldsymbol{\sigma} : \frac{d\boldsymbol{\varepsilon}}{dt} + \sum_{\alpha} g_\alpha \frac{dm_\alpha}{dt} - S\frac{dT}{dt} - \frac{d\Psi}{dt} \tag{6.16}$$

Now let Ψ_s and S_s be the skeleton free energy and entropy defined by:

$$\Psi_s = \Psi - m_\alpha \psi_\alpha; \qquad S_s = S - m_\alpha s_\alpha \tag{6.17}$$

Definitions (6.4) and (6.17), together with the fluid state equations (see §3.1.2) applied to each fluid α, allow us to express Φ_s in the form:

$$\Phi_s = \boldsymbol{\sigma} : \frac{d\boldsymbol{\varepsilon}}{dt} + \sum_{\alpha} p_\alpha \frac{d\phi_\alpha}{dt} - S_s\frac{dT}{dt} - \frac{d\Psi_s}{dt} \tag{6.18}$$

The same analysis as the one performed in §3.2.3 eventually allows us to identify Φ_s as the mechanical dissipation associated with the skeleton. In thermoporoelasticity this dissipation is zero and we write:

$$\sigma_{ij} d\varepsilon_{ij} + \sum_{\alpha} p_\alpha d\phi_\alpha - S_s dT - d\Psi_s = 0 \tag{6.19}$$

so that the skeleton state equations of unsaturated thermoporoelasticity are:

$$\boldsymbol{\sigma} = \frac{\partial \Psi_s}{\partial \boldsymbol{\varepsilon}}; \quad p_\alpha = \frac{\partial \Psi_s}{\partial \phi_\alpha}; \quad S_s = -\frac{\partial \Psi_s}{\partial T} \tag{6.20}$$

Since $\Psi d\Omega_0$ and $S d\Omega_0$ stand respectively for the free energy and for the entropy of the whole matter contained in volume $d\Omega_t$, Ψ_s and S_s are the free energy and the entropy of the skeleton, provided that the latter, in addition to the solid matrix, includes the interfaces between the different components. Indeed the fluid–fluid interfaces and the solid–fluid interfaces possess their own proper interfacial energy and entropy. Consider the usual case where the porous space is filled by two fluids, one being the wetting fluid, $\alpha = w$, and the other the non-wetting fluid, $\alpha = nw$. The wetting fluid is the fluid showing a preponderant affinity for the solid, with a contact angle θ less than 90° (see Fig. 6.1). Perfect wettability corresponds to $\theta = 0^\circ$ and is often met for clean porous materials (rock, glass, etc.) and a liquid/gas saturating mixture where the gas is the non-wetting fluid. Let $\gamma_{s,\alpha}$ be the free energy per unit of surface of the interface $\mathrm{A}_{s,\alpha}$ between the solid matrix and either the wetting fluid, that is $\alpha = w$, or the non-wetting fluid, that is $\alpha = nw$. The surface entropy of the interface $\mathrm{A}_{s,\alpha}$ is $S_{s,\alpha} = -d\gamma_{s,\alpha}/dT$. Similarly let $\gamma_{w,nw}$ be the free energy related to the interface $\mathrm{A}_{w,nw}$ between the wetting fluid and the non-wetting fluid, and whose entropy is $S_{w,nw} = -d\gamma_{w,nw}/dT$. When the surface free energy of an interface depends only on temperature, it is eventually identified with the surface tension exerted along the interface. Owing to the above remarks and to the additive character of energy, we now write:

$$\Psi_s = \psi_s + \phi U \tag{6.21}$$

where ψ_s represents the free energy of the solid matrix per unit of macroscopic volume $d\Omega_0$, while ϕU represents the overall interfacial energy per unit of porous volume $d\Omega_0$, that is:

$$\phi U = \gamma_{s,w}\, a_{s,w} + \gamma_{s,nw}\, a_{s,nw} + \gamma_{w,nw}\, a_{w,nw} - \gamma_{s,w} a_s \tag{6.22}$$

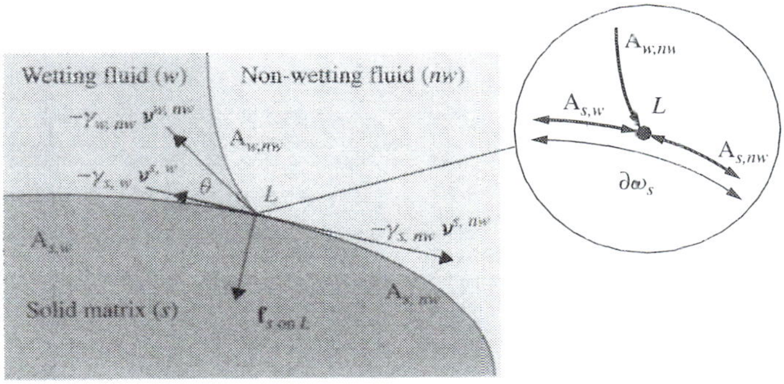

Figure 6.1: Contact angle θ between the wetting fluid ($\alpha = w$) and the non-wetting fluid ($\alpha = nw$) and surface tensions $\gamma_{\alpha,\beta}$. Forces $-\gamma_{\alpha,\beta}\nu^{\alpha,\beta}$ and force $f_{s \text{ on } L}$ are the forces exerted conjointly by the interfaces and the solid matrix on the triple junction line L.

In (6.22), $a_{s,\alpha}$ or $a_{w,nw}$ denotes the current aera of interfaces $A_{s,\alpha}$ or $A_{w,nw}$ while $a_s = a_{s,w} + a_{w,nw}$ is the total aera of the solid internal walls of the porous network per unit of volume $d\Omega_0$. According to expression (6.22), the complete saturation with respect to the wetting fluid is conventionally adopted as a zero reference state for the overall interfacial energy so that $U(S_w = 1) = 0$.

Since the porous space is filled only by a wetting fluid and a non-wetting fluid, that is $\alpha = w, nw$, definition (6.2) of saturation degree S_α allows us to rewrite (6.19) in the equivalent form:

$$\sigma_{ij} d\varepsilon_{ij} + p^* d\phi - \phi p_c dS_w - S_s dT - d\Psi_s = 0 \tag{6.23}$$

where p^* is the averaged fluid pressure defined by:

$$p^* = S_w \, p_w + S_{nw} \, p_{nw} \tag{6.24}$$

while p_c stands for the macroscopic capillary pressure exerted on the interface between the two fluids, that is:

$$p_c = p_{nw} - p_w \tag{6.25}$$

The state equations can be now written in the form:

$$\boldsymbol{\sigma} = \frac{\partial \Psi_s}{\partial \boldsymbol{\varepsilon}}; \quad p^* = \frac{\partial \Psi_s}{\partial \phi}; \quad \phi p_c = -\frac{\partial \Psi_s}{\partial S_w}; \quad S_s = -\frac{\partial \Psi_s}{\partial T} \tag{6.26}$$

6.2.3 Thermodynamics of Porous Media with Phase Change

Consider now the case examined in §6.1.3, where a liquid–vapour phase change can occur. Substitution of (6.10) into expression (6.14a) of dissipation Φ_{int} yields:

$$\Phi_{int} = \Phi_s + \Phi_{\rightarrow} \tag{6.27}$$

where Φ_s is still identified from (6.18), while $\Phi_{\rightarrow}$ accounts for the possible dissipation related to the phase change:

$$\Phi_{\rightarrow} = (g_l - g_v)\overset{\circ}{m}_{l\rightarrow v} \tag{6.28}$$

Since the air and the vapour occupy the same porosity ϕ_g, (6.19) reduces to:

$$\sigma_{ij} d\varepsilon_{ij} + p_l d\phi_l + p_g d\phi_g - S_s dT - d\Psi_s = 0 \tag{6.29}$$

where p_g is the gas pressure, that is the pressure of the vapour–air mixture:

$$p_g = p_v + p_a \tag{6.30}$$

Once the wetting fluid (e.g. the liquid water, $w = l$) and the non-wetting fluid (e.g. the wet air, $nw = g$) are identified state equations (6.26), which are relative to the sole skeleton, apply.

The local thermodynamic equilibrium between the liquid and its vapour requires the dissipation $\Phi_{\rightarrow}$ to be zero. We write:

$$g_l(p_l, T) = g_v(p_v, T) \tag{6.31}$$

Assuming that the liquid–vapour equilibrium is maintained throughout the evolution and taking into account the fluid state equations (see §3.1.2), we differentiate (6.31), yielding:

$$\frac{dp_l}{\rho_l} - \frac{dp_v}{\rho_v} = (s_l - s_v)\, dT \tag{6.32}$$

Let us assume that the liquid undergoes incompressible transformations and that the vapour is an ideal gas, so that (4.54) applies with $f = l$, as do (4.55) and (4.57) when identifying in the latter index f as v and pressure p as p_v. Accordingly, the integration of (6.32) gives the celebrated Kelvin's law:

$$p_l - p_{atm} = \frac{\rho_l RT}{\mathcal{M}_v} \ln \frac{p_v}{p_{vs}(T)} \tag{6.33}$$

where $p_{vs}(T)$ is the pressure of the saturating vapour at temperature T and for the liquid at atmospheric pressure p_{atm}, and whose expression is:

$$\frac{p_{vs}(T)}{p_{vs}(T_0)} = \exp\left\{\frac{\mathcal{M}_v}{RT}\left[\frac{\mathcal{L}_0}{T_0}(T - T_0) + \left(C_p^l - C_p^v\right)\left(T - T_0 - T \ln \frac{T}{T_0}\right)\right]\right\} \tag{6.34}$$

where:

$$\mathcal{L}_0 = T_0(s_{v0} - s_{l0}) \tag{6.35}$$

stands for the latent heat of the phase change at temperature T_0 and $p_l = p_{atm}$ (for water and $T_0 = 373°$K, $\mathcal{L}_0 = 2.26\ 10^6$ J/K while $p_{vs} = p_{atm}$). If the liquid is water, the ratio $p_v/p_{vs}(T)$ is eventually identified with the relative humidity h_r:

$$h_r = \frac{p_v}{p_{vs}(T)} \tag{6.36}$$

6.3 Capillary Pressure Curve

6.3.1 Energy Approach to the Capillary Pressure Curve

In the non-deformable and isothermal case where $\varepsilon_{ij} = 0$, $\phi = \phi_0$ and $T = T_0$, state equations (6.26) reduce to:

$$\phi_0 p_c = -\frac{d\Psi_s}{dS_w} \tag{6.37}$$

so that the capillary pressure depends on saturation degree only, that is:

$$p_c = p_c(S_w) \tag{6.38}$$

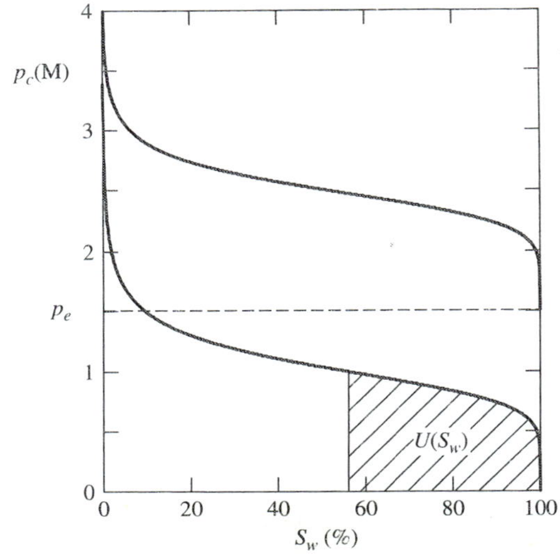

Figure 6.2: Typical aspect of a capillary curve when taking the capillary modulus M as the unit for the capillary pressure. The lower curve is obtained by letting $m = 0.85$ in (6.40) and adopting a zero value for the entry pressure p_e, that is the pressure threshold required to make the non-wetting fluid actually enter the porous material when starting the drainage process from complete saturation. The hashed zone corresponds to the interface energy. The upper curve is obtained by letting $p = 1 - m$ in (6.41) with the same value of for m, while adopting a value of $p_e = 1.5$M for the entry pressure. With the same entry pressures both curves would have practically coincided. Division of the capillary pressure values by $\rho_w g$ gives the saturation profile $z = z(S_w)$ resulting from natural imbibition (see (6.49)).

The curve p_c as a function of S_w is called the capillary pressure curve. Its general aspect is given in Fig. 6.2. The capillary pressure decreases from a maximum value, which can turn out to be infinite for the zero saturation $S_w = 0$, to a possible threshold value, the 'entry pressure' p_e, which is the pressure required for the saturation degree S_w actually to decrease from the complete saturation value $S_w = 1$. In a drainage process the entry pressure is the minimum pressure required for the non-wetting fluid actually to enter the material and, in fact, within the pores having the largest access radius according to the Laplace equation (for further details see §6.6). In order to account for a possible non-zero entry pressure p_e we rewrite (6.38) in the form:

$$\langle p_c - p_e \rangle = \mathsf{M}\pi_c(S_w); \quad \pi_c(1) = 0 \tag{6.39}$$

where $\langle p_c - p_e \rangle$ stands for the positive part of $p_c - p_e$ and M for a capillary modulus or a reference pressure. Various expressions have been proposed for capturing the dimensionless function $\pi_c(S_w)$. When the wetting fluid is water and the non-wetting fluid is wet air, an expression often used is:[3]

$$\pi_c(S_w) = \left(S_w^{-\frac{1}{m}} - 1 \right)^{1-m} \qquad 0 < m < 1 \tag{6.40}$$

[3]See van Genuchten M.Th. (1980), 'A closed-form equation for predicting the hydraulic conductivity of unsaturated soils', *Soil Science Society of America Journal*, **44**, 892–898.

or, alternatively:

$$\pi_c(S_w) = \left(S_w^{-1} - 1\right)^p \tag{6.41}$$

For a value of m close to one, the latter expression conveniently approaches (6.40) by letting $p = 1 - m$ (see Fig. 6.2).

In the non-deformable and isothermal case, the skeleton energy Ψ_s reduces to the interfacial energy $\phi_0 U$, and we rewrite (6.37) in the form:

$$p_c = -\frac{dU}{dS_w} \tag{6.42}$$

As sketched in Fig. 6.2, the interfacial energy U as a function of saturation S_w is determined from the capillary curve according to:

$$U(S_w) = \int_{S_w}^{1} p_c(S)\, dS \tag{6.43}$$

In the still non-deformable, but non-isothermal, case, state equations (6.26) entail:

$$\phi_0 p_c = -\frac{\partial \Psi_s}{\partial S_w}; \quad S_s = -\frac{\partial \Psi_s}{\partial T} \tag{6.44}$$

Since the solid matrix is non-deformable we let $\phi = \phi_0$ in (6.21) so that (6.44) gives:

$$p_c = -\frac{\partial U}{\partial S_w} \tag{6.45a}$$

$$S_s = -\frac{\partial \psi_s}{\partial T} - \phi_0 \frac{\partial U}{\partial T} = S_s^0 + (1 - \phi_0) C_s \ln \frac{T}{T_0} - \phi_0 \frac{\partial U}{\partial T} \tag{6.45b}$$

where C_s is the volumetric heat capacity of the solid matrix.[4] The capillary pressure now depends on both S_w and T, that is:

$$p_c = p_c(S_w, T) \tag{6.46}$$

with the experimental identification:

$$U(S_w, T) = \int_{S_w}^{1} p_c(S, T)\, dS \tag{6.47}$$

The usual expressions of the capillary pressure curve, for example (6.40)–(6.41), can be conveniently extended by letting:

$$p_c = \mathsf{M}(T)\, \pi_c(S_w)$$

[4]Since we conventionally adopted $U(S_w = 1) = 0$ (see (6.22)) a more accurate expression than (6.45b) would have to include the term $-(d\gamma_{s,w}/dT)a_s$.

6.3.2 Capillary Pressure, Natural Imbibition and Isotherm of Sorption

For sufficiently permeable materials, such as rocks and sands, the capillary pressure curve can be directly determined in a drainage experiment where, starting from complete saturation, the capillary pressure is increased to make the non-wetting fluid progressively invade the porous space. Alternatively, the capillary pressure curve can be captured from a natural imbibition experiment. In natural imbibition a dry sample lying in the upper half-space $z > 0$ comes into contact with the wetting fluid along the plane $z = 0$, the latter remaining at atmospheric pressure. The capillary pressure acts upwards against the vertical gravity forces, resulting in an equilibrium imbibition profile that matches the capillary pressure curve. More precisely, the vertical equilibrium of the wetting fluid in the z direction requires:[5]

$$-\frac{\partial p_w}{\partial z} - \rho_w g = 0 \tag{6.48}$$

At equilibrium the pressure of the non-wetting fluid, assumed to be the air, is equal to atmospheric pressure everywhere. The height h of the saturated zone is then determined by the integration of (6.48) with boundary conditions $p_w = p_{atm}$ at $z = 0$ and $p_w = -p_e + p_{atm}$ at $z = h$, where p_e is the entry pressure involved in (6.39). For $z > h$ the remaining capillary fringe of imbibition is determined by the integration of (6.48), when letting in the latter $p_w = -p_c(S_w) + p_{atm}$ and requiring the solution to match the condition $p_w = -p_e + p_{atm}$ at $z = h$. The procedure eventually gives:

$$\begin{aligned} 0 < z < h = \frac{p_e}{\rho_w g} &: \quad S_w = 1 \\ h = \frac{p_e}{\rho_w g} < z &: \quad p_c(S_w) - p_e = \rho_w g\,(z - h) \end{aligned} \tag{6.49}$$

According to (6.49), the saturation profile of natural imbibition $z = z(S_w)$ is identified with that of the capillary pressure $p_c = p_c(S_w)$ represented in Fig. 6.2, provided that the capillary pressure values are divided by $\rho_w g$. For $z > h$ the capillary fringe of imbibition is scaled by the characteristic length $\ell_c = \mathsf{M}/\rho_w g$ where M is the capillary modulus (see (6.39)). Characteristic length ℓ_c quantifies the strength of capillary effects with regard to gravity forces.

For weakly permeable materials, such as cement-based materials and clays, the capillary pressure that is required in the drainage experiment actually to decrease the saturation rapidly turns out to be too important. In practice it is difficult to reach values of saturation degree lower than 90% and, for a water liquid–wet air mixture, the capillary pressure curve is then indirectly determined through the sorption isotherm. In a sorption experiment the sample is maintained at thermodynamic equilibrium with the outer atmosphere whose relative humidity h_r is controlled by way of saturated salt solutions. Varying the latter turns out to vary the vapour pressure of the outer atmosphere in equilibrium with the liquid water forming the solutions. Between two successive values of the relative humidity, the

[5] Equilibrium equation (6.48) can be recovered by letting $\mathcal{V}_w = 0$ and $\mathbf{f} = -g\mathbf{e}_z$ in Darcy's law (6.70a). It is worthwhile to note that the equilibrium condition is the same in the saturated and the unsaturated situations.

change in liquid water saturation is recorded by weighing the sample so that the relative humidity can eventually be plotted as a function of the liquid water saturation S_l, yielding the so-called sorption isotherm:

$$h_r = h_r(S_l) \tag{6.50}$$

Since the wet air forming the inner gas mixture is constantly maintained at equilibrium with the outer atmosphere, its pressure p_g remains equal to the atmospheric pressure p_{atm}. The capillary pressure p_c is eventually identified with $p_{atm} - p_l$, where pressure p_l stands for the liquid water pressure, so that (6.33) and (6.36) allow us to write:

$$p_c = -\frac{\rho_l RT}{\mathcal{M}_v} \ln h_r \tag{6.51}$$

Use of (6.50) and (6.51) finally leads to the indirect determination of the capillary pressure curve $p_c = p_c(S_l)$. Figure 6.3 illustrates the resulting curves when proceeding in such a way for a cement paste. Letting $p_{atm} \simeq 0.1$ MPa, the experimental data reported in Fig. 6.3 show that the liquid water pressure $p_l = p_{atm} - p_c$ within cement-based materials becomes negative for a saturation degree still close to 100%. Since the liquid water pressure in the sorption experiment is indirectly controlled via the liquid–vapour equilibrium by Kelvin's law (6.33), such a negative pressure has to be interpreted as a thermodynamic effective pressure accounting for the chemical and physical interactions existing between the liquid water and the internal walls of the cement matrix delimiting the porous space

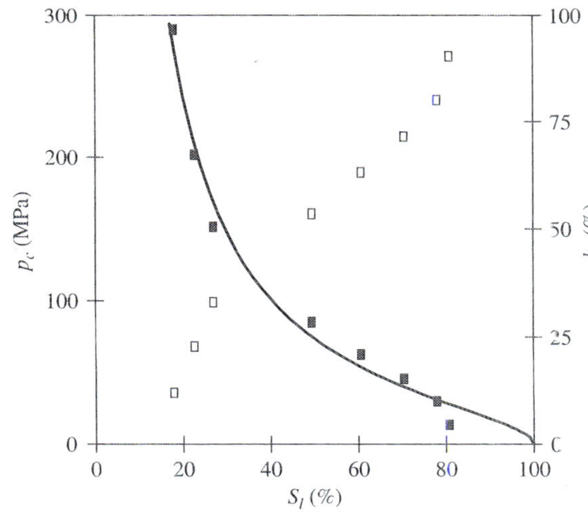

Figure 6.3: Capillary curve relative to a cement paste as obtained from a desorption experiment. The solid line results from the fitting of the experimental data by using (6.40) with $m = 0.46$, M = 37.55 MPa and adopting a zero entry pressure (from Mainguy M., Coussy O., Baroghel-Bouny V. (2001), 'The role of air pressure in the drying of weakly permeable materials', *Journal of Engineering Mechanics*, ASCE, **127**, (6), 582–592). See also Černy R., Rovnanikova P. (2002), *Transport Processes in Concrete*, Spon Press, for data relative to sorption isotherms of cement-based materials.

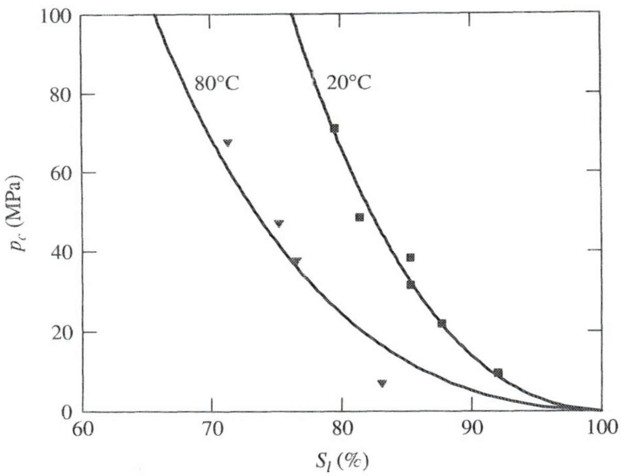

Figure 6.4: Capillary curve relative to a compacted artificial clay as determined from the sorption isotherm for two different values of the temperature. The solid line results from the fitting of the experimental data by using (6.41) with $p = 1.92$, $M(20°C) = 950$ MPa, $M(80°C) = 350$ MPa and a zero entry pressure (from Dangla P., Coussy O., Olchitzky E., Imbert C. (2000), 'Non-linear thermo-mechanical couplings in unsaturated clay barriers', in *Proceedings of IUTAM Symposium on Theoretical and Numerical Methods in Continuum Mechanics of Porous Materials*, ed. Ehlers W., Kluwer Academic, Dordrecht).

(see §3.6.3 for the general meaning of a thermodynamic effective pressure). In addition, the dependence on temperature T of the capillary curve can be obtained when varying the temperature at which the sorption experiment is performed. Figure 6.4 illustrates the resulting curves for a compacted artificial clay. Note that the pressure p_l of the liquid water again becomes negative for a saturation degree still close to 100%. In the context of unsaturated soils, owing to this negativeness, the capillary pressure is also called suction.

6.4 Unsaturated Thermoporoelastic Constitutive Equations

6.4.1 Energy Separation and the Equivalent Pore Pressure Concept

In the general deformable case state equations (6.26) show that the total stress σ_{ij}, the averaged fluid pressure p^* and the capillary pressure p_c act as independent state variables with regard to the state equations of the skeleton. Nevertheless, as analysed in the previous section, we saw in the non-deformable case that the capillary curve depended only on the saturation degree S_w and possibly on temperature T according to (6.46). A first approach to unsaturated thermoporoelasticity consists in extending (6.46) to the deformable case. Accordingly, the substitution of (6.46) into the third of the state equations (6.26), followed by integration, allows us to express $\Psi_s(\varepsilon_{ij}, S_w, T)$ in the form:

$$\Psi_s(\varepsilon_{ij}, S_w, T) = \psi_s(\varepsilon_{ij}, \phi, T) + \phi U(S_w, T) \tag{6.52}$$

Assuming that (6.46) still holds in the deformable case, this turns out to be equivalent to a hypothesis of energy separation. Indeed, according to (6.52), the solid matrix energy

ψ_s is expressed separately from the interfacial energy U whose expression involves state variables ε_{ij}, ϕ and T only and not saturation degree S_w.

Substitution of (6.52) into (6.23) and use of the third of the state equations (6.26) give:

$$\sigma_{ij}d\varepsilon_{ij} + \pi d\phi - \mathcal{S}_s dT - d\psi_s = 0 \tag{6.53}$$

where the entropy $\mathcal{S}_s$ is defined by:

$$\mathcal{S}_s = S_s + \phi \frac{\partial U}{\partial T} \tag{6.54}$$

so that the latter stands for the entropy of the solid matrix since the interface entropy $-\phi\partial U/\partial T$ has been removed from S_s. Furthermore the pressure π in (6.53) is defined by:

$$\pi = p^* - U \tag{6.55}$$

Alternatively, use of (6.24) and (6.47) provides a differential definition of π:

$$d\pi = S_w\, dp_w + S_{nw}\, dp_{nw} = -S_w dp_c + dp_{nw} \tag{6.56}$$

Comparing (4.2) and (6.53), we conclude that the pressure π, with regard to the solid matrix, acts the same as the pressure p does in the saturated situation. For this reason π is called the equivalent pore pressure.[6] Note that the minus sign affecting U in the expression (6.55) of π accounts for the tensile character of the surface stresses acting along the interfaces.

6.4.2 Equivalent Pore Pressure and Averaged Fluid Pressure

The averaged fluid pressure p^* defined by (6.24) is often invoked to play the role of an overall equivalent pore pressure instead of π. Comparing (6.24) and (6.55), we conclude that letting p^* play the role of an equivalent pore pressure instead of π turns out to ignore the tensile surface stresses acting along the interfaces and their subsequent effects upon the deformation of the solid matrix. For instance, p^* can be conveniently used to approach the constitutive equations of porous materials exhibiting a double porous network. As more lengthily analysed in the advanced analysis sections of the first four chapters, the double porous network concept is aimed at capturing the effects of the two very distinct porous networks exhibited by some materials as rocks or concrete materials. Roughly speaking, the first is formed of rounded pores, while the second is formed of penny-shaped cracks. Although both networks are saturated by the same and unique wetting fluid, their quite different geometries lead us to distinguish the fluid pressure p_α and the saturation degree S_α attached to each network referred to by index $\alpha = 1$ or 2. However, with regard to the wetting fluid viewed as a whole, the porous material remains saturated. Accordingly, we let $S_w = S_1 + S_2 = 1$ in (6.47), resulting in $U = 0$, so that (6.55) leads us to identify the averaged pressure p^* with the equivalent pore pressure π.

[6]The equivalent pore pressure concept was initially introduced in the differential form (6.56) in Coussy O. (1995), *Mechanics of Porous Continua*, John Wiley & Sons, Chichester. See also Coussy O., Eymard R., Lassabatère T. (1998), 'Constitutive modelling of unsaturated drying deformable materials', *Journal of Engineering Mechanics*, **124**, (6), 658–667.

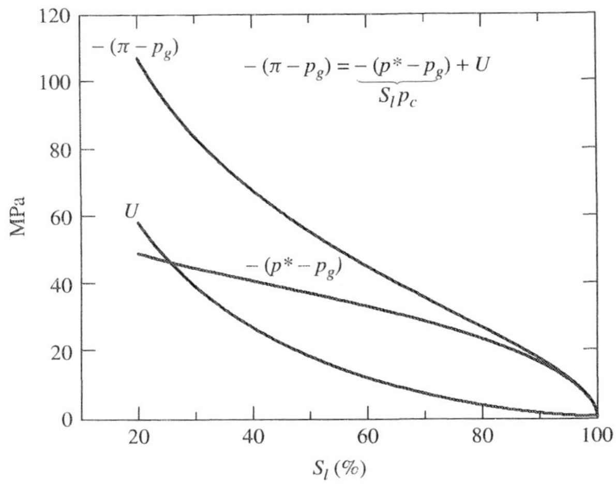

Figure 6.5: Respective contributions of $S_l p_c = -\left(p^* - p_g\right)$ and surface energy U to $-\left(\pi - p_g\right)$ for the cement paste whose capillary pressure has been reported in Fig. 6.3. In practice the gas pressure p_g remains close to atmospheric pressure $p_{atm} \simeq 0.1$ MPa so that $S_l p_c$ can be identified with a good accuracy to $-p^*$ and $-\left(\pi - p_g\right)$ to $-\pi \simeq -p^* + U$. (Coussy O., Dangla T., Lassabatère V., Baroghel-Bouny (2003)).

Extending the analysis, the averaged pore pressure p^* can possibly be used as an equivalent pore pressure provided that the wetting fluid and the non-wetting fluid occupy well-separated pores so that the contact area between them turns out to be negligible. This assumption is generally not supported by experimental evidence for porous materials such as cement-based materials or clays. According to definitions (6.24) and (6.55) of p^* and π, we write:

$$\pi - p_g = p^* - p_g - U = -S_w p_c - U \tag{6.57}$$

In most cases the wetting fluid is liquid water, $w = l$, while the non-wetting fluid is wet air, $nw = g$. The pressure p_g of the latter remains generally close to atmospheric pressure p_{atm} and can be neglected with respect to the terms appearing on the right hand side of (6.57). This results in a negative equivalent pressure π, that is a suction, which can provoke significant shrinkage. As illustrated in Fig. 6.5 for the cement paste whose capillary pressure has been reported in Fig. 6.3, the respective contributions of the averaged pore pressure $p^* \simeq -S_l p_c$ and of the opposite of interfacial energy $-U$ to the equivalent pore pressure π can be separately assessed from the capillary curve (6.46) and from the relation (6.47). As soon as the liquid water saturation degree S_l departs significantly from 100%, the contribution U of the interfacial energy can no longer be neglected with respect to the contribution of the averaged pore pressure p^* to π.

6.4.3 Equivalent Pore Pressure and Thermoporoelastic Constitutive Equations

Letting G_s be defined by:

$$G_s = \psi_s - \pi\,\phi \tag{6.58}$$

from (6.53) we derive the alternative energy balance:

$$\sigma_{ij}\, d\varepsilon_{ij} - \phi\, d\pi - S_s\, dT - dG_s = 0 \tag{6.59}$$

From (6.59) we finally derive the state equations of unsaturated thermoporoelasticity in the form:

$$G_s = G_s(\varepsilon_{ij}, \pi, T): \quad \sigma_{ij} = \frac{\partial G_s}{\partial \varepsilon_{ij}}; \quad \phi = -\frac{\partial G_s}{\partial \pi}; \quad \mathcal{S}_s = -\frac{\partial G_s}{\partial T} \tag{6.60}$$

which are written exactly the same as state equations (4.5) of saturated thermoporoelasticity, provided that the pressure p in the latter is replaced by the equivalent pore pressure π. As a consequence all the developments of Chapter 4 apply For instance, constitutive equations (4.32)–(4.33) are extended in the form:

$$\sigma - \sigma^0 = K\epsilon - b(\pi - \pi_0) \tag{6.61}$$

$$\phi - \phi_0 = b\epsilon + \frac{\pi - \pi_0}{N} \tag{6.62}$$

where b and N still stand for Biot's coefficient and Biot's modulus satisfying (4.35). In the very special case of the double porous network, we saw in the previous section that $\pi = p^*$. Substitution of $\pi = p^*$ into (6.61) and use of definition (6.24) of p^* allow us eventually to recover (4.113a) by letting $b_\alpha = bS_\alpha$.

6.4.4 Equivalent Pore Pressure, Wetting and Free Swelling of Materials

Consider a stress-free porous material subjected to an outer relative humidity higher than its inner relative humidity. In response the porous material absorbs water vapour from the outer atmosphere so that the higher outer relative humidity is progressively established within the inner layers. Simultaneously the liquid water condenses within the latter in order to maintain the internal liquid–vapour equilibrium. Accordingly the liquid pressure increases, while the capillary pressure and, consequently, the equivalent pore pressure decrease so that the porous material eventually swells according to (6.61). While the wetting kinetics (or conversely, the drying kinetics) is governed by the transport phenomena (see §6.6.4), the asymptotic state is governed by the outer relative humidity only, since, asymptotically, the air pressure recovers the atmospheric pressure value. Accordingly, we combine (6.50), (6.51), (6.56) and (6.61) to get the asymptotic free swelling due to wetting in the form:

$$\epsilon = \frac{b}{nK} \times \frac{RT}{\mathcal{M}_v} \int_{h_r^0}^{h_r} \frac{w(h)}{h}\, dh \tag{6.63}$$

where h_r^0 denotes the initial relative humidity to which the swelling due to wetting is referred, while $w = \rho_l n S_l$ is the liquid water mass content and is expressed as a function of the relative humidity h_r by inverting the relation (6.50) provided by the sorption experiment.

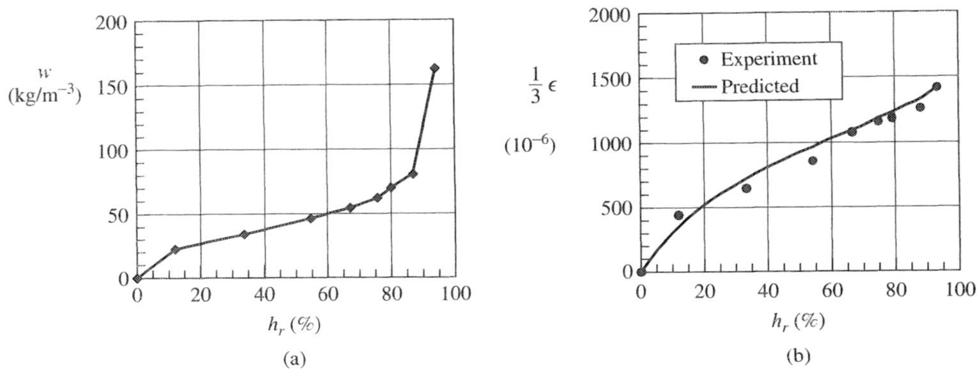

Figure 6.6: Checking of the energy separation hypothesis (6.52): (i) the water mass content $w = \rho_l n S_l$ is plotted against relative humidity h_r by means of an adsorption experiment (a); (ii) the predicted values of the axial swelling $\epsilon/3$ of the sample due to hydration through the use of (6.63) are compared with the observed ones (b). The experimental data are related to a cellulose fibre cement composite with $K = 8.25 \times 10^3$ MPa while the value of Biot's coefficient is assessed to $b = 0.79$ (from Carmeliet (1999), with kind permission of Kluwer Academic Publishers, see footnote).

Expression (6.63) of the swelling due to wetting has been derived by using the equivalent pore pressure concept and, consequently, by assuming the separation of energies according to (6.52). Conversely, as illustrated[7] in Fig. 6.6, the relevance of the latter can be experimentally checked by determining $w(h_r)$ from a sorption experiment (Fig. 6.6a), and comparing the predicted values (6.63) for the free swelling ϵ due to wetting to the observed ones (Fig. 6.6b).

For materials such as clays whose solid grains undergo negligible changes, the swelling strain is equivalently captured by the increase of the void ratio e. In the saturated case the constitutive equation linking the void ratio to Terzaghi's effective stress $\sigma + p$ is well approached through a relation of the form (see Fig. 6.7a):

$$de = -\kappa \frac{d\,(\sigma + p)}{\sigma + p} \tag{6.64}$$

where p is the pore pressure. According to (6.64) and to the equivalent pore pressure concept, the changes in void ratio e in the free swelling test must be governed by:

$$de = -\kappa \frac{d\pi}{\pi} \tag{6.65}$$

As illustrated in Fig. 6.7, the comparison between the value of the coefficient κ measured in an experiment performed on a saturated sample (i.e. (6.64) and Fig. 6.7a) and

[7]The checking of the energy separation hypothesis reported in Fig. 6.6 is from Carmeliet J. (1999), 'Coupling of damage and fluid-solid interactions in quasi-brittle nonsaturated porous materials', in *Proceedings of the IUTAM Symposium on Theoretical and Numerical Methods in Continuum Mechanics of Porous Materials*, 307–312, ed. Ehlers W., Kluwer Academic, Dordrecht. The checking procedure is based upon the earlier work of Coussy O., Eymard R., Lassabatère T. (1998), 'Constitutive modelling of unsaturated drying deformable materials', *Journal of Engineering Mechanics*, **124**, (6), 658–667.

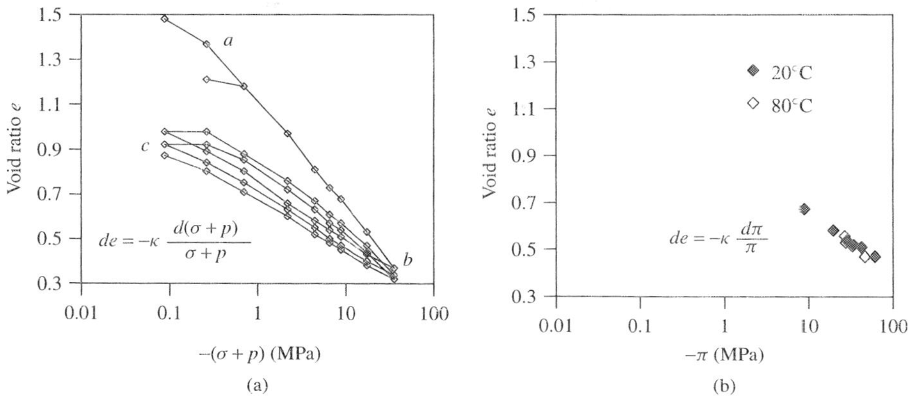

Figure 6.7: Checking the validity of the equivalent pore pressure concept: (i) the coefficient κ involved in constitutive equation (6.64) is measured in a test performed on a saturated sample along the elastic unloading path bc (a); (ii) the validity of the equivalent pore pressure concept requires the value of coefficient κ so measured to be the same as the one delivered by the free swelling test (b). The experimental data are related to the compacted artificial clay whose capillary pressure curve leading to the determination of π has been reported in Fig. 6.4. The validity of the equivalent pore pressure concept is confirmed here since the two values of κ so determined are both found to be close to 0.1. This value is even found to be irrespective of the temperature 20°C or 80°C at which the free swelling test is determined, showing that the thermal dilation of the solid grains forming the matrix is actually negligible (see Fig. 8.11 and §8.4.4 for the analysis of the plastic loading path ab).

the value of the same coefficient κ measured in a free swelling test (i.e. (6.65) and Fig. 6.7b) eventually provides the means to check the validity of the equivalent pore pressure concept.

6.5 Heat and Mass Conduction

6.5.1 Fourier's Law, Thermal Equation and Phase Change

Dissipation φ_{th} associated with the heat transfer and given by (6.14b) is expressed in the same way as in the saturated case so that Fourier's law can be formulated as in (3.57). Making the same assumptions as there made in §5.3.1, but starting now from (6.15) instead of (5.117), we derive:

$$T\left(\frac{\partial S}{\partial t} + \sum_{\alpha} s_{\alpha} \nabla \cdot \mathbf{w}^{\alpha}\right) = \kappa \nabla^2 \theta \tag{6.66}$$

Considering a possible phase change by letting $\alpha = l, v$ or g, substitution of (6.17) and (6.10) into (6.66) leads us to rewrite the thermal equation in the form:

$$T\left(\frac{\partial S_s}{\partial t} + \sum_{\alpha=l,v,g} m_{\alpha} \frac{\partial s_{\alpha}}{\partial t} + (s_v - s_l)\overset{\circ}{m}_{l\to v}\right) = \kappa \nabla^2 \theta \tag{6.67}$$

6.5.2 Darcy's Law

Considering the case where the porous space is filled by a wetting fluid and a non-wetting fluid, we let $\alpha = w, nw$. Use of fluid state equations (3.10) then allows us to write the positiveness of dissipation (6.14c) associated with the fluid transport in the form:

$$\varphi_f = (-\nabla p_w + \rho_w \mathbf{f}) \cdot \mathcal{V}^w + (-\nabla p_{nw} + \rho_{nw} \mathbf{f}) \cdot \mathcal{V}^{nw} \geq 0 \tag{6.68}$$

where $\mathcal{V}^\alpha$ is the filtration vector associated with fluid α:

$$\mathcal{V}^\alpha = \frac{\mathbf{w}^\alpha}{\rho_\alpha} = n S_\alpha (\mathbf{V}^\alpha - \mathbf{V}^s) \tag{6.69}$$

Accordingly, Darcy's law (3.46) extends to the unsaturated case in the form:

$$\mathcal{V}^w = \frac{\varkappa k_{rw}(S_w)}{\eta_w} (-\nabla p_w + \rho_w \mathbf{f}) \tag{6.70a}$$

$$\mathcal{V}^{nw} = \frac{\varkappa k_{rnw}(S_{nw})}{\eta_{nw}} (-\nabla p_{nw} + \rho_{nw} \mathbf{f}) \tag{6.70b}$$

where $k_{rw}(S_w)$ and $k_{rw}(S_{nw})$ are the relative permeabilities related to, respectively, the wetting fluid and the non-wetting fluid and satisfying:

$$k_{r\alpha}(0) = 0 \leq k_{r\alpha}(S_\alpha) \leq k_{r\alpha}(1) = 1 \tag{6.71}$$

When the wetting fluid and the non-wetting fluid are liquid water and wet air, respectively, expressions commonly used in association with expression (6.40) for the capillary curve, that is referring to the same m, are[8]:

$$k_{rw}(S_w) = \sqrt{S_w} \left(1 - \left(1 - S_w^{\frac{1}{m}}\right)^m\right)^2 \tag{6.72a}$$

$$k_{rnw}(1 - S_w) = \sqrt{1 - S_w} \left(1 - S_w^{\frac{1}{m}}\right)^{2m} \tag{6.72b}$$

In Fig. 6.8 we plot k_{rw} and k_{rnw} against S_w for various values of m.

6.5.3 Fick's Law

When the fluid is a vapour–air mixture, the vapour is transported both by advection, as a component of the mixture, and by molecular diffusion through the gas. The former is governed by Darcy's law (6.70b) applied to the filtration vector $\mathcal{V}^g$ related to the mixture, while the latter is governed by Fick's law. Their explicit formulation requires us to express the filtration vector $\mathcal{V}^g$ related to the vapour–air gaseous mixture considered as a whole as a function of the filtration vectors $\mathcal{V}^v$ and $\mathcal{V}^a$ related to the vapour and to the air considered separately.

[8] See Luckner L., van Genutchen M.Th., Nielsen D.R. (1989), 'A consistent set of parametric models for the two-phase flow of immiscible fluids in the subsurface', *Water Resources Research*, **25**, (10), 2187–2193.

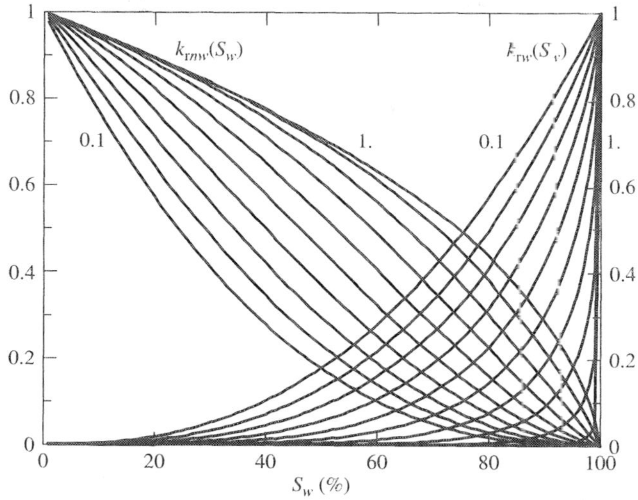

Figure 6.8: Relative permeabilities k_{rw} and k_{rnw} plotted against S_w when varying m from 0.1 to 1 in (6.72).

Let then x_v and x_a be the mole fractions related to the vapour and to the air, that is the ratio of their molar concentration $c_{v \text{ or } a}$ to the total concentration $c = c_v + c_a$:

$$x_v = \frac{c_v}{c}; \quad x_a = \frac{c_a}{c} \quad \text{with} \quad x_v + x_a = 1 \tag{6.73}$$

The filtration vector $\mathcal{V}^g$ is conveniently defined as the molar average:

$$\mathcal{V}^g = x_v \mathcal{V}^v + x_a \mathcal{V}^\alpha \tag{6.74}$$

so that:

$$\mathcal{V}^v = \mathcal{V}^g + x_a(\mathcal{V}^v - \mathcal{V}^a) \tag{6.75a}$$

$$\mathcal{V}^a = \mathcal{V}^g - x_v(\mathcal{V}^v - \mathcal{V}^a) \tag{6.75b}$$

Recalling that $(\nabla g_\alpha)_T = (1/\rho_\alpha)\nabla p_\alpha$ and $p_\alpha = x_\alpha p_g$ for an ideal mixture formed of ideal gases while using (6.30), (6.69), (6.73)–(6.75), the positiveness (6.14c) of the dissipation φ_f associated with the transport of the vapour and of the air can be expressed in the form:

$$-\nabla p_g \cdot \mathcal{V}^g - \nabla x_v \cdot (\mathcal{V}^v - \mathcal{V}^a) \geq 0 \tag{6.76}$$

where, for the sake of simplicity, we did not consider body forces $\mathbf{f}$. The first term in (6.76) is similar to (6.68). It accounts for the dissipation related to the advective transport of the vapour–air mixture considered as a whole. The latter is governed by Darcy's law by letting $nw = g$ in (6.70b). The second term is the dissipation related to the molar diffusion of the vapour through the air: it is the product of the driving force $-\nabla x_v$ by the relative filtration vector $\mathcal{V}^v - \mathcal{V}^a$ it produces. Consequently the law governing the diffusion of vapour through the air has to relate $\mathcal{V}^v - \mathcal{V}^a$ to $-\nabla x_v$.

According to the kinetic theory of gases, the molar diffusion of the vapour in the air occurring in free space is governed by Fick's law, that is:

$$x_a x_v(\mathcal{V}^v - \mathcal{V}^a) = -D_{va}\nabla x_v \tag{6.77}$$

In most cases the vapour is the water vapour. Combining the kinetic theory of gases and experiments, it can be experimentally shown that the diffusion coefficient D_{va} involved in Fick's law (6.77) can be well accounted for through the following expression:[9]

$$D_{va} = D_0 \frac{p_{atm}}{p_g}; \quad D_0 = \delta_0 \left(\frac{T}{T_0}\right)^{1.88} \tag{6.78}$$

where $\delta_0 = 2.17 \times 10^{-5}$ m^{2}s, $T_0 = 273$ K and $p_{atm} = 101325$ Pa.

The law governing the molar diffusion of the vapour through the air, but now within a porous material, and guaranteeing the positiveness of the associated dissipation, can be consistently extended from (6.77) in the form:

$$x_a x_v(\mathcal{V}^v - \mathcal{V}^a) = -D\nabla x_v; \quad D = (n_g \times \tau)D_{va} \tag{6.79}$$

With respect to the formulation (6.77) of Fick's law in free space, the diffusion coefficient D_{va} is now affected by the factor $n_g \times \tau$. The latter firstly accounts for the reduction of the volume offered to the diffusion by means of the gas porosity n_g. It secondly accounts for the actual length of the diffusion path of molecules through the porous space by means of the so-called tortuosity τ. The usual expression for τ, which at the least has to depend on the overall porosity n and on the gas saturation degree S_g, is:[10]

$$\tau(n, S_g) = n^{1/3} S_g^{7/3} \tag{6.80}$$

Combining Darcy's and Fick's laws and using (6.75), we eventually have:

$$\mathcal{V}^v = -\frac{\varkappa k_{rg}}{\eta_g}\nabla p_g - D\nabla \ln x_v \tag{6.81a}$$

$$\mathcal{V}^a = -\frac{\varkappa k_{rg}}{\eta_a}\nabla p_g - D\nabla \ln x_a \tag{6.81b}$$

Using $x_v = p_v/p_g$ and (6.78)–(6.79), we finally write (6.81) in the more operationally explicit form:

$$\mathcal{V}^v = -\frac{\varkappa k_{rg}}{\eta_g}\nabla p_g - (nS_g\tau)D_0\frac{p_{atm}}{p_v}\nabla\left(\frac{p_v}{p_g}\right) \tag{6.82a}$$

$$\mathcal{V}^a = -\frac{\varkappa k_{rg}}{\eta_g}\nabla p_g + (nS_g\tau)D_0\frac{p_{atm}}{p_a}\nabla\left(\frac{p_v}{p_g}\right) \tag{6.82b}$$

---

[9] See de Vries D.A., Kruger A.J. (1966), 'On the value of the diffusion coefficient of water vapour in air', *Phénomènes de transport avec changement de phase dans les milieux poreux ou colloïdaux*, éd. CNRS, 561–572.

[10] See Millington R.J. (1959), 'Gas diffusion in porous media', *Science*, **130**, 100–102.

6.6 Advanced Analysis

6.6.1 The Stress Partition Theorem in the Unsaturated Case

For the sake of clarity we consider here a porous material whose porous space is filled by only two fluids, referred to by index 1 or 2. Similar to the saturated case (see §2.5.1 for the notation), we define a macroscopic partial stress σ^{Π} related to the skeleton particle, $\Pi = S$, and to the fluid particles, $\Pi = F_\alpha$, $\alpha = 1, 2$, and an intrinsic partial stress σ^{π} related to the solid matrix, $\pi = s$, and to the fluids, $\pi = f_\alpha$:

$$\boldsymbol{\sigma}^{S} = \langle f_{\omega_s}(\mathbf{z})\,\boldsymbol{\sigma}^{s}(\mathbf{z})\rangle = (1-n)\,\boldsymbol{\sigma}^{s} \tag{6.83a}$$

$$\boldsymbol{\sigma}^{F_\alpha} = \left\langle f_{\omega_{f\alpha}}(\mathbf{z})\,\boldsymbol{\sigma}^{f_\alpha}(\mathbf{z})\right\rangle = nS_\alpha\boldsymbol{\sigma}^{f_\alpha} \tag{6.83b}$$

where $\sigma^{s}(\mathbf{z})$ and $\sigma^{f_\alpha}(\mathbf{z})$ are the actual stress fields related to the matrix and to the fluids at the microscopic point located at $\mathbf{z}$ within the volume ω. The equilibrium of each continuum $\Pi = S, F_\alpha$ considered separately is (see (2.30)):

$$\nabla\cdot\boldsymbol{\sigma}^{\Pi} + \rho_{\Pi}\mathbf{f} + \mathbf{f}_{int}^{\rightarrow\Pi} = 0 \tag{6.84}$$

where the volume force $\mathbf{f}_{int}^{\rightarrow\Pi}$ accounts for the macroscopic mechanical interaction exerted on the Π continuum. The same developments as those treated in §2.5.1 allow us to express (6.84) in the form (see (2.67)):

$$\nabla_x\cdot\boldsymbol{\sigma}^{\Pi} + \rho_{\Pi}\mathbf{f} + \frac{1}{\omega}\int_{\partial\omega_\pi}\boldsymbol{\sigma}^{\pi}(\mathbf{z})\cdot\mathbf{n}^{\pi}\,f(\mathbf{z}-\mathbf{x})\,da_a = 0 \tag{6.85}$$

where $\mathbf{n}^{\pi}$ is the outward unit normal to surface $\partial\omega_\pi$ delimiting the volume actually occupied by the particle $\Pi = S$ or F_α and separating the latter from the volume occupied by the other two particles.

Along the interface $A_{\alpha,\beta}$ between two components α and β, either fluid-fluid ($\alpha = 1$, $\beta = 2$) or solid–fluid ($\alpha = s$, $\beta = 1$ or 2), surface tension exerts a strength $\gamma_{\alpha,\beta}$ per unit of surface (see Fig. 6.1). The union of all these interfaces can be viewed as forming the 'fourth component'. The local intrinsic partial stress $\boldsymbol{\gamma}^{\alpha,\beta}(\mathbf{z})$ related to interface $A_{\alpha,\beta}$ can be expressed in the form:

$$\boldsymbol{\gamma}^{\alpha,\beta}(\mathbf{z}) = \gamma_{\alpha,\beta}\delta_{ab}\mathbf{t}^{(a)}\otimes\mathbf{t}^{(b)} \tag{6.86}$$

In (6.86), $\mathbf{t}^{(1)}$ and $\mathbf{t}^{(2)}$ are the vectors forming a curvilinear orthonormal basis in the plane tangent to $A_{\alpha,\beta}$ such as:

$$\mathbf{t}^{(a)} = \frac{\partial z_i(s_1,s_2)}{\partial s_a}\mathbf{e}_i;\quad \sum_{a=1}^{a=2} ds_a^2 = \sum_{i=1}^{i=3} dz_i^2 \tag{6.87}$$

where $\mathbf{e}_{i=1,2,3}$ forms a fixed orthonormal basis, while s_a is the curvilinear abscissa related to the $\mathbf{t}^{(a)}$ direction. Recall then that:

$$\frac{d\mathbf{t}^{(a)}}{ds_a} = \frac{\mathbf{n}^{\alpha,\beta}}{R_a};\qquad \mathbf{n}^{\alpha,\beta} = \mathbf{t}^{(1)}\times\mathbf{t}^{(2)} \tag{6.88}$$

where $\mathbf{n}^{\alpha,\beta}$ is the unit normal to interface $\mathrm{A}_{\alpha,\beta}$, while R_a is its radius of curvature in the $\mathbf{t}^{(a)}$ direction. Now let σ^{IV} be the macroscopic partial stress related to the fourth component. It can be defined as:

$$\boldsymbol{\sigma}^{IV} = \left\langle \delta_{\cup \mathrm{A}_{\alpha,\beta}} \boldsymbol{\gamma}^{\alpha,\beta}(\mathbf{z}) \right\rangle = \frac{1}{\omega} \int_{\cup \mathrm{A}_{\alpha,\beta}} \boldsymbol{\gamma}^{\alpha,\beta}(\mathbf{z})\, f(\mathbf{z}-\mathbf{x})\, da_\omega \tag{6.89}$$

where $\delta_{\cup \mathrm{A}_{\alpha,\beta}}$ stands for the characteristic or Dirac δ-function attached to the fourth component in so far as a surface immersed in the volume ω is concerned. Partial stress σ^{IV} can be split into the contributions related to each interface $\mathrm{A}_{\alpha,\beta}$ according to:

$$\boldsymbol{\sigma}^{IV} = \boldsymbol{\Gamma}^{s,f_1} + \boldsymbol{\Gamma}^{s,f_2} + \boldsymbol{\Gamma}^{f_1,f_2} \tag{6.90}$$

where:

$$\boldsymbol{\Gamma}^{\alpha,\beta} = \frac{1}{\omega} \int_{\mathrm{A}_{\alpha,\beta}} \gamma_{\alpha,\beta} \delta_{ab} \mathbf{t}^{(a)} \otimes \mathbf{t}^{(b)} f(\mathbf{z}-\mathbf{x})\, ds_1 ds_2 \tag{6.91}$$

Using (6.87), Cartesian components $\Gamma_{ij}^{\alpha,\beta}$ of $\boldsymbol{\Gamma}^{\alpha,\beta}$ can be expressed in the form:

$$\Gamma_{ij}^{\alpha,\beta} = \frac{1}{\omega} \int_{\mathrm{A}_{\alpha,\beta}} \gamma_{\alpha,\beta} \delta_{ab} \frac{\partial z_i}{\partial s_a} \frac{\partial z_j}{\partial s_b} f(\mathbf{z}-\mathbf{x})\, ds_1 ds_2 \tag{6.92}$$

so that:

$$\frac{\partial \Gamma_{ij}^{\alpha,\beta}}{\partial x_j} = -\frac{1}{\omega} \int_{\mathrm{A}_{\alpha,\beta}} \gamma_{\alpha,\beta} \delta_{ab} \frac{\partial z_i}{\partial s_a} \frac{\partial f}{\partial s_b} (\mathbf{z}-\mathbf{x})\, ds_1 ds_2 \tag{6.93}$$

or, equivalently:

$$\nabla_x \cdot \boldsymbol{\Gamma}^{\alpha,\beta} = -\frac{1}{\omega} \int_{\mathrm{A}_{\alpha,\beta}} \gamma_{\alpha,\beta} \delta_{ab} \mathbf{t}^{(a)} \frac{\partial f}{\partial s_b} (\mathbf{z}-\mathbf{x})\, ds_1 ds_2 \tag{6.94}$$

Use of (6.88) allows us to rewrite the previous equation in the form:

$$\begin{aligned} \nabla_x \cdot \boldsymbol{\Gamma}^{\alpha,\beta} = {} & \frac{1}{\omega} \int_{\mathrm{A}_{\alpha,\beta}} \frac{2\gamma_{\alpha,\beta}}{R} \mathbf{n}^{\alpha,\beta} f(\mathbf{z}-\mathbf{x})\, ds_1 ds_2 \\ & - \frac{1}{\omega} \int_{\mathrm{A}_{\alpha,\beta}} \frac{\partial}{\partial s_b} \left[\gamma_{\alpha,\beta} \delta_{ab} \mathbf{t}^{(a)} f(\mathbf{z}-\mathbf{x}) \right] ds_1 ds_2 \end{aligned} \tag{6.95}$$

where $2/R = 1/R_1 + 1/R_2$ is the mean curvature of the interface. The integrand in the second integral is eventually identified as:

$$\frac{\partial}{\partial s_b} \left[\gamma_{\alpha,\beta} \delta_{ab} \mathbf{t}^{(a)} f(\mathbf{z}-\mathbf{x}) \right] = \nabla_s \cdot \left[\boldsymbol{\gamma}^{\alpha,\beta} f(\mathbf{z}-\mathbf{x}) \right] \tag{6.96}$$

where ∇_s stands for the surface divergence operator related to surface $A_{\alpha,\beta}$. Applying the divergence theorem to (6.95) and summing over all interfaces $A_{\alpha,\beta}$, we derive the averaged macroscopic momentum balance equation relative to the fourth component, that at is:

$$\nabla_x \cdot \boldsymbol{\sigma}^{IV} + \mathbf{f}_{int}^{\to IV} = 0 \tag{6.97}$$

where the interaction force $\mathbf{f}_{int}^{\to IV}$ exerted on the fourth component can be expressed in the form:

$$\begin{aligned}\mathbf{f}_{int}^{\to IV} &= -\sum_{A_{\alpha,\beta}} \frac{1}{\omega} \int_{A_{\alpha,\beta}} \frac{2\gamma_{\alpha,\beta}}{R} \mathbf{n}^{\alpha,\beta} f(\mathbf{z}-\mathbf{x})\, da_\omega \\ &+ \sum_{A_{\alpha,\beta}} \frac{1}{\omega} \int_L \gamma_{\alpha,\beta} \boldsymbol{\nu}^{\alpha,\beta} f(\mathbf{z}-\mathbf{x})\, dL\end{aligned} \tag{6.98}$$

In (6.98), $\boldsymbol{\nu}^{\alpha,\beta}$ denotes the unit vector lying in the tangent plane to interface $A_{\alpha,\beta}$ and normal to its border. The latter is eventually the triple junction line L, that is the line shared by all the interfaces (see the detail in Fig. 6.1).

The overall equilibrium of interfaces $A_{\alpha,\beta}$ and of triple junction line L requires that:

$$\begin{aligned}&-\frac{1}{\omega} \sum_{p=s,f_1,f_2} \int_{\partial\omega_\pi} \boldsymbol{\sigma}^\pi(\mathbf{z}) \cdot \mathbf{n}^\pi f(\mathbf{z}-\mathbf{x})\, da_\omega \\ &+ \sum_{A_{\alpha,\beta}} \frac{1}{\omega} \int_{A_{\alpha,\beta}} \frac{2\gamma_{\alpha,\beta}}{R} \mathbf{n}^{\alpha,\beta} f(\mathbf{z}-\mathbf{x})\, da_\omega \\ &- \sum_{A_{\alpha,\beta}} \frac{1}{\omega} \int_L \gamma_{\alpha,\beta} \boldsymbol{\nu}^{\alpha,\beta} f(\mathbf{z}-\mathbf{x})\, dL = 0\end{aligned} \tag{6.99}$$

Indeed, the last term on the left hand side of (6.99) represents the resulting force exerted by all the interfaces $A_{\alpha,\beta}$ on the tripple junction line L. Besides, on the surface shared by $\partial\omega_\pi$ and $A_{\alpha,\beta}$ the membrane equilibrium of the interface $A_{\alpha,\beta}$ can be expressed by means of the Laplace equation according to:

$$\boldsymbol{\sigma}^{\pi I}(\mathbf{z}) \cdot \mathbf{n}^{\pi I} + \boldsymbol{\sigma}^{\pi J}(\mathbf{z}) \cdot \mathbf{n}^{\pi J} = \frac{2\gamma_{\alpha,\beta}}{R} \mathbf{n}^{\alpha,\beta}; \quad \mathbf{n}^{\pi I} = -\mathbf{n}^{\pi J} = -\mathbf{n}^{\alpha,\beta} \tag{6.100}$$

Once (6.100) is substituted into (6.99), the remaining non-zero contribution of the first two integrals in (6.99) represents the force $\mathbf{f}_{s \text{ on } L}$ which is exerted by the solid matrix on the triple junction line L (see Fig. 6.1). Therefore the left hand side of (6.99) represents the overall force exerted on L, whose equilibrium is required to be zero. Equation (6.99) eventually turns out to be the explicit formulation of the action–reaction law stating the equilibrium of the 'interface' L shared by the four components, that is:

$$\mathbf{f}_{int}^{\to S} + \sum_\alpha \mathbf{f}_{int}^{\to F_\alpha} + \mathbf{f}_{int}^{\to IV} = 0 \tag{6.101}$$

Indeed, summing (6.85) for $\Pi = S, F_1, F_2$ and taking into account (6.99), we finally retrieve the macroscopic momentum equation (6.8), provided that the total stress σ is identified with the sum of the macroscopic partial stresses:

$$\boldsymbol{\sigma} = \boldsymbol{\sigma}^S + \sum_\alpha \boldsymbol{\sigma}^{F_\alpha} + \boldsymbol{\sigma}^{IV} \tag{6.102}$$

Selecting the sliding average operator associated with the weighting function $f = f_{\omega 0}$ (see (2.57)), we finally obtain:

$$\boldsymbol{\sigma} = (1-n)\,\boldsymbol{\sigma}^s - \sum_\alpha n S_\alpha p_\alpha \mathbf{1} + \frac{1}{\omega}\int_{\cup \mathsf{A}_{\alpha,\beta}} \gamma_{\alpha,\beta}\delta_{ab}\mathbf{t}^{(a)} \otimes \mathbf{t}^{(b)}\, ds_1 ds_2 \tag{6.103}$$

which extends the partition theorem to the unsaturated case.[11]

6.6.2 Capillary Hysteresis. Porosimetry

When a sample of porous material is subjected to a drainage–imbibition cycle, a hysteresis loop is generally observed in the $(S_w \times p_c)$ plane so that the link between the capillary pressure and the saturation degree cannot reduce to a one-to-one relation between p_c and S_w as in (6.38). More precisely, when starting from a complete saturation and progressively increasing the capillary pressure, the saturation degree S_w progressively decreases and the corresponding point (S_w, p_c) follows the drainage curve defined by:

$$p_c = p_c^{DR}(S_w) \tag{6.104}$$

At the end of the previous drainage process, when again decreasing the capillary pressure, saturation S_w recovers higher values and the point (S_w, p_c) follows the imbibition curve defined by:

$$p_c = p_c^{IM}(S_w) \tag{6.105}$$

The latter differs from the former in such a way that the saturation S_w related to the same capillary pressure is lower in the imbibition process than in the drainage process. Both curves conjointly form a reproducible hysteresis loop: when the direction of variation of the capillary pressure is inverted during a drainage process, the point (S_w, p_c) leaves the drainage curve to rejoin the imbibition curve at nearly the same saturation S_w; conversely, when the direction of variation of the capillary pressure is inverted during an imbibition process, the point (S_w, p_c) leaves the imbibition curve to rejoin the drainage curve at nearly the same saturation S_w. This rough description of capillary hysteresis is sketched in Fig. 6.9.

The existence of an hysteresis implies a dissipative process so that (6.23) has to become an inequality, that is:

$$\sigma_{ij} d\varepsilon_{ij} + (S_w p_w + S_{nw} p_{nw}) d\phi - \phi p_c dS_w - S_s dT - d\Psi_s \geq 0 \tag{6.106}$$

[11] The partition stress theorem can be alternatively derived by carrying out an energy approach based on the principle of virtual work (see Chateau X. Dormieux L. (1995), 'Homogénéisation d'un milieux poreux non saturé: lemme de Hill et applications', Comptes Rendus de l'Academie des Sciences, **320**, Série II b, 627–634).

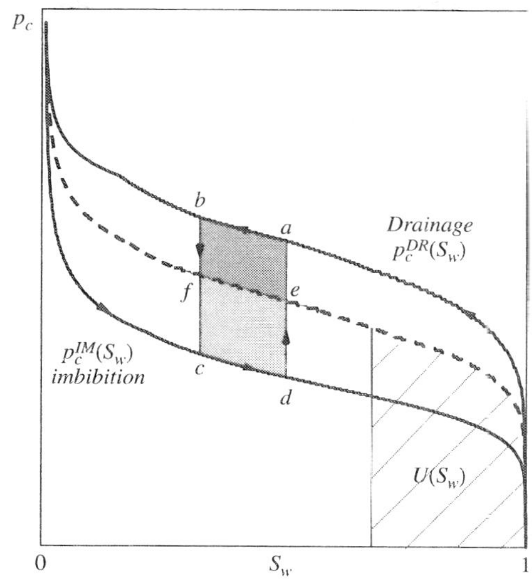

Figure 6.9: Hysteresis loop formed by the drainage curve and the imbibition curve in the $(S_w \times p_c)$ plane. Curve $-\partial U/\partial S_w$ necessarily lies within the loop. Along the drainage path ab the dissipated energy equals the area $abef$, whereas along the imbibition path cd it equals the area $cdfe$.

Taking into account the energy separation hypothesis (6.52) in (6.106) allows us to obtain the positiveness of the dissipation associated with capillary hysteresis in the form:

$$-\phi\left(p_c + \frac{\partial U}{\partial S_w}\right) dS_w \geq 0 \tag{6.107}$$

where we used (6.53) stating the nullity of the dissipation associated with the deformation of the solid elastic matrix. The above definitions of the drainage-imbibition curves combined with (6.107) result in the inequality:

$$p_c^{IM} \leq -\frac{\partial U}{\partial S_w} \leq p_c^{DR} \tag{6.108}$$

so that p_c necessarily differs from $-\partial U/\partial S_w$. Accordingly, the point $(S_w, -\partial U/\partial S_w)$ follows in the $(S_w \times p_c)$ plane a curve lying between the drainage curve and the imbibition curve. During a capillary pressure cycle such as *abfcdea* as sketched in Fig. 6.9, the energy dissipated through capillary hysteresis can now be identified. Along the drainage path ab the dissipated energy per unit of porous volume is the area between ab and the path ef followed by the point $(S_w, -\partial U/\partial S_w)$ during the drainage. It corresponds to the part of the mechanical work $-p_c dS_w$, where $dS_w < 0$, which is not stored in interfacial energy form dU (per unit of porous volume) but is dissipated in heat form. Conversely, along the imbibition path cd the dissipated energy is the area between the path cd and the path fe followed by point $(S_w, -\partial U/\partial S_w)$ during the imbibition It corresponds to the part of the interfacial energy $-dU$ (per unit of porous volume) which is not returned to the outer system in the form of mechanical work $-p_c dS_w$, where $dS_w > 0$, but is dissipated in heat form. During the whole cycle *abfcdea* the dissipated energy is finally the area between the part *ab* and the part *cd* of respectively the drainage curve and the imbibition curve. Nevertheless, the above macroscopic approach does not allow us to identify either the interfacial energy $U(S_w)$, or the origin of the capillary hysteresis. This requires specification of the microstructure of the porous space.

Indeed a standard interpretation[12] of the capillary hysteresis turns out to consider the imbibition process – or the drainage process – as the filling up – or the pouring out – of spheroid-shaped microscopic pores with distinct radii. The pores are linked to each other by fine capillary tubes of negligible volume so that the latter do not significantly contribute to the saturation level but play a key role in the hysteresis analysis. Consider for instance the current imbibition state c of Fig. 6.9 and let R be the upper radius bounding the radius of pores which are already filled up and contribute to the current saturation degree S_w. The value of radius R is given by Laplace's law for the membrane interface equilibrium. Indeed, assuming a zero contact angle between the wetting fluid and the solid matrix, R identifies the curvature of the interface between the wetting fluid and the non-wetting fluid (see Fig. 6.10). Accordingly, we write:

$$p_c^{IM} = \frac{2\gamma_{w,nw}}{R} \tag{6.109}$$

Laplace's law eventually guarantees the very existence of R since, at given capillary pressure p_c^{IM}, according to (6.109) there is no possible membrane equilibrium between the wetting fluid and the non-wetting fluid within pores having a radius less than R or greater than R. This latter remark is at the origin of the methods of porosimetry aimed at determining the pore radius distribution through the fraction $\varphi(R)$ of the porous volume consisting of pores with a radius greater than R. Indeed, we may write:

$$\varphi(R) = 1 - S_w \tag{6.110}$$

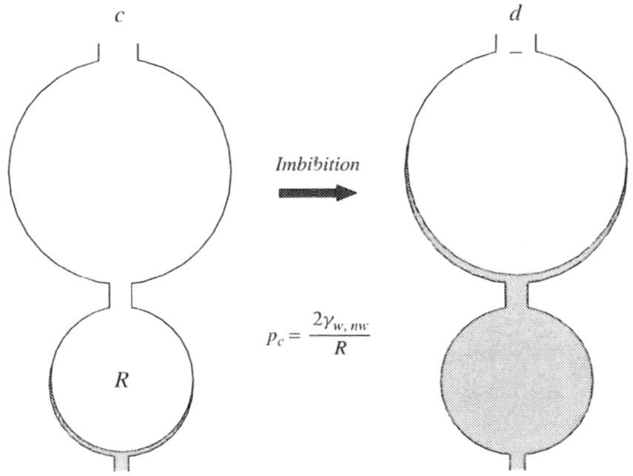

Figure 6.10: The filling up of a given pore occurs when capillary pressure p_c fits the radius R of the pore according to Laplace's law.

[12]For a comprehensive presentation see Dullien F.A.L. (1979), *Porous media: fluid transport and pore structure*, Academic Press, New York.

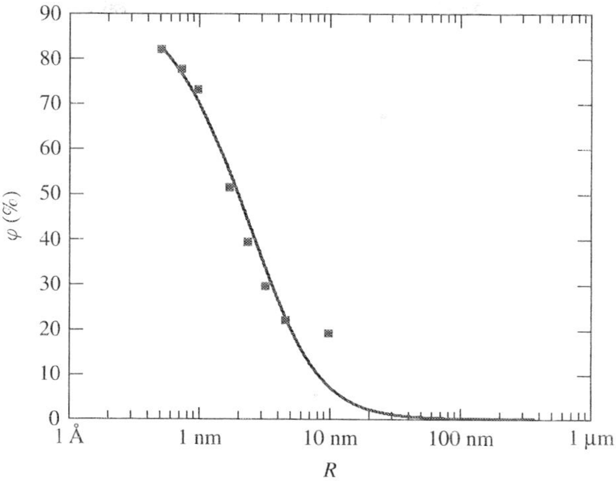

Figure 6.11: Pore distribution function $\varphi(R)$ as determined from the imbibition capillary pressure curve (6.105), the definition (6.110) of $\varphi(R)$ and Laplace's law (6.109). The curve relates to the experimental data reported in Fig. 6.3, while adopting the surface energy of a liquid water–air interface, that is $\gamma_{w,nw} = 73$ mJ/m.

The pore distribution function $\varphi(R)$ is then determined from the imbibition capillary pressure curve (6.105), the definition (6.110) of $\varphi(R)$ and Laplace's law (6.109) (see Fig. 6.11).

Consider now an infinitesimal decrease in capillary pressure. This decrease can be carried out either by releasing the external pressure of the non-wetting fluid as in the imbibition experiment, or by increasing the outer relative humidity as in the adsorption experiment (see §6.3.2). The greatest value $R + dR$ of the pores filled up consecutively to the drop in capillary pressure has to fit the new capillary pressure according to (6.109), resulting in the saturation increase dS_w. In the actual imbibition experiment the filling up is achieved by invasion of the wetting fluid. In the adsorption experiment, when the wetting fluid is water, the filling up is eventually achieved by condensation of the water vapour of the wetted air, according to both Kelvin's law (6.33) and Laplace's law (6.109), the left hand member of the former identifying the opposite of the current capillary pressure. Denoting dV_w as the infinitesimal volume per unit of volume which is newly filled up and neglecting the volume of the capillary tubes linking the pores between them, we write:

$$dV_w = -\phi \frac{d\varphi}{dR} dR \tag{6.111}$$

Let $da_{s,w}$ be the newly wetted surface per unit of volume relative to dV_w. Since the pores are assumed to be spheroids, $da_{s,w}$ can be expressed in the form:

$$da_{s,w} = \frac{3}{R} dV_w = \frac{3}{R} \phi dS_w \tag{6.112}$$

In the infinitesimal filling process variation $da_{w,nw}$ of the interface between the wetting and the non-wetting fluid is a second-order term when compared with $da_{s,w}$. In addition

$da_{s,nw} = -da_{s,w}$ so that (6.22) gives:

$$\phi dU = (\gamma_{s,w} - \gamma_{s,nw})\, da_{s,w} \tag{6.113}$$

Besides, owing to the nullity of the contact angle, the equilibrium of triple junction line L (see Fig. 6.1) requires that $\gamma_{s,nw} = \gamma_{s,w} + \gamma_{w,nw}$. Combining (6.109), (6.112) and (6.113) finally leads to the following identification of $-\partial U/\partial S_w$:

$$-\frac{\partial U}{\partial S_w} = \frac{3}{2} p_c^{IM}(S_w) \tag{6.114}$$

so that U can be obtained by a simple integration. However, it must be pointed out that such an identification of the interfacial energy relies deeply on the initial assumption of spheroid-shaped microscopic pores.

If the capillary pressure is now again increased from state d of Fig. 6.10, the drainage of the pore of radius R can start only for a capillary pressure equal to the radius r of the fine capillary tube allowing access to the pore (see Fig. 6.12). Consequently, as actually observed, the saturation S_w remains constant along path da, while Laplace's law gives the value of the capillary pressure at which the drainage actually starts, that is:

$$p_c^{DR}(S_w) = \frac{2\gamma_{w,nw}}{r} \tag{6.115}$$

6.6.3 Capillary Pressure Curve, Deformation and Equivalent Pore Pressure

An hypothesis less restrictive than the energy separation assumption (6.52) consists in assuming that the interfacial energy U per unit of porous volume also depends on the current porous volume. Instead of (6.52) we now write:

$$\Psi_s(\varepsilon_{ij}, S_w, T) = \psi_s(\varepsilon_{ij}, \phi, T) + \phi U(S_w, \phi, T) \tag{6.116}$$

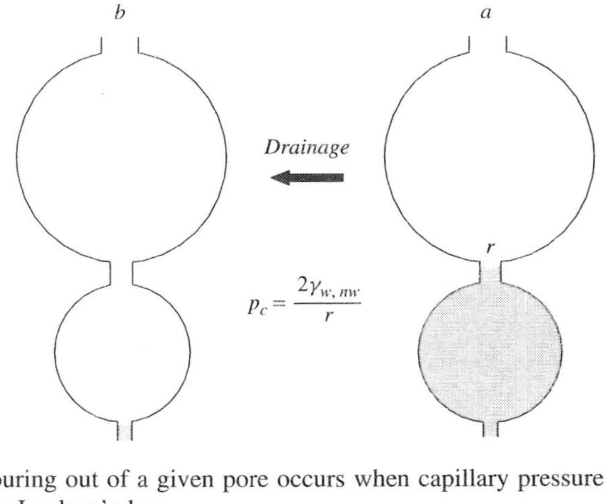

Figure 6.12: The pouring out of a given pore occurs when capillary pressure p_c equals the access radius r according to Laplace's law.

In order to identify a possible candidate for $U(S_w, \phi, T)$ we first introduce the length ℓ scaling the current porous volume material according to:

$$\phi d\Omega_0 = \ell^3 \tag{6.117}$$

so that:

$$\frac{d\phi}{\phi} = 3\frac{d\ell}{\ell} \tag{6.118}$$

Besides, dimensional analysis gives:

$$U = \frac{\gamma_{w,nw}}{\ell} f\left(\frac{\ell_i}{\ell}\right) \tag{6.119}$$

where we have omitted other variables than those relative to the current geometry, ℓ_i standing for all the relevant characteristic lengths, $i = 1, \ldots$ required for the fine description of the geometry of the porous network. Differentiation of (6.119) gives:

$$\frac{dU}{U} = -\frac{d\ell}{\ell} \tag{6.120}$$

which is where we let $d(\ell_i/\ell) = 0$, equivalent to assuming that the porous space undergoes a homothetic transformation. Getting rid of ℓ between (6.118) and (5.120) and integrating the resulting equation, we derive:

$$U = \phi^{-1/3}\Gamma(S_w, T) \tag{6.121}$$

so that (6.116) gives:

$$\Psi_s(\varepsilon_{ij}, S_w, T) = \psi_s(\varepsilon_{ij}, \phi, T) + \phi^{2/3}\Gamma(S_w, T) \tag{6.122}$$

A combination of (6.23), the third of the state equations (6.26) and (6.122) allows us to recover (6.53) provided that the definition (6.55) of the equivalent pore pressure π is replaced by:

$$\pi = S_w p_w + S_{nw} p_{nw} - \frac{2}{3}U \tag{6.123}$$

whereas a substitution of (6.122) into the third of the state equations (6.26) leads us to express the capillary pressure p_c in the form:

$$p_c = -\phi^{-1/3}\frac{\partial\Gamma}{\partial S_w} \tag{6.124}$$

The factor 2/3 affecting U in (6.123) accounts for the two-dimensional nature of the interfaces compared with the three-dimensional nature of the porous material volume in which the former are immersed. Similarly, the exponent $-1/3$ applying to ϕ in (6.124) accounts for the 1D nature of the curvature of the interface, while the minus sign relates to the one that arises from Laplace's law (6.100) governing the membrane equilibrium of interfaces. This can be more rigorously inferred by inspecting the stress partition theorem (6.103) related to the unsaturated case. The very concept of an equivalent pore pressure representing the overall effects of both the fluid pressure and the surface tension implies that the partial stress $\boldsymbol{\sigma}^{IV}$ related to the fourth phase formed by all the interfaces and corresponding to the last term on the right hand side of (6.103) must be a spherical tensor whose diagonal components are equal to a third of its trace. Restricting consideration here to the case of a wetting fluid and a non-wetting fluid, we write:

$$\frac{1}{\omega}\int_{\cup \mathsf{A}_{\alpha,\beta}} \gamma_{\alpha,\beta}\delta_{ab}\mathbf{t}^{(a)}\otimes\mathbf{t}^{(b)}\,ds_1 ds_2$$
$$= \frac{2}{3}(\gamma_{s,w}\,a_{s,w} + \gamma_{s,nw}\,a_{s,nw} + \gamma_{w,nw}\,a_{w,nw})\mathbf{1} \tag{6.125}$$

Combining (6.22), (6.103) and (6.125) gives:

$$\boldsymbol{\sigma} = (1-n)\,\boldsymbol{\sigma}^{s} - n\left(S_w p_w + S_{nw} p_{nw} + \gamma_{s,w} a_s - \frac{2}{3}U\right)\mathbf{1} \tag{6.126}$$

In the saturated case $S_w = 1$ and $S_{nw} = 0$ so that (6.126) reduces to:

$$\boldsymbol{\sigma} = (1-n)\,\boldsymbol{\sigma}^{s} - n(p_w + \gamma_{s,w} a_s)\mathbf{1} \tag{6.127}$$

Expression (6.123) is again retrieved if we require the equivalent pore pressure π to play in (6.126) the same role as the one played by p_w in (6.127). It is eventually worthwhile to note that the hypothesis of an homothetic transformation for the porous space in the former simple approach turns out in the latter refined approach to assume the sphericity of the stress tensor attached to the surface tension.

6.6.4 Isothermal Drying of Weakly Permeable Materials

The drying of materials is a field of investigation in itself and this section is limited to give a very first insight into the continuum approach to the modelling of such a process when the temperature T can be assumed to remain constant.[13] In addition, the analysis will be restricted to the drying of weakly permeable materials. For these weakly permeable materials the molecular diffusion that allows the dry air to enter the porous material preponderates when compared with the opposite process, that is the advective transport which, in response to any gas pressure increase, would enforce the vapour–air mixture considered as a whole to leave the porous material. The condition ensuring $\mathcal{V}^{g} \simeq 0$, and,

[13]The assumption of isothermal conditions is relevant provided that (i) the same temperature as the initial one is imposed on the external surface of the material; (ii) the characteristic time related to thermal diffusion is much smaller than the ones associated with fluid transport (for further details see for instance Coussy O., Eymard R., Lassabatère T. (1998), 'Constitutive modelling of unsaturated drying deformable materials', *Journal of Engineering Mechanics*, **124**, (6), 658–667.

consequently, allowing us to neglect the first term in regard to the second term on the right hand side of (6.82b), is roughly expressed in the form:

$$\frac{\varkappa}{\eta_g} \times \frac{p_g}{\tau D_0} \ll 1 \tag{6.128}$$

Letting $\eta_g = 1.8 \times 10^{-5}$ kg/(m/s), $p_g \leq 10^6$ Pa, $\tau D_0 \sim 10^{-3}$ m^2/s^1, condition (6.128) leads us to require:

$$\varkappa \ll 10^{-19} \text{ m}^2 \tag{6.129}$$

The permeability and the capillary curve are macroscopic properties that both result from the detailed geometry of the porous space. As a consequence, weakly permeable materials such that condition (6.129) is met are equally porous materials such that the capillary pressure rapidly reaches values much greater than atmospheric pressure. The cement paste and the artificial clay of Figs. 6.3 and 6.4 constitute such examples of porous materials. Besides, within a drying material the gas pressure of the vapour–air mixture cannot be much greater than several times atmospheric pressure. For weakly permeable materials, in addition to $\mathcal{V}_g \simeq 0$, we can therefore conclude that $\|\nabla p_g\| \ll |\nabla p_c\|$, yielding:

$$\nabla p_l \simeq -\nabla p_c \tag{6.130}$$

Consider now a 1D drying experiment. A sample of porous material, lying between $x = -L$ and $x = L$, is in uniform thermodynamic equilibrium and its initial liquid saturation degree is denoted S_l^0. At time $t = 0$ the sample end surfaces $x = \pm L$ come into contact with the outer atmosphere which imposes there its relative humidity $h_r^{out} = p_v^{out}/p_{vs}(T)$ and provokes a thermodynamic imbalance due to the higher relative humidity, say h_r^0, initially prevailing within the porous material. As a consequence the porous material exchanges water vapour with the outer atmosphere in order to install progressively the outer relative humidity within the layers close to the surface. In turn the liquid water of the latter simultaneously evaporates in order to maintain the vapour–liquid equilibrium. The vapour so supplied by the evaporation of the skin layers leads to a decrease in the liquid pressure. As a result, the gradient in liquid pressure progressively builds up and forces the liquid of the inner layers to move towards the surface and, eventually, to evaporate when coming into contact with the outer atmosphere. Consequently the saturation degree of the liquid progressively decreases within the material and the whole process stops when the equilibrium saturation associated with the outer relative humidity is reached.

So depicted, the drying process involves mainly the motion of the water in liquid form and its subsequent evaporation within the layer close to the surface so that, by comparison, the inner evaporation is assumed to be negligible. This actually holds for weakly permeable materials satisfying conditions (6.128)–(6.129) and allowing the dry air actually to enter the porous material in order to replace progressively the water leaving the inner layers of the porous material in both vapour and liquid form. Therefore the some what paradoxical conclusion is that the drying of such weakly permeable materials is eventually achieved through Darcean advective transport of the water in liquid form, from the inner layers to the skin layers where it finally evaporates on contact with the outer atmosphere.

Neglecting the velocity and the deformation of the skeleton,[14] so that $m_\alpha \simeq nS_\alpha\rho^\alpha$ where $n \simeq \phi_0$, the mass balance for the liquid water (index l), the vapour (index v) and the dry air (index a) are:

$$\frac{\partial(nS_l\rho_l)}{\partial t} + \nabla\cdot\left(\rho_l \mathcal{V}^l\right) = -\overset{\circ}{m}_{l\to v} \tag{6.131a}$$

$$\frac{\partial(n(1-S_l)\rho_v)}{\partial t} + \nabla\cdot\left(\rho_v \mathcal{V}^v\right) = \overset{\circ}{m}_{l\to v} \tag{6.131b}$$

$$\frac{\partial(n(1-S_l)\rho_a)}{\partial t} + \nabla\cdot\left(\rho_a \mathcal{V}^a\right) = 0 \tag{6.131c}$$

Using Darcy's law (6.70a) with $w = l$ and neglecting body forces, (6.131a) can be specified in the 1D form:

$$\frac{\partial(nS_l\rho_l)}{\partial t} + \frac{\mathrm{M}\rho_l\varkappa}{\eta_l}\frac{\partial}{\partial x}\left(k_{rl}(S_l)\frac{\partial\pi_c(S_l)}{\partial x}\right) = -\overset{\circ}{m}_{l\to v} \tag{6.132}$$

To derive (6.132) we assumed incompressibility of the liquid flow and we used assumption (6.130), while letting $p_c = \mathrm{M}\pi_c$, where M stands for the capillary modulus (see (6.39) with $p_e = 0$). Furthermore, under assumption (6.129) there is no significant Darcean advective transport of the vapour–air mixture considered as a whole. Neglecting the first term of the right hand member in (6.82), we substitute the resulting simplified transport laws into (6.131b) and (6.131c) and, subsequently, we add (6.131b) and (6.132). The procedure gives:

$$\begin{aligned}&\frac{\partial S_l}{\partial t} + \frac{\mathcal{M}_v}{\rho_l RT}\frac{\partial\left[(1-S_l)p_v\right]}{\partial t} - \frac{\mathcal{M}_v}{\rho_l RT}D_0 p_{atm}\frac{\partial}{\partial x}\\ &\times\left[(1-S_l)\tau(n,S_l)\frac{\partial}{\partial x}\left(\frac{p_v}{p_g}\right)\right] - \frac{\mathrm{M}\varkappa}{\eta_l}\frac{\partial}{\partial x}\left(D_l(S_l)\frac{\partial S_l}{\partial x}\right) = 0\end{aligned} \tag{6.133a}$$

$$\times\ \frac{\partial\left[(1-S_l)(p_g-p_v)\right]}{\partial t} + D_0 p_{atm}\frac{\partial}{\partial x}\left[(1-S_l)\tau\frac{\partial}{\partial x}\left(\frac{p_v}{p_g}\right)\right] = 0 \tag{6.133b}$$

where the (dry) air and the vapour have both been assumed to be ideal gases and where we note:

$$D_l(S_l) = -\frac{1}{n}k_{rl}(S_l)\frac{d\pi_c}{dS_l} \tag{6.134}$$

Equation (6.133a) accounts for the transport of water, in both liquid and vapour form, while (6.133b) accounts for the transport of dry air. Among the three thermodynamic variables p_v, S_l and p_g, only two are actually independent. Indeed, the differential system (6.133) must be solved when ensuring in addition the thermodynamic equilibrium between

[14]The effect of the kinematics and the deformation of the skeleton on the drying and the associated desaturation process of a porous material generally turns out to be negligible. Indeed, the only possible effect concerns the variation of the intrinsic permeability due to the change in porosity. By contrast, the effect of the drying or wetting on the skeleton deformation cannot generally be neglected (see §6.4.4).

the liquid water and its vapour. The latter is accounted for through Kelvin's law (6.33), which we write in the form:

$$-\mathrm{M}\pi_c(S_l) + p_g - p_{atm} = \frac{\rho_l RT}{\mathcal{M}_v} \ln \frac{p_v}{p_{vs}(T)} \tag{6.135}$$

Finally, considering half the sample, we add the boundary and symmetry conditions:

$$x = L: \; p_g = p_{atm}; \quad p_v = p_v^{out}; \qquad x = 0: \; \frac{\partial S_l}{\partial x} = \frac{\partial}{\partial x}\left(\frac{p_v}{p_g}\right) = 0 \tag{6.136}$$

while the initial conditions are expressed in the form:

$$t = 0: \quad p_g = p_{atm}; \quad p_v = p_v^0 \tag{6.137}$$

Owing to the thermodynamic equilibrium between the vapour and the liquid water, Kelvin's law (6.135) and initial conditions (6.137) provide the uniform initial saturation S_l^0 as the saturation satisfying:

$$-\mathrm{M}\pi_c\left(S_l^0\right) = \frac{\rho_l RT}{\mathcal{M}_v} \ln \frac{p_v^0}{p_{vs}(T)} \tag{6.138}$$

Inspection of the set of equations (6.133)–(6.136) leads us to identify two characteristic times τ_F and τ_D, respectively defined by:

$$\tau_F = \frac{\rho_l RT}{\mathcal{M}_v p_{atm}} \times \frac{L^2}{D_0}; \qquad \tau_D = \frac{\eta_l L^2}{\mathrm{M}\varkappa} \tag{6.139}$$

The characteristic time τ_F scales the rate at which the vapour and the dry air are transported within the porous material through the molecular (or Fickean) diffusion, while the characteristic time τ_D scales the overall rate at which the advective (or Darcean) transport of the liquid water takes place. The two transport phenomena will occur at well-separated time scales provided that the ratio of their characteristic times satisfies:

$$\varepsilon = \frac{\tau_F}{\tau_D} = \frac{\rho_l RT}{\mathcal{M}_v p_{atm}} \times \frac{\mathrm{M}\varkappa}{\eta_l D_0} \ll 1 \tag{6.140}$$

Letting $\rho_l = 1000\ \mathrm{kg/m^3}$, $R = 8.314\ \mathrm{J/(molK)}$, $T = 293$ K, $\mathcal{M}_v = 18$ g, $p_{atm} =$ 101325 Pa, $\eta_l = 1 \times 10^{-3}\ \mathrm{kg/(ms)}$, $D_0 = 2.17 \times 10^{-5}\ \mathrm{m^2/s}$[1] and adopting $\mathrm{M} =$ 37.55 MPa (see Fig. 6.3), condition (6.140) of time scale separation requires the intrinsic permeability $\varkappa$ to fall in the same range as the one imposed by condition (6.129).

The time scale separation condition (6.140) leads to a two-step drying process. The first step involves mainly a transient molecular diffusion process while, in the second subsequent step, the water is mainly transported in liquid form through a Darcean advection process. In order to explore the first step of the drying process, we let:

$$\bar{x} = \frac{x}{L}; \qquad \bar{t} = \frac{t}{\tau_F} \tag{6.141}$$

Substituting (6.141) into (6.133) and disregarding the ε-order term in (6.133a) we derive:

$$\frac{\mathcal{M}_v}{\rho_l RT}\frac{\partial\left[(1-S_l)p_v\right]}{\partial\bar{t}} - \frac{\partial}{\partial\bar{x}}\left[(1-S_l)\tau(n,S_l)\frac{\partial}{\partial\bar{x}}\left(\frac{p_v}{p_g}\right)\right] = -\frac{\partial S_l}{\partial\bar{t}} \tag{6.142a}$$

$$\frac{\partial\left[(1-S_l)p_g\right]}{\partial\bar{t}} = -\frac{\rho_l RT}{\mathcal{M}_v}\frac{\partial S_l}{\partial\bar{t}} \tag{6.142b}$$

Taking into account (6.137) and (6.138), (6.142b) can be integrated in the form:

$$\frac{p_g}{p_{atm}} = 1 + (\gamma - 1)\frac{S_l^0 - S_l}{1 - S_l} \quad \text{where } \gamma = \frac{\rho_l RT}{p_{atm}\mathcal{M}_v} \tag{6.143}$$

The steady state related to the molecular diffusion is obtained by integrating (6.142a) after nullifying the time derivatives. In addition, taking into account boundary conditions (6.136), we finally obtain:

$$\bar{t} \gg 1: \quad x_v = \frac{p_v}{p_g} = \frac{p_v^{out}}{p_{atm}} \tag{6.144}$$

Asymptotic values S_l^∞ and p_g^∞ related to the first step of the drying process must simultaneously satisfy (6.135), (6.143) and (6.144), providing S_l^∞ as the solution of the non-linear equation:

$$-\frac{\mathrm{M}}{p_{atm}}\pi_c\left(S_l^\infty\right) + (\gamma - 1)\frac{S_l^0 - S_l^\infty}{1 - S_l^\infty} = \gamma \ln\left[\frac{p_v^{out}}{p_{vs}(T)}\left(1 + (\gamma - 1)\frac{S_l^0 - S_l^\infty}{1 - S_l^\infty}\right)\right] \tag{6.145}$$

In order to assess the drop in liquid water saturation and the maximum excess of gas pressure to be expected at the end of the first drying step, we adopt the sorption isotherm of Fig. 6.3. We set $T = 293$ K, with $p_{vs}(T) = 2333$ Pa, and $h_r^0 = p_v^0/p_{vs}(T) = 87\%$, while, for the outer relative humidity, we choose $h_r^{out} = p_v^{out}/p_{vs}(T) = 50\%$. Using successively (6.138), (6.145) and (6.142), we numerically obtain:

$$S_l^0 = 0.892975; \quad S_l^\infty = 0.892915; \quad \frac{p_g^\infty}{p_{atm}} = 1.741 \tag{6.146}$$

Although the difference between the initial saturation S_l^0 and the asymptotic saturation S_l^∞ turns out to be quite negligible, it is worthwhile to note that the asymptotic gas pressure is nearly twice the initial atmospheric pressure. This is mainly due to the high contrast existing between the mass density of the liquid water and the mass density of the amount of vapour produced by the slight evaporation of the former (see (6.143) where $\gamma \simeq 1330$).[15]

[15] For a numerical investigation of the differential system (6.133) supporting the above asymptotic analysis of the drying process of weakly permeable materials, see Mainguy M., Coussy O., Baroghel-Bouny V. (2001), 'The role of air pressure in the drying of weakly permeable materials', *Journal of Engineering Mechanics*, ASCE, **127**, (6), 582–592.

Kelvin's law (6.135) and the first two boundary conditions of (6.136) provide the value S_l^L of the liquid saturation prevailing on the border $x = \pm L$ of the sample that is:

$$-M\pi_c\left(S_l^L\right) = \frac{\rho_l RT}{\mathcal{M}_v} \ln \frac{p_v^{out}}{p_{vs}(T)} \tag{6.147}$$

Obviously, owing to a boundary layer effect not yet accounted for the value of the asymptotic liquid water saturation S_l^∞ of the first drying step given by (6.146) does not match the value of the saturation at the border S_l^L. For instance, when using the same experimental data as the ones which led to the assessment (6.146) of S_l^∞, assessment (6.147) of S_l^L furnishes:

$$S_l^L = 0.4243 \tag{6.148}$$

In fact, the boundary condition at $x = \pm L$ is eventually met with the help of the second step of the drying process. The latter starts immediately in the skin layer close to the end surfaces at $x = \pm L$ where the capillary pressure gradient concentrates at very early times. In order to explore the second drying step involving the inner layers we now let:

$$\overline{x} = \frac{x}{L}; \qquad \tilde{t} = \frac{t}{\tau_D} = \frac{\overline{t}}{\varepsilon} \tag{6.149}$$

which, substituted into (6.133a), give:

$$\frac{\partial S_l}{\partial \tilde{t}} + \frac{\mathcal{M}_v}{\rho_l RT}\frac{\partial\left[(1-S_l)p_v\right]}{\partial \tilde{t}} - \frac{1}{\varepsilon}\frac{\partial}{\partial \overline{x}}\left[(1-S_l)\tau(S_l)\frac{\partial}{\partial \overline{x}}\left(\frac{p_v}{p_g}\right)\right] - \frac{\partial}{\partial \overline{x}}\left(D_l(S_l)\frac{\partial S_l}{\partial \overline{x}}\right) = 0 \tag{6.150}$$

When the leading ε^{-1}-order term is equal to zero, the asymptotic regime (6.144) of the early drying step is recovered. Indeed, the molecular Fickean diffusion acts at a time scale much shorter than the one related to the advective Darcean transport. Therefore, when the second step of the drying process becomes significantly active within the material, the molecular diffusion has already made the vapour fraction $x_v = p_v/p_g$ uniform within the porous material down to the outer value p_v^{out}/p_{atm}. The diffusion equation governing the second step of the drying process is finally obtained by nullifying the ε^0-order term in (6.150). Combining (6.135) and (6.144) gives:

$$-M\pi_c(S_l) + p_{atm}\left(\frac{p_v}{p_v^{out}} - 1\right) = \frac{\rho_l RT}{\mathcal{M}_v} \ln \frac{p_v}{p_{vs}(T)} \tag{6.151}$$

so that the second term on the left hand side of (6.150) turns out to be negligible with regard to the first term. We finally obtain:

$$\frac{\partial S_l}{\partial \tilde{t}} - \frac{\partial}{\partial \overline{x}}\left(D_l(S_l)\frac{\partial S_l}{\partial \overline{x}}\right) = 0 \tag{6.152}$$

to which we add the boundary and initial conditions:

$$\overline{x}=1:\ S_l=S_l^L;\ \overline{x}=0:\ \frac{\partial S_l}{\partial \overline{x}}=0;\ \widetilde{t}=0:\ S_l=S_l^{\infty}\simeq S_l^0 \tag{6.153}$$

Since the mass loss is mainly due to the change in water content in liquid form, the relative mass loss history $\Delta M\,(t)$ during the drying process of the sample can be expressed in the form:

$$\Delta M\,(t)=n\rho_l\Sigma\int_{-L}^{+L}\big(S_l^0-S_l\,(x,t)\big)dx \tag{6.154}$$

where Σ is the cross section of the sample. Owing to the symmetry to the problem and to boundary conditions (6.153), integration of (6.152) allows us to rewrite $\Delta M\,(t)$ in the form:

$$\Delta M\,(t)=-2n\rho_l\Sigma L D_l\big(S_l^L\big)\int_0^{\widetilde{t}}\frac{\partial S_l}{\partial \overline{x}}\,(\overline{x}=1,s)\,ds \tag{6.155}$$

Proceeding as in §5.2.2, the early time solution for $t\ll\tau_D$ is determined by letting:

$$S_l=f\left(\eta=\frac{1-\overline{x}}{\sqrt{\widetilde{t}}}\right) \tag{6.156}$$

which, when substituted into (6.152)–(6.153) with $\widetilde{t}\ll 1$, leads to the ordinary differential equation:

$$\eta\frac{df}{d\eta}+2\frac{d}{d\eta}\left(D_l\,(f)\,\frac{df}{d\eta}\right)=0 \tag{6.157}$$

and to boundary conditions:

$$f\,(0)=S_l^L;\qquad f\,(\eta\to\infty)=S_l^0 \tag{6.158}$$

Collecting the above results, the early time mass loss is finally found to depend on the square root of time according to:

$$\Delta M\,(t)=\alpha\times\sqrt{t};\qquad \alpha=4\sqrt{\frac{M\varkappa}{\eta^l}}\rho_l\Sigma D_l\big(S_l^L\big)\frac{df}{d\eta}\,|_{\eta=0} \tag{6.159}$$

The previous result can be used to assess the value of the intrinsic permeability $\varkappa$ of weakly permeable materials from the experimental early mass loss history. This requires numerically solving the non-linear differential system (6.157)–(6.158) in order to derive $df/d\eta\,|_{\eta=0}$, whose value eventually depends on function $D_l(S_l)$ (see (6.134)) and the values of the liquid saturation degree S_l^0 and S_l^L. An example of such a determination is given in Fig. 6.13 for a cement paste. Such a low value found for the intrinsic permeability, $\varkappa=0.98\times10^{-21}$ m^2, leads us to conclude that the transported liquid water in the drying process is the water creeping along the solid internal walls of the porous network. In

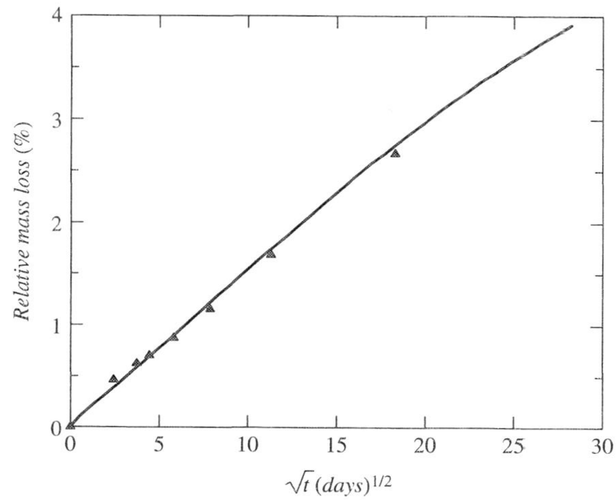

Figure 6.13: Assessment of the intrinsic permeability $\varkappa$ from the early time history of the mass loss due to drying. The triangles correspond to the mass loss recorded during the drying of a cylindrical sample (diameter $\varnothing = 16$ cm, length $2L = 10$ cm, initial mass $\mathcal{M}_0 = 41954$ g) formed from the cement paste whose capillary curve is given in Fig. 6.3 and subjected to initial and boundary conditions (6.146) and (6.148). The solid line corresponds to the numerical solution of (6.152)–(6.153), when letting $\varkappa = 0.98 \times 10^{-21}$ m^2, the latter value being assessed from a linear regression applied to the early time experimental points and from the resulting value found for coefficient α involved in (6.159) (after Mainguy *et al.* (2001), see footnote 15).

turn the question arises about the actual relevance of Darcy's law for describing such a transport since the law implicitly assumes a viscous flow at the scale of the porous network. However, a careful analysis shows that Darcy's law is only apparent. Indeed, the driving force of the liquid water motion is eventually the gradient of its chemical potential involving the complex tensile forces exerted by the solid internal walls on the creeping water (see Fig. 6.3 for the order of magnitude of these tensile stresses). The resulting transport law can only be written in the form of a Darcy-like law involving an effective, apparent, intrinsic permeability $\varkappa^{eff}$ conveniently expressed in m^2 for comparison with the usual values (see §3.6.3 and (3.124)–(3.125)).

Chapter 7

Penetration Fronts

An analysis of the durability performance of porous materials requires in addition an analysis of the penetration of aggressive agents through the pore solution and of the way in which this penetration affects the mechanical properties. For instance, the kinetics of the decrease in elastic properties and strength of geomaterials subjected to leaching mainly depends on the underlying diffusion–dissolution process. Another example is provided by the penetration of chloride ions in concrete structures inducing a high risk of steel corrosion. The decontamination of concrete in energy facilities or the remediation of soils is another major concern in environmental engineering.

In this chapter, motivated by such durability issues, we focus attention on the kinetics of penetration fronts. Whatever the special physics at work, the latter is mainly governed by the same vanishing diffusion process operating at the penetration front. This allows a general unified analysis of the penetration of fronts associated with apparently quite distinct phenomena, for example dissolution, ionic diffusion and migration, forced imbibition, etc. If the physico-chemical mechanisms have a strong influence on the mechanical behaviour of porous materials, the influence of the material deformation on physico-chemical mechanisms generally turns out to be negligible. Accordingly, in the analysis of the kinetics of penetration fronts the deformation of the porous medium can be ignored so that the Lagrangian approach and the Eulerian approach merge. Penetration fronts constitute propagating surfaces of discontinuity and we conclude this chapter by giving a general approach to the propagation of surfaces of discontinuity, extending the analysis to the propagation of poroelastic waves.

7.1 Dissolution Fronts

The leaching of geomaterials that are partly formed by calcium-based constituents form a good example of the deterioration governed by the penetration of dissolution fronts. When the geomaterials are naturally wetted by deionized streaming water, the gradient of concentration causes the diffusion of the calcium ions from the inner solution towards the fresh water at the geomaterial surface. This leads to a decrease in the initial concentration

Poromechanics O. Coussy
 ISBN 0-470-84920-7

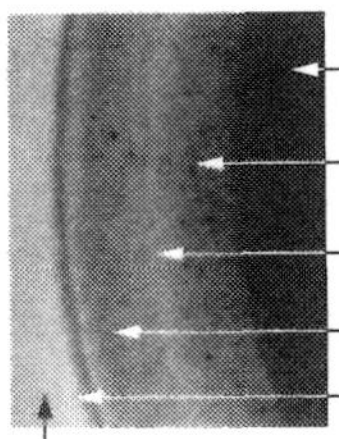

Figure 7.1: Successive dissolution fronts during the leaching process of a cement paste associated with the various calcium-based constituents. C-S-H stands for Calcium Silicate Hydrates with variable composition (from Adenot F. (1992), 'Durabilité du Béton: Caractérisation et Modélisation des Processus Physiques et Chimiques de Dégradation du Ciment', *PhD Thesis*, Université d'Orléans, France).

that breaks the initial equilibrium prevailing between the calcium ions belonging to the inner solution and the calcium ions bound to the constituents of the solid matrix. The restoration of the initial equilibrium is obtained through the dissolution of the calcium-based constituents, resulting in the progressive penetration of a dissolution front within the geomaterial, as shown in Fig. 7.1 in the particular case of a cement-based material. We analyse below the kinetics of the penetration front resulting from the coupling between the diffusion and the dissolution processes.

7.1.1 Mass Balance and Fick's Law for the Solute

The mass balance equation relative to the solid mineral bound to the matrix and the one relative to the mineral that diffuses in solute form through the porous space are expressed respectively in the form:

$$\frac{\partial\left[(1-n)\,\rho_b\right]}{\partial t}+r_{\rightarrow}=0 \tag{7.1a}$$

$$\frac{\partial\,(n\rho)}{\partial t}+\nabla\cdot\mathbf{w}-r_{\rightarrow}=0 \tag{7.1b}$$

where n is the material porosity whose variations induced by dissolution are assumed to be negligible in the following; ρ_b is the mass density per unit of matrix volume of the solid mineral bound to the matrix, while ρ is the mass density of solute per unit of porous volume; and $\mathbf{w}$ is the mass flux of solute diffusing through the porous medium. Finally $r_{\rightarrow}$ is the rate of the solid constituent dissolving between time t and $t+dt$. It acts as a sink in the skeleton continuity equation (7.1a), and as a source in the solute continuity equation (7.1b). The continuity equation for the mineral viewed as a whole, that is in both solid and solute forms, is obtained by adding (7.1a) and (7.1b), giving:

$$\frac{\partial\,(n\rho)}{\partial t}+\frac{\partial\left[(1-n)\,\rho_b\right]}{\partial t}+\nabla\cdot\mathbf{w}=0 \tag{7.2}$$

or, equivalently:

$$\frac{\partial m}{\partial t} + \nabla \cdot \mathbf{w} = 0 \tag{7.3}$$

where m denotes the mineral total mass density:

$$m = (1 - n)\,\rho_b + n\rho \tag{7.4}$$

The continuity equations have to be completed by the law governing the diffusion of the solute through the solution. Disregarding here any electrical activity or multi-species effects, the diffusion law reduces to Fick's first law applied to the solute:

$$\mathbf{w} = -D\,\nabla\rho \tag{7.5}$$

where D is the effective diffusion coefficient. In the general formulation of Fick's law, $\mathbf{w}$ must be replaced by the mass flux of solute with respect to the solution. Substitution of (7.5) into (7.2) gives:

$$\frac{\partial\,(n\rho)}{\partial t} + \frac{\partial\,[(1-n)\,\rho_b]}{\partial t} - \nabla \cdot (D\,\nabla\rho) = 0 \tag{7.6}$$

or, equivalently:

$$\frac{\partial m}{\partial t} - \nabla \cdot (D\,\nabla\rho) = 0 \tag{7.7}$$

7.1.2 Instantaneous Dissolution and the Formation of a Penetration Front

The solid–solute equilibrium requires the chemical potential $\mu_{sol}\,(\rho)$ of the mineral in solute form to be equal to the chemical potential μ_s of the solid mineral bound to the matrix. Letting ρ^{Eq} be the equilibrium value of the solute mass density accounting for such a solid–solute equilibrium condition (for further details see §3.6.3), the kinetics law ruling the dissolution process can be expressed in the form of Kuhn–Tucker conditions:

$$\rho_b \geq 0; \quad \rho - \rho^{Eq} \leq 0; \quad \rho_b(\rho^{Eq} - \rho) = 0 \tag{7.8}$$

When the mineral involved in the dissolution process belongs to several distinct constituents of the solid matrix, for instance in the dissolution pattern reproduced in Fig. 7.1, equilibrium equation (7.8) must be replaced by a relation having the form $\rho_b = f\,(\rho)$, in order to account for the equilibrium of the solute with the solid mineral of the constituents successively dissolving. Nevertheless the approach will remain mainly the same.[1]

Equation (7.8) allows us to express the solute mass density ρ as a non-decreasing function of the total mass density m with:

$$\rho = f\,(m) \quad \begin{cases} 0 \leq m \leq n\rho^{Eq} : \; f\,(m) = m \\ n\rho^{Eq} \leq m \; : \; f\,(m) = n\rho^{Eq} \end{cases} \tag{7.9}$$

[1] See Mainguy M., Coussy O. (2000), 'Propagation fronts during calcium leaching and chloride penetration', *Journal of Engineering Mechanics*, **126**, (3), 250–257.

Substitution of (7.9) into (7.7) leads to:

$$\frac{\partial m}{\partial t} - \nabla \cdot (D_m\,(m)\ \nabla m) = 0 \tag{7.10}$$

where the overall diffusion coefficient $D_m\,(m)$ is expressed in the form:

$$D_m(0 \le m < n\rho^{Eq}) = D; \qquad D_m(m > n\rho^{Eq}) = 0 \tag{7.11}$$

When m becomes greater than $n\rho^{Eq}$, the diffusion coefficient D_m vanishes so that the diffusion process stops. This results in the formation of a front penetrating at finite speed within the material and separating a zone where the mineral bound to the matrix in solid form is entirely dissolved from a zone where the matrix is still unaltered with constant mass density $\rho_b = \rho_b^0$. When passing across the front, m and $\mathbf{w}$ are discontinuous. With the notation of §1.5.2 the discontinuities undergone by the latter can be expressed respectively in the form:

$$[[m]] = (1-n)\,\rho_b^0; \quad [[\mathbf{w}]] = D\,\nabla\rho \tag{7.12}$$

The jump condition associated with continuity equation (7.3) which has to be satisfied across the front can be derived from (1.78), that is:

$$[[\mathbf{w} - m\mathbf{V}]] \cdot \mathbf{n} = 0 \tag{7.13}$$

where $\mathbf{n}$ is the unit normal to the front while $\mathbf{V}$ is the front speed. Substitution of (7.12) into (7.13) finally provides the condition governing the motion of the penetration front in the form:

$$D\,\nabla\rho \cdot \mathbf{n} = (1-n)\,\rho_b^0 \mathbf{V} \cdot \mathbf{n} \tag{7.14}$$

7.1.3 Stefan-like Problem

Consider a semi-infinite porous medium lying in the region $x \ge 0$, the solute and the solid constituent being in an equilibrium state. At time $t = 0$ the border $x = 0$ is suddenly subjected to renewed fresh water at zero solute concentration so that a dissolution front progressively penetrates the semi-infinite porous medium. This diffusion–dissolution problem is similar to the diffusion–melting problem of a semi-infinite solid space subjected on its border to a temperature higher than the melting temperature. In 1891 Stefan obtained the solution to this problem that can be adapted to the dissolution problem.

The initial and boundary conditions for the solute mass density ρ are expressed in the form:

$$\rho\,(x, t = 0) = \rho^{Eq}; \quad \rho\,(x = 0, t) = 0 \tag{7.15}$$

In the entirely degraded zone the dissolution rate $r_{\rightarrow}$ is zero so that (7.1b) and (7.5) combine to give the unidimensional diffusion equation:

$$\frac{\partial\,(n\rho)}{\partial t} - D\frac{\partial^2\rho}{\partial x^2} = 0 \tag{7.16}$$

At time t the entirely degraded zone extends from the border $x = 0$ to the location $x_d(t)$ of the dissolution front where:

$$\rho(x = x_d, t) = \rho^{Eq} \tag{7.17}$$

Noting:

$$\overline{\rho} = \frac{\rho}{\rho^{Eq}}; \qquad \xi = \frac{x}{2\sqrt{(D/n)\,t}} \tag{7.18}$$

the standard solution of (7.15)–(7.17) is written:

$$\overline{\rho} = \frac{\operatorname{erf}(\xi)}{\operatorname{erf}(\xi_d)} \tag{7.19}$$

where erf stands for the error function (see (5.39)) and where ξ_d is linked to the location $x_d(t)$ of the dissolution front according to:

$$x_d(t) = 2\xi_d\sqrt{(D/n)\,t} \tag{7.20}$$

Jump condition (7.14) here is specified in the form:

$$D\frac{\partial \rho}{\partial x}\,|_{x=x_d} = (1-n)\,\rho_b^0 \frac{dx_d}{dt} \tag{7.21}$$

Using (7.17)–(7.21) and definition (5.39) of the error function, the dimensionless location ξ_d of the dissolution front is finally determined by solving the implicit equation:

$$\varepsilon \exp\left(-\xi_d^2\right) = \sqrt{\pi}\,\xi_d \operatorname{erf}(\xi_d) \tag{7.22}$$

where ε is the mass density ratio defined by:

$$\varepsilon = \frac{n\rho^{Eq}}{(1-n)\,\rho_b^0} \tag{7.23}$$

In most cases the solid mass to dissolve is much greater than the solute mass present in the inner solution at equilibrium. This large mass density ratio approximation turns out to let $\varepsilon \ll 1$. Accordingly, (7.22) gives $\xi_d \ll 1$ so that the approximation $\operatorname{erf}(\xi_d \ll 1) \simeq 2\xi_d/\sqrt{\pi} + O\left(\xi_d^3\right)$ substituted into (7.22) finally provides:

$$\varepsilon \ll 1: \quad \xi_d \simeq \sqrt{\frac{\varepsilon}{2+\varepsilon}} \tag{7.24}$$

In Fig. 7.2 we plot $\overline{\rho}$ against ξ for various values of ε. For large values of mass density ratio ε, the dissolution front penetrates so slowly within the porous material that the diffusion regime in regions $x < 2\xi_d\sqrt{(D/n)\,t}$ eventually reduces to a succession of steady state diffusion processes represented by straight lines in Fig. 7.2.

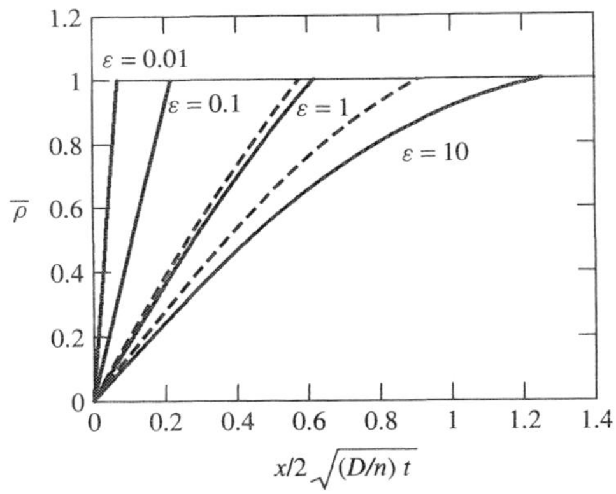

Figure 7.2: Normalized solute mass density $\overline{\rho} = \rho/\rho^{Eq}$ plotted against $\xi = x/2\sqrt{(D/n)t}$ for the Stefan-like dissolution problem and for various values of mass density ratio $\varepsilon = n\rho^{Eq}/\left(1-n\right)\rho_b$. The dashed lines represent the large mass density ratio approximation (7.24). For $\varepsilon = 0.01, 0.1$ they cannot be distinguished from the exact (numerical) solution.

7.2 Solute Penetration with Non-linear Binding

In the previous section we analysed the penetration of a dissolution front resulting from the extraction of the solute from the interstitial pore solution achieved by contact with an outer solution at lower concentration. Conversely, here we analyse the penetration of a solute within a porous material with the simultaneous binding of the diffusing particles onto the solid matrix. The resulting diffusion–binding process of aggressive particles can significantly impair the durability of materials and the structures they constitute. For instance, reinforced concrete structures exposed to seawater or deicing salt exhibit a high risk of steel corrosion and the structural performance relies strongly on the position of the penetration front of the chloride ions with respect to the structure reinforcement. We give below a general analysis of the penetration of a solute in a porous material, associated with a non-linear binding process.[2]

7.2.1 The Binding Process and the Formation of a Penetration Front

The penetration of a solute through a porous material, associated with the binding of the latter onto the solid matrix, is still governed by (7.6), where now ρ stands for the mass density of the free particles forming the solute, while ρ_b refers to the bound particles of the same constituent. The underlying complex chemical–physical sorption processes at the origin of binding are generally not well identified. However, in comparison with diffusion, the particle binding generally occurs so fast that the solute can be considered as remaining constantly in equilibrium with the particles that fix onto the solid matrix. As a result, the local kinetics law can be fortunately and appropriately replaced by an equilibrium condition linking the mass density of the free particles to that of the bound ones:

[2]The main results are taken from Coussy O., Eymard R. (2003), 'Non-linear binding and the diffusion-migration test', *Transport in Porous Media*, **1770**, 1–24.

$$\rho_b = f(\rho) \tag{7.25}$$

Several laws are commonly used:[3]

$$\begin{array}{lll} \text{Linear } (lin) & f(\rho) = K\rho & K > 0 \\ \text{Langmuir } (Lan) & f(\rho) = C\dfrac{\rho}{1+\alpha\rho} & C > 0,\ \alpha > 0 \\ \text{Freundlich } (Fre) & f(\rho) = \mu\rho^{\gamma} & \mu > 0,\ 0 < \gamma < 1 \end{array} \tag{7.26}$$

Use of (7.25) in (7.4)–(7.7) leads again to the non-linear diffusion equation (7.10) provided that the overall diffusion coefficient $D_m(m)$ is now expressed in the form:

$$D_m(m) = \frac{D}{n + (1-n)\frac{df}{d\rho}} \tag{7.27}$$

The solute diffusion is slowed down by the binding of the particles onto the solid matrix, resulting in the division of the diffusion coefficient D by the factor $1 + ((1-n)/n))(df/d\rho)$. Since $(d^2 f/d\rho^2) \leq 0$, the lower the solute concentration, the slower the diffusion process. Provided that the overall diffusion coefficient D_m never vanishes, as in the case of linear interaction $(7.26)_{lin}$ or in the case of a non-linear interaction governed by Langmuir's isotherm $(7.26)_{Lan}$, the diffusion–binding process eventually reduces to a standard diffusion process. By contrast, for sufficiently strong non-linear interactions, such as those captured by Freundlich's interaction isotherm $(7.26)_{Fre}$, we get:

$$\rho \to 0: \quad \frac{df}{d\rho} \to \infty; \quad D_m \to 0 \tag{7.28}$$

Therefore, in the limit of a zero concentration the interaction is so intense that the ratio $df/d\rho = (\partial\rho_b/\partial t)/(\partial\rho/\partial t)$ of the rate of particles binding to the solid matrix to the rate of particles diffusing through the porous space becomes infinite. The overall diffusion coefficient D_m vanishes and the diffusion stops. Accordingly, for a zero initial concentration of solute in the pore solution, a penetration front propagates at finite speed in the material bulk.

To analyse the propagation of the penetration front let us consider a semi-infinite half-space $x \geq 0$ at inner zero initial concentration, which is suddenly subjected to an outer solution at concentration ρ_0 maintained constant on the border $x = 0$. We write:

$$\rho(x, t = 0) = 0; \quad \rho(x = 0, t) = \rho_0 \tag{7.29}$$

For the 1D problem at hand, when adopting Freundlich's isotherm $(7.26)_{Fre}$, diffusion equation (7.6) can be specialized in the form:

$$\frac{\partial}{\partial t}\left[\rho + \frac{1-n}{n}\mu\rho^{\gamma}\right] - (D/n)\frac{\partial^2\rho}{\partial x^2} = 0 \tag{7.30}$$

[3]Other laws are the quadratic law, $f(\rho) = k_1\rho - k_2\rho^2$, generalized Langmuir's law, $f(\rho) = C(\rho^m/(1 + \alpha\rho^m))$, and the exponential law, $f(\rho) = C\exp(-\alpha/\rho)$.

By letting:

$$\overline{\rho} = \frac{\rho}{\rho_0}; \quad \xi = \frac{x}{\sqrt{(D/n)\,t}}; \tag{7.31}$$

the previous problem reduces to finding the solution $\overline{\rho}\,(\xi)$ of the ordinary differential equation:

$$\xi\,\frac{d}{d\xi}\left[\overline{\rho} + r\overline{\rho}^{\gamma}\right] + 2\frac{d^2\overline{\rho}}{d\xi^2} = 0 \tag{7.32}$$

with boundary conditions:

$$\overline{\rho}\,(\xi = 0) = 1; \quad \overline{\rho}\,(\xi \to \infty) = 0 \tag{7.33}$$

In (7.32) r is the parameter defined by:

$$r = \frac{1-n}{n}\frac{\mu}{\rho_0^{1-\gamma}} \tag{7.34}$$

As time flows, the ions diffuse within the semi-infinite half-space with a penetration front located at:

$$x_0\,(t) = \xi_0\,(\gamma, r)\sqrt{(D/n)\,t} \tag{7.35}$$

where $\xi_0\,(\gamma, r)$ is the solution of $\overline{\rho}\,(\xi_0) = 0$ and depends on r and γ. As ξ tends towards ξ_0 the asymptotic profile for $\overline{\rho}$ can be looked for in the form:

$$\xi \to \xi_0; \quad \overline{\rho} \simeq A\,(\xi_0 - \xi)^{\beta} \tag{7.36}$$

Substitution of (7.36) into (7.32) gives:

$$-\xi_0 r\gamma A^{\gamma-1}\,(\xi_0 - \xi)^{1-\beta(1-\gamma)} + 2\,(\beta - 1) \simeq 0 \tag{7.37}$$

Equation (7.37) allows the identification of β and A, resulting in:

$$\overline{\rho}\,(\xi \to \xi_0) \simeq \left[\frac{1}{2}r\xi_0\,(1-\gamma)\,(\xi_0 - \xi)\right]^{\frac{1}{1-\gamma}} \tag{7.38}$$

A crude assessment of $\xi_0\,(\gamma, r)$ can be achieved by imposing boundary condition $\overline{\rho} = 1$ for $\xi = 0$, yielding:

$$\xi_0 \simeq \sqrt{\frac{2}{r\,(1-\gamma)}} \tag{7.39}$$

We find that $\xi_0 \to \infty$ as $r \to 0$ (no binding) or $\gamma \to 1$ (linear binding). The closer the coefficient γ is to $\frac{1}{2}$, the closer the exact solution is to the approximate solution defined by (7.38) and (7.39) (see Fig. 7.3).

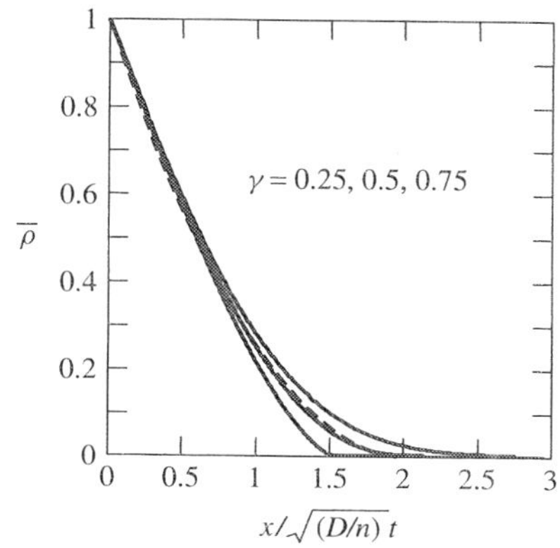

Figure 7.3: Normalized penetration profiles of the free solute mass density $\overline{\rho} = \rho/\rho_0$ numerically obtained and plotted against $\xi = x/\sqrt{(D/n)t}$, for $r = 1$ and three values of coefficient γ defining Freundlich's binding isotherm $(7.26)_{Fre}$. The dashed line is the approximate solution defined by (7.38) and (7.39) when adopting $\gamma = \frac{1}{2}$.

7.2.2 The Time Lag and the Diffusion Test

Material properties γ and μ are key durability properties with respect to the intrusion of an aggressive solute. Using (7.39), it is tempting to assess their values from the experimental determination of ξ_0, that is from the location of the penetration front. Unfortunately, asymptotic solution (7.38) reveals that $\partial\rho/dx \simeq 0$ while $\partial\rho_b/\partial x$ behaves as $(x_0 - x)^{(2\gamma-1)/(1-\gamma)}$ as $x \to x_0$. Hence, an accurate assessment of the location of the penetration front can be carried out only from the profiles of bound particles, especially for $0 < \gamma < \frac{1}{2}$. This would require the crushing of the sample.

An alternative method to determine the binding isotherm is afforded by the diffusion test. In the diffusion test sketched out in Fig.7.4 a sample of length L, with zero initial concentration, is placed between two compartments: an upstream compartment where various concentrations of diffusing ions can be imposed; and a downstream compartment where the cumulative mass Q of arriving ions per unit of sample section is measured as a function of time t by means of regular samplings. At large times the corresponding function $Q(t)$ becomes a straight line whose abscissa at the origin defines the so-called time lag τ (see Fig. 7.4). The time lag τ, as a function of upstream concentration ρ_0, quantifies the slowdown of the diffusion process due to particle binding. Therefore a key to determine the binding isotherm $f(\rho)$ is to derive an expression linking τ, $f(\rho)$ and ρ_0.

The downstream compartment volume is always large compared with the volume sample so that the concentration imposed downstream can be assumed to be zero. We write:

$$\rho(x = 0, t) = \rho_0; \quad \rho(x = L, t) = 0 \tag{7.40}$$

According to Fick's first law (7.5), $Q(t)$ is:

$$Q(t) = -D \int_0^t \frac{\partial\rho}{\partial x}(x = L, t)\, dt \tag{7.41}$$

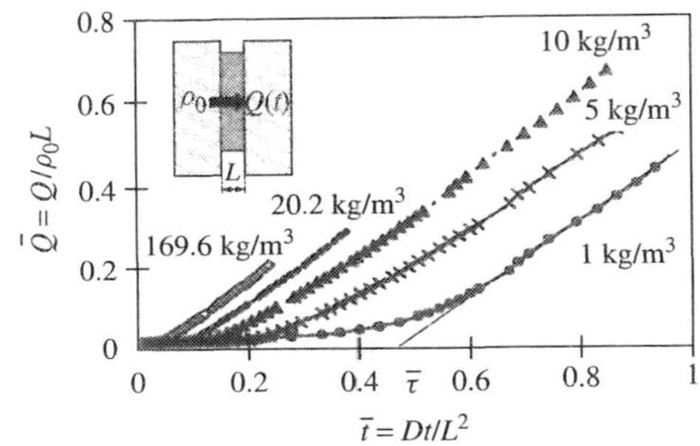

Figure 7.4: In the diffusion test the cumulative mass Q of particles arriving in the downstream compartment is measured as a function of time t, for various upstream concentrations ρ_0. At large times the corresponding function $Q(t)$ becomes a straight line whose abscissa τ at the origin is the time lag which quantifies the slowdown of the diffusion process due to particle binding. The experimental results, reported here in dimensionless form, are relative to a cement mortar, while the particles are chloride ions. The porosity of the samples is $n \simeq 0.14$. The experimental data are from Bigas J.-P. (1994), 'Diffusion des ions chlorures dans les mortiers', *PhD Thesis*, Génie civil—INSA Toulouse.

We multiply (7.30) by x and integrate by parts the resulting equation between $x = 0$ and $x = L$, while using boundary conditions (7.40) We obtain:

$$-DL\frac{\partial \rho}{\partial x}(x = L, t) - D\rho_0 = -\frac{\partial}{\partial t}\int_0^L x\left[n\,\rho + (1-n)\,f(\rho)\right]dx \tag{7.42}$$

The time integration of (7.42) with a zero initial condition for ρ finally yields:

$$Q(t) = \frac{\rho_0 D}{L}t - \frac{1}{L}\int_0^L x\left[n\,\rho + (1-n)\,f(\rho)\right]dx \tag{7.43}$$

At large times the time derivatives can be neglected in (7.30). The space integration of the resulting equation, together with boundary conditions (7.40) provides the asymptotic concentration profile $\rho_\infty(x)$ in the form:

$$\rho_\infty(x) = \rho_0\left(1 - \frac{x}{L}\right) \tag{7.44}$$

Substitution of (7.44) into (7.43) furnishes the cumulative mass of particles arriving in the asymptotic regime in the downstream compartment:

$$Q(t \to \infty) = \frac{\rho_0 D}{L}(t - \tau) \tag{7.45}$$

or, in dimensionless form:

$$\overline{Q}(\bar{t} \to \infty) \simeq \bar{t} - \bar{\tau} \tag{7.46}$$

where we note:

$$\overline{Q} = \frac{Q}{\rho_0 L}; \quad \bar{t} = \frac{t}{t_0}; \quad t_0 = \frac{L^2}{D} \tag{7.47}$$

whereas $\overline{\tau} = \tau / t_0$ is the dimensionless time lag defined by:

$$\overline{\tau} = \frac{n}{6} + \frac{1-n}{\rho_0} \int_0^1 \overline{x}\, f\left(\rho_0 \left(1 - \overline{x}\right)\right) d\overline{x} \tag{7.48}$$

Equation (7.48) constitutes the new key to determine the non-linear binding isotherm from the time lag. Indeed, by progressively increasing upstream concentration ρ_0 from zero, diffusion coefficient D can firstly be measured from the successive slopes $\rho_0 D/L$ of $Q\,(t \rightarrow \infty)$. Once D is determined, the determination of the successive dimensionless time lags $\overline{\tau}$ provides an assessment of the unknown interaction isotherm $f\,(\rho)$ with prerequisite knowledge only of the porosity.

Adopting Freundlich's isotherm $(7.26)_{Fre}$ we derive:

$$\ln\left(\overline{\tau} - \frac{n}{6}\right) = \ln \frac{(1-n)\,\mu}{(\gamma+1)\,(\gamma+2)} - (1-\gamma)\ln \rho_0 \tag{7.49}$$

Accordingly, the quantity $\ln\left(\overline{\tau} - \frac{n}{6}\right)$ plotted against $\ln \rho_0$ must be a straight line. The measurement of both the slope of the latter and its abscissa at the origin provides the identification of exponent γ and constant μ (see Fig. 7.5). Another time of interest is the breakthrough time, τ_b, matching the first arrival of particles in the downstream compartment. It differs from zero as soon as a penetration front exists, that is when condition (7.28) is fulfilled. An assessment of time τ_b is provided by (7.35), when letting the location x_0 of the penetration front be equal to L and adopting expression (7.39) for ξ_0. The procedure gives:

$$\overline{\tau}_b = \frac{\tau_b}{t_0} \simeq \frac{1}{2}(1-n)(1-\gamma)\frac{\mu}{\rho_0^{1-\gamma}} \tag{7.50}$$

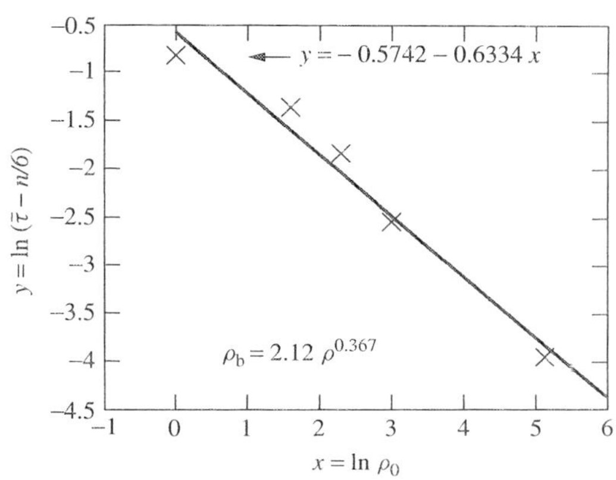

Figure 7.5: Identification of Freundlich's binding isotherm $\rho_b = \mu \rho^{\gamma}$ from the successive measurements of the time lag and its theoretical expression (7.49). The experimental data represented by crosses match the results reported in Fig. 7.4 with $n \simeq 0.14$.

The relevance of the values of γ and μ previously obtained by exploiting (7.49) can then be independently checked. We can compare the predicted values of the dimensionless breakthrough time $\overline{\tau}_b$ they deliver by using (7.50) with the experimental order of magnitude of $\overline{\tau}_b$ that the diffusion test can provide (see Fig. 7.4).

7.3 Ionic Migration with Non-linear Binding

7.3.1 Ionic Migration in the Advection Approximation

In most cases the solute is formed of ions. The solute transport is sensitive to the local electrical field $\mathbf{E}$ and Fick's law (7.5) must be modified according to the Nernst–Planck–Einstein relation:

$$\mathbf{w} = -D\,\nabla\rho - D\rho\frac{F}{RT}z\mathbf{E} \tag{7.51}$$

where z stands for the valency of the ions forming the solute while F is Faraday's constant, R is the ideal gas constant[4] and T is the absolute temperature. To analyse the effects of an external electrical field on the propagation of the penetration front let us again consider a semi-infinite half-space $x \geq 0$ subjected to initial and boundary conditions (7.29). The pre-established external electrical field before the migration process starts is assumed to be sufficiently intense in order that the migrating ions do not significantly affect it. Letting $\mathbf{E} = E_0\mathbf{e}_x$, where the constant E_0 is the algebraic strength of the external electrical field in the $\mathbf{e}_x$ direction, (7.51) can be specified in the form:

$$\mathbf{w} = w_x\mathbf{e}_x:\ w_x = -D\,\frac{\partial\rho}{\partial x} - \operatorname{sgn}\,(zE_0)\,D\frac{\rho}{\ell} \tag{7.52}$$

where sgn $(\cdot)$ stands for the sign of the quantity in parentheses. In (7.52) ℓ is the length defined by:

$$\ell = \frac{RT}{F\,|zE_0|} \tag{7.53}$$

and scales the range of the diffusion strength compared with that of the advection caused by the external electrical field. In the following we assume $\operatorname{sgn}\,(zE_0) = -1$ in order that the externally applied electrical field contributes positively to the ionic transport in the $\mathbf{e}_x$ direction. Substitution of (7.52) into (7.2) and use of equilibrium condition (7.25) give the advection–diffusion equation governing the migration of the ionic solute:

$$\frac{\partial}{\partial t}\left[n\,\rho + (1-n)\,f\,(\rho)\right] - D\frac{\partial}{\partial x}\left(\frac{\partial\rho}{\partial x} - \frac{\rho}{\ell}\right) = 0 \tag{7.54}$$

Let us scale x and t according to:

$$\overline{x} = \frac{x}{X};\ \ \overline{t} = \frac{t}{\Theta} \tag{7.55}$$

[4] $F = 9.6486\ 10^4$ J/ (mol V), $R = 8.31$ J/ (mol K).

so that the advection–diffusion equation (7.54) can be rewritten in the form:

$$\frac{\partial}{\partial \bar{t}}\left[n\,\rho+(1-n)\,f\,(\rho)\right]-\left(\frac{D\Theta}{X^2}\right)\frac{\partial^2\rho}{\partial \bar{x}^2}+\left(\frac{D\Theta}{X\ell}\right)\frac{\partial\rho}{\partial \bar{x}}=0 \tag{7.56}$$

Consider now times t and locations x of respective order of magnitude $\Theta = O(t)$ and $X = O(x)$ such as:

$$O\left(\frac{Dt}{x\ell}\right)=\frac{D\Theta}{X\ell}=O\,(1) \tag{7.57}$$

Under assumption (7.57), the advection approximation consists in exploring the region of space and time such as:[5]

$$\frac{D\Theta}{X^2}\ll\left(\frac{D\Theta}{X\ell}\right);\quad \frac{D\Theta}{X^2}\ll 1\Rightarrow\ x\gg\ell;\quad t\gg\frac{\ell^2}{D} \tag{7.58}$$

In the advection approximation, when returning to the original variables, the advection–diffusion equation (7.56) reduces to the advection equation:

$$\frac{\partial}{\partial t}\left[n\,\rho+(1-n)\,f\,(\rho)\right]+\frac{D}{\ell}\frac{\partial\rho}{\partial x}=0 \tag{7.59}$$

With boundary and initial conditions (7.29) a solution of the advection equation (7.59) is:

$$\rho\,(x,t)=\rho_0\,[1-H\,(x-ct)] \tag{7.60}$$

where H is the Heaviside function, with $H(x-ct>0)=1$ and $H(x-ct<0)=0$, and where c is defined by:

$$c=\frac{c_0}{\lambda}\quad\text{with}\quad\lambda=1+\frac{1-n}{n}\frac{f\,(\rho_0)}{\rho_0}\quad\text{and}\quad c_0=\frac{D/n}{\ell} \tag{7.61}$$

Solution (7.60) is a step function of intensity ρ_0 which propagates at speed c. Hence, at $x = ct$ the total ionic mass density, that is $m = (1-n)\,\rho_b + n\rho$, and the mass flux $\mathbf{w}$ undergo the respective jumps:

$$[[m]]=-\left[n\rho_0+(1-n)\,f\,(\rho_0)\right];\quad [[\mathbf{w}]]=-D\frac{\rho_0}{\ell} \tag{7.62}$$

and expression (7.61) of speed c satisfies the Rankine–Hugoniot jump condition (7.13). For the usual binding isotherms (see (7.26)), the explicit expressions of $\lambda = c_0/c$ are:

$$\begin{aligned}
&lin && \lambda=c_0/c=1+\frac{1-n}{n}K\\
&Lan && \lambda=c_0/c=1+\frac{1-n}{n}\frac{C}{1+\alpha\rho_0}\\
&Fre && \lambda=c_0/c=1+\frac{1-n}{n}\frac{\mu}{\rho_0^{1-\gamma}}
\end{aligned} \tag{7.63}$$

[5]The diffusion approximation would consist in exploring the region of spacetime $x \ll \ell$ and $t \ll \ell^2/D$, and neglecting the advection term $-(D/\ell)(\partial\rho/\partial x)$ in (7.54) so that the penetration profiles would match those of Fig. 7.3.

7.3.2 The Travelling Wave Solution

The step function solution derived in the advection approximation can only be viewed as the limit solution as $\ell \to 0$. Indeed, when passing across the head of a migration front accounted for by a step function, the mass density gradient $\nabla\rho$ becomes infinite and diffusion effects will inexorably spread out the migration front profile. In order to capture the structure of the transition layer we note:

$$\chi = \frac{x - ct}{\ell} \tag{7.64}$$

where the speed c does not need to be specified at this step of analysis.

Considering first a linear binding process, we substitute $(7.26)_{lin}$ and (7.64) into (7.54) and we get:

$$\ell\left(1 + \frac{1-n}{n}K\right)\frac{\partial\rho}{\partial t} + \left[c_0 - \left(1 + \frac{1-n}{n}K\right)c\right]\frac{\partial\rho}{\partial\chi} - c_0\frac{\partial^2\rho}{\partial\chi^2} = 0 \tag{7.65}$$

Adopting for c the expression $(7.63)_{lin}$ we found in the advection approximation, the second term in (7.65) vanishes. Adding the boundary conditions:

$$\rho\left(\chi \to -\infty\right) = \rho_0; \quad \rho\left(\chi \to +\infty\right) = 0 \tag{7.66}$$

the solution of (7.65) is shown to be:

$$\rho\left(x, t\right) = \frac{\rho_0}{2}\left(1 - \operatorname{erf}\frac{x - ct}{4\sqrt{\{D/\left[n + (1-n)K\right]\}t}}\right) \tag{7.67}$$

Accordingly, the current 'thickness' $e\left(t\right)$ of the transition layer can be assessed in the form:

$$e\left(t\right) = 4\sqrt{\{D/\left[n + (1-n)K\right]\}t} \tag{7.68}$$

In the case of linear binding, as time flows, according to (7.67) the diffusion effects inexorably spread out the transition layer so that the migration front eventually cannot propagate without deforming.

For more intense non-linear binding effects, in particular those captured by Freundlich's isotherm, we showed in §7.2.1 that, in the absence of any external electrical field, the interaction of the ions with the solid matrix could result in a front propagating at a finite speed. Hence, in the presence of an external electrical field and of non-linear binding effects, the propagation of a steady migration front can also be expected in spite of the diffusion process. In order to explore such an issue we substitute (7.64) into (7.54) while using definitions (7.61):

$$\ell\frac{\partial}{\partial t}\left[\rho + \frac{1-n}{n}f\left(\rho\right)\right] + \left(c_0 - c\right)\frac{\partial\rho}{\partial\chi} - c\frac{1-n}{n}\frac{\partial f\left(\rho\right)}{\partial\chi} - c_0\frac{\partial^2\rho}{\partial\chi^2} = 0 \tag{7.69}$$

A time-independent advection–diffusion profile has to match (7.69) when setting the time derivative to zero. Integration of the resulting equation with respect to χ gives:

$$\left(c_0 - c\right)\rho - c\frac{1-n}{n}f\left(\rho\right) - c_0\frac{d\rho}{d\chi} = C \tag{7.70}$$

where C is an integration constant so that the continuity of ρ and $f(\rho)$ requires continuity of $d\rho/d\chi$. According to the second boundary condition (7.66), ρ and $d\rho/d\chi$ have to vanish as χ goes to infinity so that the integration constant C turns out to be zero. Using in addition the first boundary condition (7.66) and noting that $(d\rho/d\chi)(\chi \to -\infty) = 0$, we retrieve for c the expression (7.61) we found in the advection approximation.[6] Accordingly, substitution of (7.61) into (7.70), where we now let $C = 0$, leads to:

$$\frac{\rho}{\rho_0} - \frac{f(\rho)}{f(\rho_0)} - \frac{\lambda}{\lambda - 1}\frac{1}{\rho_0}\frac{d\rho}{d\chi} = 0 \tag{7.71}$$

Returning to the initial space and time variables, the solution of (7.71) is then found to be:

$$-\int_{\rho_0/2}^{\rho} \frac{d\tilde{\rho}/\rho_0}{f(\tilde{\rho})/f(\rho_0) - \tilde{\rho}/\rho_0} = \frac{\lambda - 1}{\lambda}\frac{x - ct}{\ell} \tag{7.72}$$

where the origin $\chi = 0$ has been chosen where the concentration is half the upstream concentration. A steady migration front can actually form provided that the previous equation can be inverted in order to give ρ as a function of $x - ct$, requiring $d\chi/d\rho$ to be non zero everywhere and consequently ρ to be a decreasing function of χ. Inspecting (7.71), we finally require the binding function $f(\rho)$ to be strictly concave as the usual binding functions are and therefore to satisfy:

$$\forall \rho_0\colon\ 0 < \rho < \rho_0\colon\ \frac{f(\rho)}{\rho} > \frac{f(\rho_0)}{\rho_0} \tag{7.73}$$

According to (7.72), non-linear binding effects then allow the formation of a so-called 'travelling wave' whose 'thickness' e is now independent of time and is given by:

$$e = \frac{\lambda}{1 - \lambda}\ell \tag{7.74}$$

Condition $\rho(\chi \to -\infty) = \rho_0$ in (7.66) eventually means that $\rho(x - ct \ll -\ell) = \rho_0$ so that the travelling wave solution will be physically relevant as soon as the characteristic length ℓ is small compared with the extent of the porous medium.[7]

Owing to the strength of the binding effects two situations can eventually be met, depending on the behaviour of the binding function as the concentration goes to zero. Let us first consider:

$$\frac{f(\rho)}{\rho}\Big|_{\rho \to 0} < +\infty \tag{7.75}$$

The integral in (7.72) then diverges as ρ goes to zero. As a result the binding effects are too weak to prevent the diffusion process from spreading out the travelling wave over the whole x axis so that no wavefront where $\rho = 0$ can actually be observed. For instance,

[6] Indeed, step function (7.60), where c is given by (7.61), is the standard 'entropy weak solution' towards which the solution of (7.54) converges when the diffusion term vanishes.

[7] At room temperature $T = 293$ K, the order of magnitude of ℓ is $1/(40\,|zE_0|)$, where the strength E_0 of the electrical field is expressed in volts per length unit.

the case of Langmuir's binding isotherm we substitute $(7.26)_{Lan}$ into (7.72) to obtain finally:

$$2\left(1-\frac{\rho}{\rho_0}\right)^{\frac{1}{\alpha\rho_0}+1}\left(\frac{\rho}{\rho_0}\right)^{-\frac{1}{\alpha\rho_0}} = \exp\left(\frac{x-ct}{e}\right) \tag{7.76}$$

Instead of (7.75) we now explore stronger binding effects by considering:

$$\frac{f(\rho)}{\rho}\,|_{\rho\to 0} = +\infty \tag{7.77}$$

The integral in (7.72) then converges as ρ goes to zero. The binding effects are strong enough to counterbalance the diffusion process so that an actual wavefront where $\rho = 0$ can be observed at a finite distance along the x-axis. For instance, in the case of Freundlich's binding isotherm we substitute $(7.26)_{Fre}$ into (7.72), resulting in:

$$-\int_{1/2}^{\rho/\rho_0} \frac{d\overline{\rho}}{\overline{\rho}^{\gamma}-\overline{\rho}} = \frac{x-ct}{e} \tag{7.78}$$

For the usual value $\gamma = \frac{1}{2}$ the previous solution can be written explicitly as:

$$\frac{\rho}{\rho_0} = \left[1-\exp\left(\frac{x-ct}{2e} - \ln\left(\sqrt{2}+2\right)\right)\right]^2 \tag{7.79}$$

In Fig. 7.6 we compare the travelling wave profile (7.79) with (7.76) when letting $\alpha\rho_0 = 1$. The travelling waves significantly depart from each other only close to the actual front $x - ct = 2e\ln(\sqrt{2}+2)$ of the travelling wave relative to Freundlich's isotherm.

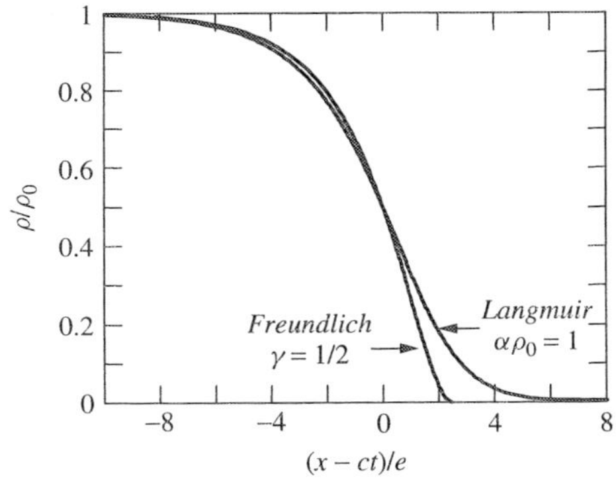

Figure 7.6: Comparison of the travelling wave profiles (7.76) and (7.79), related respectively to Langmuir's binding isotherm $(7.26)_{Lan}$ when letting $\alpha\rho_0 = 1$ and Freundlich's binding isotherm $(7.26)_{Fre}$ when letting $\gamma = \frac{1}{2}$. The profiles are adjusted by imposing the value $\rho = \rho_0/2$ at $x = ct$ for both travelling waves and by adopting for both binding isotherms the same value for λ defined in (7.61).

7.3.3 The Time Lag and the Migration Test

The duration of the diffusion test is scaled at the lower end by the time lag $t_0 = nL^2/6D$ that would be recorded in the absence of binding (see (7.47) and (7.48)). The order of magnitude of the diffusion coefficient D in porous materials such as the cement mortar corresponding to the experimental results reported in Fig. 7.4 is about 10^{-12} m^2/s. A standard thickness sample is $L \simeq 1$cm. Adopting in addition $n \simeq 0.15$, the order of magnitude of t_0 is found to be about 30 days. It becomes prohibitive for routine tests aiming at the determination of D from the asymptotic slope of Q ($t \to \infty$). Furthermore, when the solute is formed of ions, the electrical interactions existing between all the ionic species present in the material cannot be neglected in the absence of an applied external electrical field. Indeed, owing to these multi-species electrical interactions, the diffusion coefficient D, as measured from the asymptotic slope of Q ($t \to \infty$), is generally found to depend apparently on the upstream concentration ρ_0. In order to overcome these difficulties, the migration experiment consists in applying an external electric potential of high intensity to the sample, in order to decrease drastically both the test duration and the multi-species effects. In addition the migration experiment can also provide a very convenient assessment of the binding isotherm.

More precisely the migration experiment consists in applying a difference U_0 of the electrical potential between the two sample ends, between the cathode (−) and the anode (+) (see Fig. 7.7) and in constantly renewing the solution of the upstream and the downstream compartments until the steady state electrical regime is established. In the steady state regime the local electroneutrality is restored and the electrical potential is a linear function of x:

$$U = E_0 x; \quad E_0 = \frac{U_0}{L} \tag{7.80}$$

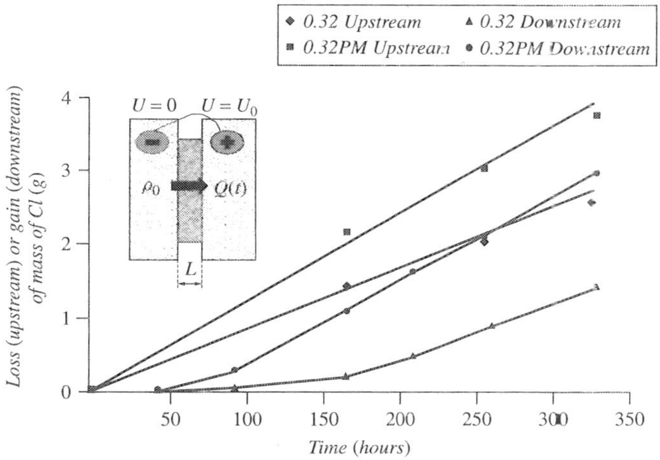

Figure 7.7: Cumulative mass of (chloride) ions leaving the upstream compartment (*Upstream*) and arriving in the downstream compartment (*Downstream*) in the migration test for two concrete samples of different mixes (32 and 32*PM*) (from Truc O., Ollivier J.-P., Carcassès L. M. (2000), 'A new way for determining the chloride diffusion coefficient in concrete from steady state migration test', *Cement and Concrete Research*, **30**, 217–226 with permission from Elsevier).

where E_0 is the (constant) strength of the electrical field. Substituting (7.80) into (7.53), we note:

$$\eta = \frac{RT}{F\,|zU_0|} = \frac{\ell}{L} \tag{7.81}$$

In the migration test the sample length L scales the x coordinate so that, letting $X = L$ in (7.56) and using (7.81), the time is found to be appropriately scaled according to the characteristic time Θ defined by:

$$\Theta = \eta t_0 = \eta \frac{L^2}{D} \tag{7.82}$$

so that the duration of the migration experiment is actually reduced by the factor η with respect to the duration of the diffusion experiment.

Using the unidimensional form of the Nernst–Planck–Einstein equation (7.52), the cumulative mass $Q\,(t)$ of arriving ions in the downstream compartment can now be written:

$$Q\,(t) = -D \int_0^t \left(\frac{\partial \rho}{\partial x} - \frac{\rho}{\ell} \right) (x = L, t)\, dt \tag{7.83}$$

We multiply (7.54) by the yet unknown test function $\psi\,(x)$ and integrate twice by parts between $x = 0$ and $x = L$. We obtain:

$$\left[-\psi D \left(\frac{\partial \rho}{\partial x} - \frac{\rho}{\ell} \right) \right]_0^L = - \left[\psi' D \rho \right]_0^L + \int_0^L D\rho \left[\psi'' + \frac{\psi'}{\ell} \right] dx$$
$$- \frac{\partial}{\partial t} \int_0^L \psi \left[n\, \rho + (1 - n)\, f\,(\rho) \right] dx \tag{7.84}$$

where ψ' and ψ'' stand respectively for the first and the second derivative of ψ. For the test function ψ we finally choose the one satisfying:

$$\psi'' + \frac{\psi'}{\ell} = 0; \quad \psi\,(0) = 0; \quad \psi\,(L) = 1 \tag{7.85}$$

resulting in:

$$\psi\,(x) = \frac{1 - \exp\left(-\frac{x}{\ell}\right)}{1 - \exp\left(-\frac{1}{\ell}\right)} \tag{7.86}$$

Substituting (7.85) into (7.84) and using boundary conditions (7.40), the subsequent integration with respect to time with zero initial condition provides an expression for $Q\,(t)$ in the form:

$$Q\,(t) = \frac{\rho_0 D}{\ell \left(1 - \exp\left(-\frac{1}{\eta}\right)\right)} t - \int_0^L \psi \left[n\, \rho + (1 - n)\, f\,(\rho) \right] dx \tag{7.87}$$

The steady state profile that establishes as time goes to infinity matches the equation obtained by setting the time derivative to zero in (7.54), that is:

$$t \to \infty: \quad \frac{d}{dx}\left(\frac{d\rho}{dx} - \frac{\rho}{\ell}\right) = 0 \tag{7.88}$$

whose solution $\rho_\infty(x)$, satisfying boundary condition (7.40), is:

$$\rho_\infty(x) = \rho_0 \frac{\exp\left(\frac{1}{\eta}\right) - \exp\left(\frac{x}{\ell}\right)}{\exp\left(\frac{1}{\eta}\right) - 1} \tag{7.89}$$

As time goes to infinity, the mass of ions arriving in the downstream compartment is finally expressed in the form:

$$Q(t \to \infty) = \frac{\rho_0 D}{\ell\left(1 - \exp\left(-\frac{1}{\eta}\right)\right)} t - \int_0^L \psi(x)\left[n\,\rho_\infty(x) + (1-n)\, f(\rho_\infty(x))\right] dx \tag{7.90}$$

The previous equation allows us to obtain an explicit expression for the time lag related to any binding isotherm $f(\rho)$. In order to limit the duration of the migration test, parameter η is chosen as the smallest possible, that is such that $\eta \ll 1$, and functions $\psi(x)$ and $\rho_\infty(x)$ can be approximated according to:

$$\eta \ll 1: \quad \psi(x) \simeq 1 - \exp\left(-\frac{x}{\ell}\right); \quad \rho_\infty(x) \simeq \rho_0 \tag{7.91}$$

Substitution of (7.91) into (7.90) gives $Q(t \to \infty)$ in the form:

$$\eta \ll 1: \quad Q(t \to \infty) \simeq \frac{cD}{\ell}\left[t - \tau_m(1 - r_i)\right] \tag{7.92}$$

where τ_m defined by:

$$\tau_m = \frac{L}{c} = n\lambda\eta t_0 \tag{7.93}$$

is the travel time through the sample of the transient solution (7.60)–(7.61) in the advection approximation. Indeed, according to (7.58) the advection approximation is relevant provided that $x \gg \ell$ and $t \gg \eta^2 t_0$. Accordingly, in the range $\eta = \ell/L \ll 1$ the step function profile (7.60) is established at early times within the sample long before it reaches the sample end $x = L$. In addition, since $\tau_m \ll t_0$, the diffusion effects cannot significantly develop at the head of the penetration front, even though no travelling wave can actually form, such as in the case of a linear binding process. When the step function profile reaches the sample end, owing to diffusion effects it progressively transforms into the asymptotic profile (7.89). Using the unidimensional form of the Nernst–Planck–Einstein equation (7.52) together with (7.89), the constant flux of ion mass associated with the asymptotic profile is:

$$w_x^\infty = -D\frac{\partial \rho_\infty(x)}{\partial x}\rho_\infty(x)\frac{D}{\ell} = \frac{\rho_0 D}{\ell}\frac{1}{1-\exp\left(-\frac{1}{\eta}\right)} \simeq \frac{\rho_0 D}{\ell} \tag{7.94}$$

In the steady state regime the migration of ions is eventually achieved by means of the electrical contribution within almost the whole sample corresponding to the second term in expression (7.94) for w_x^∞. It is only very close to the end that the electrical contribution vanishes and that the transport is finally achieved through pure diffusion corresponding to the first term in expression (7.94) for w_x^∞. According to (7.92), the time lag in the migration test is expressed in the form $\tau_m(1-\eta)$ so that the diffusion effects involved in the whole process are accounted for through the factor $(1-\eta)$ affecting the travelling time τ_m.

According to (7.92), the diffusion coefficient D can be determined from the slope $(\rho_0 D/\ell)S$ relative to the mass of ions $SQ\,(t \to \infty)$ arriving in the downstream compartment (S stands for the sample section). From the experimental data reported in Fig. 7.7 and performed at $U_0 = 12$V for discs (thickness $L = 3$ cm, diameter $\emptyset = 11$ cm), so that $\eta \simeq 0.0021 \ll 1$, the values for the diffusion coefficient and for $\eta t_0 = \eta L^2/D$ as determined from the above procedure are:

$$\begin{aligned} 0.32{:}D &= 9.12\times 10^{-13}\text{m}^2/\text{s}; \quad \eta t_0 = 576.46\ \text{h} \\ 0.32PM{:}D &= 1.30\times 10^{-12}\text{m}^2/\text{s}; \quad \eta t_0 = 404.4\ \text{h} \end{aligned} \tag{7.95}$$

In addition the time lag $\tau \simeq \tau_m$ can be extracted from the same experimental data. Accordingly, (7.93) and (7.95) can combine to deliver a first assessment $n\lambda_1 = \tau_m/\eta t_0$ of $n\lambda$. We get:

$$\begin{aligned} 0.32{:}\tau_m &= 153.13\ \text{h}, \quad n\lambda_1 = 0.2656 \\ 0.32PM{:}\tau_m &= 67.2\ \text{h}, \quad n\lambda_1 = 0.1661 \end{aligned} \tag{7.96}$$

According to definition (7.61) of λ, the mass M_∞ of ions, in both bound form and solute form, remaining trapped within the sample as time goes to infinity is expressed in the form:

$$M_\infty = \rho_0 LS\lambda \tag{7.97}$$

so that the measurement of M_∞ provides another independent assessment $n\lambda_2$ of $n\lambda$. In Fig. 7.7 the curves which are labelled *Upstream* represents the cumulative mass of ions leaving the upstream compartment. Therefore M_∞ can be assessed from the difference between the *Upstream* curve and the *Downstream* curve as time goes to infinity. For the 0.32 sample M_∞ is approximately equal to 1.245 g, and to 1 g for the $0.32PM$ sample, leading to:

$$0.32{:}\quad n\lambda_2 = 0.2181; \quad 0.32PM{:}\quad n\lambda_2 = 0.1752 \tag{7.98}$$

which match to good accuracy the first assessment (7.96) obtained for $n\lambda$.

7.4 Advanced Analysis

7.4.1 Stefan-like Problem with Non-instantaneous Dissolution

When the mass of extracted solute is experimentally found not to depend linearly on the square root of time, the dissolution process involves its proper kinetics. In order to explore the possible influence of a non-instantaneous dissolution process in the Stefan-like problem (see §7.1.3), we now assume that the kinetics of dissolution obeys a first-order law such as:

$$r_{\rightarrow} = n\frac{\rho^{Eq} - \rho}{t_d} \tag{7.99}$$

In (7.99) the characteristic time t_d of dissolution can be interpreted as the time $t_d = \ell^2/d$ scaling the microdiffusion effects involving the internal mesoscopic diffusion length ℓ and the microdiffusion coefficient d.

Substituting (7.99) into (7.1a) and adopting the notation of §7.1.3 with $\overline{\rho}_b = \rho_b/\rho_b^0$, the kinetics of dissolution of the solid mineral bound to the matrix is now governed by:

$$\frac{\partial \overline{\rho}_b}{\partial \overline{t}} = \overline{\rho} - 1 \tag{7.100}$$

where the dimensionless time $\overline{t}$ is defined by:

$$\overline{t} = \varepsilon\frac{t}{t_d} \tag{7.101}$$

In (7.101), ε is the mass density ratio as defined in (7.23), so that t_d/ε is the time eventually scaling the dissolution process of the solid matrix at the macroscopic level. Indeed, when suddenly imposing a zero solute concentration at the border of a semi-infinite porous medium, (7.100) shows that the solid skin layer at $x = 0$ dissolves according to:

$$\overline{\rho}\left(x = 0, \overline{t}\right) = 0\text{:}\quad \overline{\rho}_b\left(x = 0, \overline{t} < 1\right) = 1 - \overline{t} \tag{7.102}$$

Substituting (7.99) into (7.1b) and letting:

$$\overline{x} = \frac{x}{\sqrt{(D/n)\,t_d}} \tag{7.103}$$

we obtain the unidimensional diffusion–dissolution equation:

$$\varepsilon\frac{\partial \overline{\rho}}{\partial \overline{t}} - \frac{\partial^2 \overline{\rho}}{\partial \overline{x}^2} = -\left(\overline{\rho} - 1\right) \tag{7.104}$$

governing the solute mass density in the inner solution as long as the dissolving solid mineral bound to the matrix exists.

The skin layer at $\overline{x} = 0$ finishes dissolving at the dimensionless time $\overline{t} = 1$. For $\overline{t} = 1$ the first term in (7.104) can be neglected in the large mass density ratio approximation we adopt from now on, consisting in $\varepsilon \ll 1$. Solving the resulting equation together with (7.100) and (7.102), we eventually derive:

$$\overline{\rho}\left(\overline{x}, \overline{t}=1\right) = \overline{\rho}_b\left(\overline{x}, \overline{t}=1\right) = 1 - \exp\left(-\overline{x}\right) \tag{7.105}$$

Beyond the dissolution of the skin layer, that is for $\overline{t} > 1$, a front located at $\overline{x} = \overline{x}_d\left(\overline{t}\right)$ penetrates at finite speed within the material and separates a zone $\overline{x} < \overline{x}_d\left(\overline{t}\right)$, where the solid mineral bound to the matrix is entirely dissolved, from a zone $\overline{x} > \overline{x}_d\left(\overline{t}\right)$, where the dissolution of the matrix is still in progress. In the entirely degraded zone there is no longer a source term in the diffusion equation governing the free solute concentration so that the term on the right hand side of (7.104) disappears, yielding:

$$\overline{t} > 1;\ 0 < \overline{x} < \overline{x}_d\left(\overline{t}\right): \varepsilon\frac{\partial \overline{\rho}}{\partial \overline{t}} - \frac{\partial^2 \overline{\rho}}{\partial \overline{x}^2} = 0 \tag{7.106}$$

Analysing (7.106) in the same way as we did for (7.54) to derive the advection approximation, we obtain:

$$\varepsilon\frac{\overline{x}_d^2\left(\overline{t}\right)}{\overline{t}} \ll 1 \quad \Rightarrow \quad \varepsilon\frac{\partial \overline{\rho}}{\partial \overline{t}} \ll \frac{\partial^2 \overline{\rho}}{\partial \overline{x}^2} \tag{7.107}$$

Accordingly, we neglect the first term in (7.106) and we solve (7.106) with $\overline{\rho}\left(\overline{x}=0, \overline{t}\right) = 0$:

$$\overline{t} > 1;\ 0 < \overline{x} < \overline{x}_d\left(\overline{t}\right): \frac{\overline{\rho}\left(\overline{x}, \overline{t}\right)}{\overline{\rho}\left(\overline{x}_d\left(\overline{t}\right), \overline{t}\right)} = \frac{\overline{x}}{\overline{x}_d\left(\overline{t}\right)};\quad \overline{\rho}_b\left(\overline{x}, \overline{t}\right) = 0 \tag{7.108}$$

The evolution of $\overline{\rho}\left(\overline{x} > \overline{x}_d\left(\overline{t}\right), \overline{t} > 1\right)$ is again governed by (7.104). In the large mass density ratio approximation the dissolution kinetics is slow enough that the diffusion term (second term on the left hand side of (7.104)) efficiently ensures the transport of matter resulting from the dissolution process (right hand side of (7.104)), so that the possible local accumulation of the solute can be neglected (first term in (7.104)). In doing so, the solution of (7.104) with $\overline{\rho}\left(\overline{x} \to \infty, \overline{t}\right) = 0$ is:

$$\overline{t} > 1;\ \overline{x} > \overline{x}_d\left(\overline{t}\right): \overline{\rho}\left(\overline{x}, \overline{t}\right) = \left(\overline{\rho}\left(\overline{x}_d\left(\overline{t}\right), \overline{t}\right) - 1\right)\exp\left[-\left(\overline{x} - \overline{x}_d\left(\overline{t}\right)\right)\right] + 1 \tag{7.109}$$

The requirement of the continuity of $\partial\overline{\rho}/\partial\overline{x}$ at $\overline{x} = \overline{x}_d$ for solutions (7.108) and (7.109) gives $\overline{\rho}\left(\overline{x}_d\left(\overline{t}\right), \overline{t}\right)$ in the form:

$$\overline{\rho}\left(\overline{x}_d\left(\overline{t}\right), \overline{t}\right) = \frac{\overline{x}_d\left(\overline{t}\right)}{1 + \overline{x}_d\left(\overline{t}\right)} \tag{7.110}$$

Substitution of (7.110) into (7.109) furnishes an expression for $\overline{\rho}\left(\overline{x}, \overline{t}\right)$ that, in turn, can be substituted into (7.100). The procedure gives an equation governing $\overline{\rho}_b$ for $\overline{x} > \overline{x}_d\left(\overline{t}\right)$ and $\overline{t} > 1$ which, once integrated, leads to:

$$\overline{t} > 1;\ \overline{x} > \overline{x}_d\left(\overline{t}\right): \overline{\rho}_b\left(\overline{x}, \overline{t}\right) = 1 - \exp\left(-\overline{x}\right) - \int_1^{\overline{t}} \frac{\exp\left[-\left(\overline{x} - \overline{x}_d\left(\tau\right)\right)\right]}{1 + \overline{x}_d\left(\tau\right)}\, d\tau \tag{7.111}$$

The remaining unknown function $\overline{x}_d(\overline{t})$ must satisfy:

$$\overline{\rho}_b\left(\overline{x}_d(\overline{t}),\overline{t}\right)=0;\quad \frac{d\overline{\rho}_b\left(\overline{x}_d(\overline{t}),\overline{t}\right)}{d\overline{t}}=0 \tag{7.112}$$

which means that an observer moving at speed dx_d/dt will constantly record a zero concentration of bound ions.[8] Equation (7.112) gives the differential equation:

$$\left(1+\overline{x}_d(\overline{t})\right)\frac{d\overline{x}_d(\overline{t})}{d\overline{t}}=1 \tag{7.113}$$

Integrating the previous equation and imposing $\overline{x}_d(\overline{t}=1)=0$, we obtain:

$$\overline{x}_d(\overline{t})=\sqrt{2\overline{t}-1}-1 \tag{7.114}$$

which a posteriori ensures the fulfilment of condition (7.107) for $\overline{t}>1$. Collecting the above results, for $0<\overline{x}<\overline{x}_d(\overline{t})$ we derive:

$$\overline{\rho}\left(\overline{x},\overline{t}\right)=\frac{\overline{x}}{\sqrt{2\overline{t}-1}};\quad \overline{\rho}_b\left(\overline{x},\overline{t}\right)=0 \tag{7.115}$$

and for $\overline{x}>\overline{x}_d(\overline{t})$:

$$\overline{\rho}\left(\overline{x},\overline{t}\right)=1-\frac{\exp\left[-\left(\overline{x}-\overline{x}_d(\overline{t})\right)\right]}{\sqrt{2\overline{t}-1}};\quad \overline{\rho}_b\left(\overline{x},\overline{t}\right)=1-\exp\left[-\left(\overline{x}-\overline{x}_d(\overline{t})\right)\right] \tag{7.116}$$

In Fig. 7.8 we present the mass density profiles $\overline{\rho}\left(\overline{x},\overline{t}\right)$ and $\overline{\rho}_b\left(\overline{x},\overline{t}\right)$ for various values of the dimensionless time $\overline{t}$. When compared with experimental data, these profiles can be used to assess the order of magnitude of the characteristic time t_d of the possible non-instantaneous dissolution process.

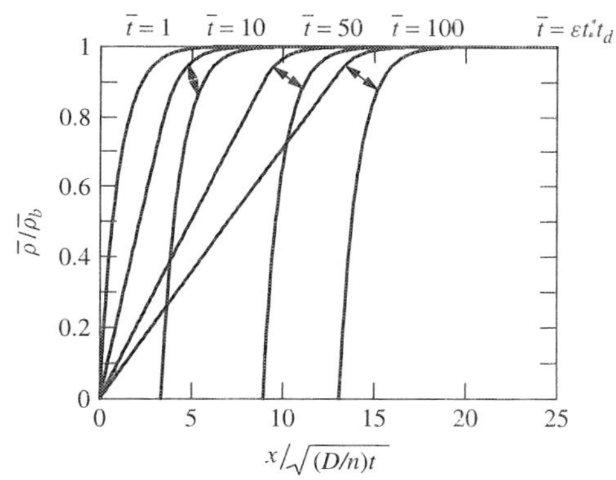

Figure 7.8: Normalized mass density profiles of solute and bound particles plotted against $\overline{x}=\sqrt{(D/n)\,t}$ for various values of the normalized time $\overline{t}=\varepsilon t/t_d$ in the case of a non-instantaneous dissolution scaled by characteristic time t_d. The profiles correspond to the large mass density ratio approximation consisting in $\varepsilon=n\rho^{Eq}/(1-n)\,\rho_b^0\ll 1$. They cannot be distinguished from the exact (numerical) solution (not represented here) for values of $\varepsilon<0.1$.

[8]Condition (7.112) can be replaced in the general framework of the kinematics of discontinuities (see §7.4.3). Indeed it is identical to (7.148) when letting $\mathcal{G}=\rho_b$ and $c=dx_d/dt$.

7.4.2 Imbibition Front

In the imbibition process a wetting fluid is injected in a porous medium and displaces a non-wetting fluid which initially saturates the porous medium. An example of great importance is the recovery of a rock reservoir of oil by means of flooding with water. With regard to durability issues, another example of concern is furnished by the imbibition of cement-based materials, initially saturated by air and invaded by seawater supplying aggressive agents such as chloride ions. In forced imbibition the volume injection rate Q of the wetting phase is imposed and leads to an advection–diffusion process that eventually relies on the same mathematical analysis we carried out for the ionic migration in §7.3.

Let us consider a semi-infinite horizontal layer $x \geq 0$ initially saturated by a non-wetting fluid (subscript nw). At time $t = 0$ a volume of wetting fluid (subscript w) is suddenly injected at constant rate Q onto the border $x = 0$. Incompressible flow is assumed for both fluids. Let S_α and $\mathcal{V}_\alpha$ be respectively the saturation degree and the component in the $\mathbf{e}_x$ direction of the filtration vector relative to fluid $\alpha = w$ or nw, so that the initial and the boundary conditions can be expressed in the form:

$$S_w\left(x, t = 0\right) = 0 \tag{7.117a}$$

$$\mathcal{V}_w\left(x = 0, t\right) = Q; \qquad \mathcal{V}_{nw}\left(x = 0, t\right) = 0 \tag{7.117b}$$

Assuming incompressible flow, the unidimensional continuity equations related to both fluids are:

$$\frac{\partial\left(nS_w\right)}{\partial t} + \frac{\partial \mathcal{V}_w}{\partial x} = 0; \qquad \frac{\partial\left(nS_{nw}\right)}{\partial t} + \frac{\partial \mathcal{V}_{nw}}{\partial x} = 0 \tag{7.118}$$

Use of relation $S_w + S_{nw} = 1$ and integration of the unique equation resulting from the addition of continuity equations (7.118) give:

$$\mathcal{V}_w + \mathcal{V}_{nw} = Q \tag{7.119}$$

where boundary conditions (7.117b) have been used in order to identify the integration constant. Owing to the horizontal character of the flow in the $\mathbf{e}_x$ direction, we write Darcy's law relative to each fluid in the form (see §6.5.2):

$$\mathcal{V}_w = -\frac{\varkappa k_{\mathrm{r}w}\left(S_w\right)}{\eta_w}\frac{\partial p_w}{\partial x}; \qquad \mathcal{V}_{nw} = -\frac{\varkappa k_{\mathrm{r}nw}\left(S_{nw}\right)}{\eta_{nw}}\frac{\partial p_{nw}}{\partial x} \tag{7.120}$$

where relative permeability $k_{\mathrm{r}\alpha}\left(S_\alpha\right)$ satisfies:

$$k_{\mathrm{r}\alpha}\left(0\right) = 0 \leq k_{\mathrm{r}\alpha}\left(S_\alpha\right) \leq k_{\mathrm{r}\alpha}\left(1\right) = 1 \tag{7.121}$$

In addition the capillary pressure as a function of the saturation degree of the wetting fluid is written in the form:

$$p_c = p_{nw} - p_w = \mathsf{M}\pi_c\left(S_w\right) \tag{7.122}$$

where M stands for a capillary modulus so that $\pi_c\left(S_w\right)$ is a dimensionless capillary pressure. Getting rid of the different unknown fields from (7.118)–(7.120) and (7.122) to the sole benefit of S_w leads us to write the continuity equation relative to the wetting fluid in the form:

$$\frac{\partial (n\, S_w)}{\partial t} - D \frac{\partial}{\partial x} \left(\delta (S_w) \frac{\partial S_w}{\partial x} - \frac{\mu (S_w)}{\ell} \right) = 0 \tag{7.123}$$

where $\delta (S_w)$ and $\mu (S_w)$ are the following diffusion and advection dimensionless functions:

$$\delta (S_w) = -\frac{d\pi_c (S_w)}{dS_w} \frac{\eta_w k_{rw} (S_w) k_{rnw} (1 - S_w)}{\eta_{nw} k_{rw} (S_w) + \eta_w k_{rnw} (1 - S_w)} \tag{7.124a}$$

$$\mu (S_w) = \frac{\eta_{nw} k_{rw} (S_w)}{\eta_{nw} k_{rw} (S_w) + \eta_w k_{rnw} (1 - S_w)} \tag{7.124b}$$

whereas D and ℓ are respectively the diffusion coefficient and the advection–diffusion length defined by:

$$D = \frac{\varkappa M}{\eta_w}; \quad \ell = \frac{D}{Q} \tag{7.125}$$

According to (7.123), the imbibition is governed by an advection–diffusion equation. The profile of the imbibition front engendered by the injection is shaped by the balance between the driving capillary pressure and the resistant viscous forces that slow down the process, so that the diffusion coefficient D results from the balance between both forces. Length $\ell = D/Q$ finally scales the range of the diffusion strength when compared with the advection one quantified by the injection rate Q: the smaller the length ℓ, the steeper the front.

In the limit of the advection approximation (7.58), (7.123) reduces to the celebrated Buckley–Leverett equation:[9]

$$\frac{\partial S_w}{\partial t} + \frac{D/n}{\ell} \frac{\partial \mu (S_w)}{\partial x} = 0 \tag{7.126}$$

The advection approximation turns out to ignore the capillary effects. Accordingly, letting $\partial p_w/\partial x = \partial p_{nw}/\partial x$, (7.119), (7.120) and (7.124b) together provide:

$$\mathcal{V}_w = -\mu (S_w)\, Q; \quad \mathcal{V}_{nw} = -\mu (S_w) \frac{\eta_w}{\eta_{nw}} \frac{k_{rnw} (1 - S_w)}{k_{rw} (S_w)} Q \tag{7.127}$$

so that boundary conditions (7.117b) when considering (7.121) can be conveniently replaced by:

$$S_w (x = 0, t) = 1 \tag{7.128}$$

[9]See for instance Bear J. (1988), *Dynamics of Fluids in Porous Media*, Dover, New York (reprint of 1972 edition published by Elsevier, New York).

A possible solution of (7.126) satisfying initial condition (7.117a) is:

$$S_w = S_w^{up} H\left(x - \upsilon\left(S_w^{up}\right)t\right); \quad \upsilon\left(S_w^{up}\right) = c_0 \mu'\left(S_w^{up}\right); \quad c_0 = \frac{D/n}{\ell} = Q/n \tag{7.129}$$

where μ' stands for the derivative of μ. Solution (7.129) represents a step function moving at speed $\upsilon\left(S_w^{up}\right)$ and admitting S_w^{up} for upstream saturation. Nevertheless, any discontinuous imbibition front moving at speed c is required to fulfil the Rankine–Hugoniot condition (7.13), specialized here in the form:

$$\left[\left[\mathcal{V}_w - ncS_w\right]\right] = 0 \tag{7.130}$$

Owing to the initial downstream condition (7.117a), the height $\left[\left[S_w\right]\right]$ of the imbibition front reduces to the upstream saturation S_w^{up} so that (7.127) and (7.130) produce the relation:

$$c = c_0 \frac{\mu\left(S_w^{up}\right)}{S_w^{up}} \tag{7.131}$$

Equating both expressions furnished by (7.129) and (7.131) for the speed, we conclude that solution (7.129) can actually develop provided only that the upstream saturation satisfies:

$$\mu'\left(S_w^{up}\right) = \frac{\mu\left(S_w^{up}\right)}{S_w^{up}} \tag{7.132}$$

However, even though a steep imbibition front such as (7.129) could actually propagate, it will be inexorably spread out owing to diffusion effects so that a transition layer will eventually form. With the aim of possibly confirming (7.132) and exploring the possibility of the formation of an autonomous transition layer, we proceed as in §7.3.2. Adopting notation (7.64), we rewrite the advection–diffusion equation (7.123) in the form:

$$\ell \frac{\partial S_w}{\partial t} + \frac{\partial\left(c_0 \mu(S_w) - cS_w\right)}{\partial \chi} - c_0 \frac{\partial}{\partial \chi}\left(\delta(S_w)\frac{\partial S_w}{\partial \chi}\right) = 0 \tag{7.133}$$

Once formed, a time-independent steady imbibition profile will have to match the equation obtained by nullifying the time derivative in (7.133), that is:

$$\frac{d\left(c_0 \mu(S_w) - cS_w\right)}{d\chi} - c_0 \frac{d}{d\chi}\left(\delta(S_w)\frac{dS_w}{d\chi}\right) = 0 \tag{7.134}$$

A first integration of (7.134) gives:

$$c_0\, \mu(S_w) - cS_w - c_0 \delta(S_w)\frac{dS_w}{d\chi} = C \tag{7.135}$$

where C is an integration constant so that the continuity of S_w, $\mu(S_w)$ and $\delta(S_w)$ requires continuity of $dS_w/d\chi$. We now add the boundary conditions:

$$S_w\,(\chi \to -\infty) = S_w^{up}; \qquad S_w\,(\chi \to +\infty) = 0 \tag{7.136}$$

In addition, (7.121) and (7.124b) give:

$$\mu(S_w = 0) = 0 \tag{7.137}$$

According to the second of the boundary conditions in (7.136) and (7.137), S_w and $dS_w/d\chi$ have to vanish as χ goes to infinity so that the integration constant C turns out to be zero. Using in addition the first of the boundary conditions in (7.136) and noting that $\partial S_w/\partial\chi\,(\chi\rightarrow-\infty)=0$, we retrieve for c the expression (7.131) we found in the advection approximation. Accordingly, substitution of (7.131) into (7.135), where we now let $C=0$, leads to:

$$\mu(S_w)-\frac{\mu\left(S_w^{up}\right)}{S_w^{up}}S_w-\delta(S_w)\frac{dS_w}{d\chi}=0 \tag{7.138}$$

Returning to the initial variables, the solution of (7.138) is:

$$\int_0^{S_w}\frac{\delta(S)}{\mu(S)-\frac{\mu\left(S_w^{up}\right)}{S_w^{up}}S}\,dS=\frac{x-ct}{\ell} \tag{7.139}$$

where the origin $\chi=0$ has been chosen at the front head where $S_w=0$. According to (7.139), length ℓ eventually scales the 'thickness' of the imbibition front. The boundary condition $S_w\,(\chi\rightarrow\infty)=S_w^{up}$ eventually means that $S_w\,(x-ct\ll-\ell)=S_w^{up}$ so that the travelling wave solution (7.139) will be physically relevant provided that ℓ is small compared with the extent of the porous medium.[10]

An imbibition front governed by the travelling wave solution (7.139) can actually form provided that the integral converges. In order to derive the corresponding condition we first remark that (7.121) and (7.124) lead to:

$$\delta\,(S_w\rightarrow 0)=-k_{rw}\,(S_w)\,\frac{d\pi_c\,(S_w)}{dS_w};\qquad \mu\,(S_w\rightarrow 0)=\frac{\eta_{nw}}{\eta_w}k_{rw}\,(S_w) \tag{7.140}$$

The relative permeability function usually satisfies $k_{rw}\,(S_w)\,/S_w\rightarrow 0$ as $S_w\rightarrow 0$ so that we require $k_{rw}\,(S_w)$ to satisfy:

$$S_w\rightarrow 0:\quad k_{rw}\,(S_w)\,\frac{d\pi_c\,(S_w)}{dS_w}\simeq aS_w^{\gamma};\quad \gamma,a>0 \tag{7.141}$$

in order that the singularity for $S_w=0$ of the integrand in (7.139) does not prevent convergence of the integral. When condition (7.141) is not fulfilled the driving capillary force is so intense that the diffusion effects spread out the imbibition profile over the whole x axis and (7.139) must be replaced by:

$$\int_{S_w^{up}/2}^{S_w}\frac{\delta(S)}{\mu(S)-\frac{\mu\left(S_w^{up}\right)}{S_w^{up}}S}\,dS=\frac{x-ct}{\ell} \tag{7.142}$$

where the origin $\chi=0$ has now been chosen where the saturation is half the upstream saturation. By contrast, when condition (7.141) is fulfilled the resistant viscous forces are strong enough to counterbalance the driving capillary force as the saturation goes to zero, resulting in the actual imbibition front (7.139) such that $S_w=0$ can be observed at a

[10] Adopting $k=1\ \mu\text{m}^2$ and $Q=1$ m/s, and considering that the wetting fluid is water, that is $\eta_w\simeq 10^{-3}\text{kg/(ms)}$, the order of magnitude of length $\ell\,(\text{m})$ is $10^{-3}\mathsf{M}$ where the capillary modulus M is expressed in MPa.

finite distance along the x axis. With order of magnitudes of $k_{rw}(S_w \to 0) \simeq O(S_w^\alpha)$ and $\pi_c(S_w \to 0) \simeq O\left(S_w^{-\beta}\right)$ condition (7.141) reduces to:

$$\gamma = \alpha - \beta - 1 \geq 0 \tag{7.143}$$

The latter condition is for instance fulfilled whatever the value of m, when adopting expressions (6.40) and (6.72) for functions $\pi_c(S_w)$, $k_{rw}(S_w)$ and $k_{rnw}(S_w)$.

In addition an autonomous imbibition front can actually form provided that (7.139) can be inverted in order to obtain S_w as a function of $x - ct$, requiring $d\chi/dS_w$ to be non-zero everywhere and consequently for S_w to be a decreasing function of χ. Accordingly (7.138) requires:[11]

$$0 \leq S_w < S_w^{up}: \quad \frac{\mu(S_w)}{S_w} < \frac{\mu\left(S_w^{up}\right)}{S_w^{up}} \tag{7.144}$$

If no such interval $\left[0, S_w^{up}\right]$ actually exists, no autonomous imbibition front can finally form, even though condition (7.141) is fulfilled. This corresponds to the limit case of viscosity ratio η_w/η_{nw} going to zero so that function $\mu(S_w)$ reduces to the step function $H(S_w)$. Conversely, as viscosity ratio η_w/η_{nw} goes to infinity, function $\mu(S_w)$ reduces to the step function $H(S_w - 1)$ so that the couple $\left(S_w^{up} = 1, c = c_0\right)$ becomes suitable with regard to condition (7.144). Accordingly, the autonomous imbibition profile (7.139) can fully develop since it meets the upstream boundary condition (7.128). For intermediary viscosity ratios η_w/η_{nw} ranging from zero to infinity, the largest value of S_w^{up} that is consistent with (7.144) is obtained by solving (7.132). This is illustrated in Fig. 7.9 for

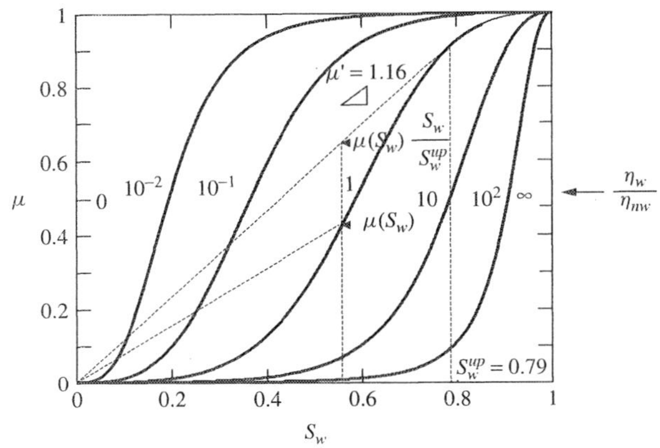

Figure 7.9: Function $\mu(S_w)$ for various values of the viscosity ratio η^w/η^{nw} resulting from expressions (6.72) for the relative permeabilities k_{rw} and k_{rnw} with $m = 0.85$. The determination of the largest upstream saturation S_w^{up} consistent with (7.144) is illustrated for $\eta^w/\eta^{nw} = 1$. The values of S_w^{up} and of the associated normalized speed $\mu'\left(S_w^{up}\right)$ respectively increase and decrease for an increasing viscosity ratio η_w/η_{nw}.

[11] Indeed, condition (7.144) turns out to be the 'entropy criterion' for admissible shocks in the mathematical sense while step function (7.129), where c is given by (7.131), is the standard 'entropy weak solution' towards which the solution of (7.133) converges when the diffusion term vanishes.

various values of the viscosity ratio η_w/η_{nw} and for the function $\mu(S_w)$ relative to expressions (6.40) and (6.72) of $\pi_c(S_w)$, $k_{rw}(S_w)$ and $k_{rnw}(S_w)$ when letting $m = 0.85$. The values of the largest suitable upstream saturation S_w^{up} and the associated normalized speed $\mu'(S_w^{up})$ respectively increase and decrease for increasing viscosity ratio η_w/η_{nw}. Indeed, the lower the viscosity ratio, the more difficult the penetration of the non-wetting fluid and, consequently, the lower the possible maximum upstream height of penetration S_w^{up}.

At early times in a region close to $x = 0$, that is for $t \ll \ell^2/D$ and $x < \ell$, when condition (7.141) is fulfilled diffusion effects spread out the profile of saturation degree S_w from one down to zero within a layer of finite extent. Subsequently the higher saturation degree contributions to the profile overtake the lower saturation degree contributions, provided that the latter belong to a range moving slower than the former owing to a lower value of related slope μ' and, consequently, associated speed $c_0\mu'$. The process progressively steepens the imbibition profile to form eventually an autonomous travelling transition layer when condition (7.132) is finally met. However, when S_w reaches the largest possible upstream height S_w^{up} for the autonomous transition layer, the integral in (7.139) no longer converges for higher values of the saturation degree. In fact, a travelling wave solution that meets boundary condition (7.128) eventually cannot fully develop: a transient regime will always occur, allowing the autonomous travelling transition layer to be matched to the upstream boundary condition (7.128). An alternative conclusion is that boundary condition (7.128) is in practice tough to carry out, the saturation degree S_w^{up} representing a threshold that is difficult to passover. All these remarks agree with the experimental observations reported in Fig. 7.10 where viscosity ratio η_w/η_{nw} is equal to unity whereas the experimental value of S_w^{up} is found to be close to $\frac{1}{2}$.

As has been recognized, the analysis of the forced imbibition process is similar to that of the migration process carried out in §7.3. Analogously the analysis of the free imbibition process turns out to be similar to that of the solute penetration process with

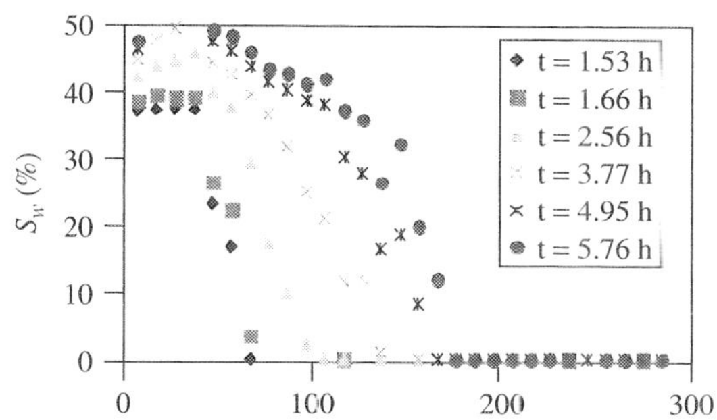

Figure 7.10: Imbibition profiles for a porous medium made of quartzite grains of radii 100 μm, packed in a cylindrical glass column. The injection rate Q is about 4 m/s. The values of the porosity and the absolute permeability are $n = 35\%$ and $k = 10\ \mu\text{m}^2$ respectively. The wetting fluid is water while the non-wetting fluid is pure n-decane with approximately the same viscosity for both fluids $\eta_w = \eta_{nw} \simeq 10^{-3}$ kg/(ms). The apparent saturation threshold occurring at the inlet $x = 0$ is equal to $\frac{1}{2}$ and can be interpreted as the largest possible upstream saturation S_w^{up} satisfying (7.144). The experimental data are from Melean Y., Broseta D., Hasmy A., Blossey R. (2003), 'Dispersion of imbibition fronts', *European Physics Letters*, **62**, (4), 505–511.

strong non-linear binding (see §7.2). In the present context free imbibition consists in suddenly imposing a complete saturation of the wetting fluid on the border of a semi-infinite medium initially saturated by the non-wetting fluid. The vanishing character of the diffusion coefficient related to the fluid invasion causes the formation of a penetration front at finite speed. Letting $L(t)$ be the imbibition length at time t, a dimensional analysis such as the one we performed in §3.3.1, applied here to the set of variables L, $\varkappa/\eta_w$, M and t, reveals that:

$$L(t) = \alpha\sqrt{\frac{\mathrm{M}\varkappa}{\eta_w}t} \tag{7.145}$$

where α is a constant coefficient whose determination requires the solution of the diffusion equation governing the saturation degree $S_w(x, t)$ of the wetting fluid (as we did in §7.2.1 for the chloride concentration). Accordingly, α depends on the viscosity ratio η_w/η_{nw}, on the relative permeability and on capillary functions $k_{rw}(S_w)$, $k_{rnw}(S_w)$ and $\pi_c(S_w)$.

7.4.3 Surfaces of Discontinuity and Wave Propagation

Kinematics of discontinuities

Let $\mathcal{G}(x_i, t)$ be any function remaining continuous across a front Σ propagating at the normal speed of displacement $c = \mathbf{c} \cdot \mathbf{n}$ in an Eulerian description (see Fig. 1.5). Owing to kinematical compatibility, the spatial discontinuities $[[\partial\mathcal{G}/\partial x_i]]$ and $[[\partial\mathcal{G}/\partial t]]$ which may affect derivatives $\partial\mathcal{G}/\partial x_i$ and $\partial g/\partial t$ when passing across Σ are not independent of each other. Indeed, consider two points infinitely close to the same geometrical point $\mathbf{x}$ of front Σ, but each of them lying on a distinct side of the latter. Since $\mathcal{G}$ is assumed to remain continuous across Σ, the infinitesimal spatial variation $d\mathcal{G}$ undergone by $\mathcal{G}$ is the same along each side of Σ. We write:

$$[[d\mathcal{G}]] = \left[\left[\frac{\partial\mathcal{G}}{\partial x_i}\right]\right] dx_i + \left[\left[\frac{\partial\mathcal{G}}{\partial t}\right]\right] dt = 0 \tag{7.146}$$

for any set of values of dx_i satisfying:

$$n_i dx_i - c dt = 0 \tag{7.147}$$

The system formed by the two previous equations has to remain undetermined with regard to dx_i and dt and results in:

$$c\left[\left[\frac{\partial\mathcal{G}}{\partial x_i}\right]\right] + \left[\left[\frac{\partial\mathcal{G}}{\partial t}\right]\right] n_i = 0 \tag{7.148}$$

which constitutes Hadamard's kinematical compatibility relation. Combining definition (1.43) of the particle derivative and (7.148) gives the alternative form:

$$C\left[\left[\frac{\partial\mathcal{G}}{\partial x_i}\right]\right] + \left[\left[\frac{d\mathcal{G}}{dt}\right]\right] n_i = 0; \quad C = c - V_i n_i \tag{7.149}$$

which can be interpreted as the Lagrangian form of Hadamard's compatibility relation (7.148) when adopting the current configuration as the reference configuration by letting $x_i = X_i$ (see (1.1)).

Consider now a quantity g whose first derivatives are continuous but whose second-order derivatives are discontinuous. Applying kinematical compatibility condition (7.148) successively to $\mathcal{G} = \partial g/\partial t$ and $\mathcal{G} = \partial g/\partial x_i$ gives the relations:

$$c\left[\left[\frac{\partial^2 g}{\partial x_i \partial t}\right]\right] + \left[\left[\frac{\partial^2 g}{\partial t^2}\right]\right] n_i = 0; \quad c\left[\left[\frac{\partial^2 g}{\partial x_i \partial x_j}\right]\right] + \left[\left[\frac{\partial^2 g}{\partial t \partial x_i}\right]\right] n_j = 0 \tag{7.150}$$

so that:

$$c^2\left[\left[\frac{\partial^2 g}{\partial x_i \partial x_j}\right]\right] - \left[\left[\frac{\partial^2 g}{\partial t^2}\right]\right] n_i n_j = 0 \tag{7.151}$$

Applying the above relations to the components of a vector field $\mathbf{g}$ gives:

$$c^2\left[\left[\nabla\left(\nabla\cdot\mathbf{g}\right)\right]\right] - \left[\left[\mathbf{n}\cdot\frac{\partial^2\mathbf{g}}{\partial t^2}\right]\right]\mathbf{n} = 0 \tag{7.152a}$$

$$c^2\left[\left[\nabla\times\nabla\times\mathbf{g}\right]\right] - \mathbf{n}\times\left[\left[\mathbf{n}\times\frac{\partial^2\mathbf{g}}{\partial t^2}\right]\right] = 0 \tag{7.152b}$$

$$c^2\left[\left[\nabla\cdot\left(\frac{\partial\mathbf{g}}{\partial t}\right)\right]\right] + \left[\left[\mathbf{n}\cdot\frac{\partial^2\mathbf{g}}{\partial t^2}\right]\right] = 0 \tag{7.152c}$$

$$c^2\left[\left[\nabla\left(\frac{\partial\mathbf{g}}{\partial t}\right)\right]\right] + \left[\left[\frac{\partial^2\mathbf{g}}{\partial t^2}\right]\right]\otimes\mathbf{n} = 0 \tag{7.152d}$$

Acceleration waves

The analysis of the propagation of acceleration waves turns out to be equivalent to the analysis of the propagation of discontinuities of the second-order time derivatives and can be carried out in the framework of the hypothesis of small movements. The latter extends the hypothesis of small perturbations (see §5.1.1) to the displacements, the velocities and their gradients, for both the skeleton and the fluid particles. Under the hypothesis of small movements, (1.61) is rewritten in the form:

$$\mathbf{w} = \rho_f^0 \mathcal{V} = \rho_f^0 \phi_0 \left(\mathbf{V}^f - \mathbf{V}^s\right) \tag{7.153}$$

Assuming uniformity of the initial porosity and of the fluid mass density through the porous medium, substitution of (7.153) into the fluid mass conservation equation (5.7) gives the relation:

$$\dot{v}_f = -\phi_0 \nabla\cdot\left(\dot{\boldsymbol{\xi}}^f - \dot{\boldsymbol{\xi}}\right) \tag{7.154}$$

where an overdot refers to the particle time derivative of the involved quantity and where we used definition (5.13) for v_f. Time integration of (7.154) leads to:

$$v_f = -\phi_0 \nabla \cdot (\boldsymbol{\xi}^f - \boldsymbol{\xi}) \tag{7.155}$$

Momentum equation (2.19) and Darcy's law (3.39) are expressed in the form:

$$\nabla \cdot \boldsymbol{\sigma} - \rho_s(1-\phi_0)\ddot{\boldsymbol{\xi}} - \rho_f \phi_0 \ddot{\boldsymbol{\xi}}^f = 0 \tag{7.156}$$

and:

$$\phi_0(\dot{\boldsymbol{\xi}}^f - \dot{\boldsymbol{\xi}}) = -k\left(\nabla p + \rho_f \ddot{\boldsymbol{\xi}}^f\right) \tag{7.157}$$

Using (1.26) and (1.28), constitutive equations (5.10) and (5.12) can be written in the form:

$$\boldsymbol{\sigma} = (\lambda + b^2 M)\nabla \cdot \boldsymbol{\xi}\,\mathbf{1} + \mu(\nabla\boldsymbol{\xi} + {}^t\nabla\boldsymbol{\xi}) - bMv_f\,\mathbf{1}; \quad v_f = b\nabla \cdot \boldsymbol{\xi} + \frac{p}{M} \tag{7.158}$$

where $\lambda = K - \frac{2}{3}\mu$ is the drained Lamé coefficient. A combination of (7.155)–(7.158) provides the equations governing the small movements of the skeleton particle and the fluid particle in the form:

$$(\lambda + 2\mu + (b-\phi_0)^2 M)\nabla(\nabla \cdot \boldsymbol{\xi}) + \phi_0(b-\phi_0)M\nabla(\nabla \cdot \boldsymbol{\xi}^f) - \mu\nabla \times (\nabla \times \boldsymbol{\xi})$$

$$+\frac{\phi_0^2}{k}(\dot{\boldsymbol{\xi}}^f - \dot{\boldsymbol{\xi}}) - \rho_s(1-\phi_0)\ddot{\boldsymbol{\xi}} = 0 \tag{7.159a}$$

$$\phi_0(b-\phi_0)M\nabla(\nabla \cdot \boldsymbol{\xi}) + \phi_0^2 M\nabla(\nabla \cdot \boldsymbol{\xi}^f)$$

$$-\frac{\phi_0^2}{k}(\dot{\boldsymbol{\xi}}^f - \dot{\boldsymbol{\xi}}) - \rho_f\phi_0\ddot{\boldsymbol{\xi}}^f = 0 \tag{7.159b}$$

For acceleration waves the velocities $\dot{\boldsymbol{\xi}}$ and $\dot{\boldsymbol{\xi}}^f$ remain continuous across the wavefront. Applying the jump operator [[ ]] to (7.159) and using relations (7.152), we conclude that discontinuities $[[\ddot{\boldsymbol{\xi}}]]$ and $[[\ddot{\boldsymbol{\xi}}^f]]$ must satisfy the dynamic relations:

$$\left(\lambda + \mu + (b-\phi_0)^2 M\right)\left[\left[\mathbf{n} \cdot \ddot{\boldsymbol{\xi}}\right]\right]\mathbf{n} + \phi_0(b-\phi_0)M[[\mathbf{n} \cdot \ddot{\boldsymbol{\xi}}^f]]\mathbf{n}$$

$$\left(\mu - c^2\rho_s(1-\phi_0)\right)\left[\left[\ddot{\boldsymbol{\xi}}\right]\right] = 0 \tag{7.160a}$$

$$\phi_0(b-\phi_0)M\left[\left[\mathbf{n} \cdot \ddot{\boldsymbol{\xi}}\right]\right]\mathbf{n} + \phi_0^2 M[[\mathbf{n} \cdot \ddot{\boldsymbol{\xi}}^f]]\mathbf{n} - c^2\rho_f\phi_0[[\ddot{\boldsymbol{\xi}}^f]] = 0 \tag{7.160b}$$

Discontinuities $[[\ddot{\boldsymbol{\xi}}]]$ and $[[\ddot{\boldsymbol{\xi}}^f]]$ decompose in the form:

$$[[\ddot{\boldsymbol{\xi}}]] = \ddot{\xi}_n\mathbf{n} + \ddot{\xi}_t\mathbf{t}; \quad \left[\left[\ddot{\boldsymbol{\xi}}^f\right]\right] = \ddot{\xi}_n^f\,\mathbf{n} + \ddot{\xi}_t^f\,\mathbf{t} \tag{7.161}$$

where $\mathbf{t}$ lies in the plane tangent to the wavefront. Substituting (7.161) into (7.160a) and (7.160b) and multiplying by $\mathbf{t}$, we derive:

$$c = c_S = \sqrt{\frac{\mu}{\rho_s(1-\phi_0)}}; \quad \ddot{\xi}_t^f = 0 \tag{7.162}$$

Indeed, only the skeleton particle can undergo a discontinuity of acceleration in the plane tangent to the wavefront. It corresponds to a transverse wave that propagates at wavespeed c_S. Substituting (7.161) into (7.160a) and (7.160b) and multiplying by $\mathbf{n}$, we alternatively derive:

$$(\mathcal{K} - c^2\mathcal{M})\begin{pmatrix} \ddot{\xi}_n \\ \ddot{\xi}_n^f \end{pmatrix} = 0 \tag{7.163}$$

where the stiffness and mass matrices $\mathcal{K}$ and $\mathcal{M}$ are defined by:

$$\mathcal{K} = \begin{pmatrix} \lambda + 2\mu + (b-\phi_0)^2 M & \phi_0(b-\phi_0)M \\ \phi_0(b-\phi_0)M & \phi_0^2 M \end{pmatrix} \tag{7.164a}$$

$$\mathcal{M} = \begin{pmatrix} \rho_s(1-\phi_0) & 0 \\ 0 & \rho_f\phi_0 \end{pmatrix} \tag{7.164b}$$

According to (7.163) two different compressional waves relative to discontinuities of acceleration along the normal to the wavefront can propagate within a saturated porous medium. Their wavespeeds c_{P_1} and c_{P_2} are solutions of:

$$\det(\mathcal{K} - c^2\mathcal{M}) = 0 \tag{7.165}$$

while the eigenmovement X_J associated with c_{P_J} satisfies:

$$\left(\mathcal{K} - c_{P_J}^2\mathcal{M}\right)X_J = 0 \tag{7.166}$$

It can be shown that the eigenmovement associated with the fastest longitudinal wave corresponds to movements of the skeleton and the fluid which are in phase, while the movements of the skeleton and the fluid are out of phase for the other longitudinal wave. In practice the latter turns out to be much slower than the former with wavespeed $c_{P_2} \ll c_{P_1}$ and is often called the slow compressional wave.[12]

Adding (7.160a) and (7.160b) we now derive:

$$(\lambda_u + 2\mu)\nabla(\nabla\cdot\boldsymbol{\xi}) + \phi_0 bM\nabla\left(\nabla\cdot(\boldsymbol{\xi}^f - \boldsymbol{\xi})\right) - \mu\nabla\times(\nabla\times\boldsymbol{\xi})$$
$$-\rho_s(1-\phi_0)\ddot{\boldsymbol{\xi}} - \rho_f\phi_0\ddot{\boldsymbol{\xi}}^f = 0 \tag{7.167}$$

where $\lambda_u = \lambda + b^2 M$ is the undrained Lamé coefficient. As retrieved by letting k go to zero in (7.159b), in the limit of a zero permeability there is no relative movement, that is $\boldsymbol{\xi}^f = \boldsymbol{\xi}$. Letting $\boldsymbol{\xi}^f = \boldsymbol{\xi}$ in (7.167) and proceeding as above to analyze the propagation

[12] The existence of the second compressional wave was first noticed by Y.I. Frenkel and subsequently fully investigated by M.A. Biot (see Biot M.A. (1956), 'The theory of propagation of elastic waves in a fluid-saturated porous solid, I lower frequency range', Journal of the Acoustical Society of America, **28**, 168–178).

of a discontinuity of acceleration along the normal to the wavefront, we find that the undrained compressional wave propagates at wavespeed c_p given by:

$$c_P = \sqrt{\frac{\lambda_u + 2\mu}{\rho}} \tag{7.168}$$

where $\rho = \rho_s (1 - \phi_0) + \rho_f \phi_0$ denotes the total mass density.

Permeability k is not involved in the expressions of wavespeeds c_{P_1} and c_{P_2} obtained by solving (7.165), so that passing continuously from a zero permeability to a very low permeability leads to a paradox. Indeed, this requires passing abruptly from the existence of only one compressional wave, with wavespeed c_P, to two compressional waves, with wavespeeds c_{P_1} and c_{P_2}. In fact the analysis of the propagation of just the discontinuities turns out to be insufficient to understand the propagation of acceleration waves in porous media. To overcome the paradox, let us consider the propagation of harmonic longitudinal waves in the $\mathbf{e}_x$ direction in the form:

$$\boldsymbol{\xi} = \xi_0 \mathrm{Re} \exp(sx + i\omega t)\, \mathbf{e}_x; \quad \boldsymbol{\xi}^f = \xi_0^f \mathrm{Re} \exp(sx + i\omega t)\, \mathbf{e}_x \tag{7.169}$$

Substitution of (7.169) into (7.159) yields:

$$(\mathcal{K}s^2 + i\omega\mathcal{N} + \omega^2\mathcal{M}) \begin{pmatrix} \xi_0 \\ \xi_0^f \end{pmatrix} = 0 \tag{7.170}$$

where the damping matrix $\mathcal{N}$ is expressed in the form:

$$\mathcal{N} = \begin{pmatrix} -\dfrac{\phi^2}{k} & \dfrac{\phi^2}{k} \\ \dfrac{\phi^2}{k} & -\dfrac{\phi^2}{k} \end{pmatrix} \tag{7.171}$$

Non-zero values obtained for ξ_0 and ξ_0^f by solving (7.170) requires s to be a solution of:

$$\det(\mathcal{K}s^2 + i\omega\mathcal{N} + \omega^2\mathcal{M}) = 0 \tag{7.172}$$

yielding:

$$\left(s^2 + \frac{\omega^2}{c_{P_1}^2}\right)\left(s^2 + \frac{\omega^2}{c_{P_2}^2}\right) - i\frac{\omega\omega_c}{c_P^2}\left(s^2 + \frac{\omega^2}{c_P^2}\right) = 0 \tag{7.173}$$

where ω_c is a characteristic circular frequency attached to the porous medium and whose expression is:

$$\omega_c = \frac{c_P^2}{c_f} \tag{7.174}$$

where c_f is the fluid diffusivity as given by (5.22). Approximate solutions of (7.173) are:

$$\omega \ll \omega_c\colon\ s^2 \simeq -\frac{\omega^2}{c_P^2} \qquad s^2 \simeq 0 \tag{7.175}$$

and:

$$\omega \gg \omega_c \begin{cases} s^2 \simeq -\dfrac{\omega^2}{c_{P_1}^2} + i\dfrac{\omega\omega_c}{c_P^2}\dfrac{1/c_P^2 - 1/c_{P_1}^2}{1/c_{P_2}^2 - 1/c_{P_1}^2} \\ \\ s^2 \simeq -\dfrac{\omega^2}{c_{P_2}^2} + i\dfrac{\omega\omega_c}{c_P^2}\dfrac{1/c_{P_2}^2 - 1/c_P^2}{1/c_{P_2}^2 - 1/c_{P_1}^2} \end{cases} \tag{7.176}$$

In the low-frequency range $\omega \ll \omega_c$ the undrained situation is recovered with only one longitudinal wave with wavespeed c_P. Conversely, in the high-frequency range $\omega \gg \omega_c$ we encounter the existence of two highly attenuated longitudinal waves, the previous wavespeeds c_{P_1} and c_{P_2} being recovered only in the limit of ratio ω/ω_c going to infinity. In the high-frequency range it could have been thought that no significant relative movement can actually occur between the skeleton and the fluid, resulting in undrained evolutions and the existence of only one longitudinal wave with wavespeed c_P. The above analysis reveals that the opposite does occur. Indeed, the higher the frequency, the higher the fluid pressure gradient, the latter being the driving force of the relative movement between the fluid and the skeleton. However, the intense relative movement conjointly implies an intense viscous dissipation of energy. The latter is actually at the origin of the wave attenuation that the analysis restricted to the sole propagation of discontinuities cannot account for. This wave attenuation provokes the wave dispersion, the wavespeed then depending on the circular frequency. Consequently, a signal with a steep front deforms when propagating, the high-frequency components of the longitudinal waves being rapidly attenuated. The slowest longitudinal wave which is the most attenuated progressively disappears, while the other is absorbed in a longitudinal wave associated with undrained evolutions with wavespeed c_P. Nevertheless, intrinsic circular frequency ω_c is very high for the usual porous materials so that any coherent excitation generally falls in the range $\omega \ll \omega_c$. Indeed, according to (7.174) the order of magnitude of the characteristic circular frequency ω_c is mainly governed by the factor $\eta_f/\rho\varkappa$ For liquid water where $\eta_f = 1 \times 10^{-3}$kg/ (ms), this order of magnitude turns out to be 500 MHz when adopting $\rho = 2 \times 10^3$ kg/m^3 and $\varkappa = 10^{-15}/\text{m}^2$ (see Table 3.1). The second compressional wave can actually be observed provided only that the porous medium is sufficiently permeable.[13] The dynamic phenomena of attenuation due to the second compressional wave have eventually to be considered when high gradients are engendered by physical discontinuities, such as wave reflection at the interface between two porous media.[14]

Acoustic tensor

From the above analysis developed for isotropic materials it turns out that wave propagation in porous media involves undrained evolutions. An alternative approach to wave propagation extended to non-isotropic and non-linear materials can be carried out as follows. Constitutive equations for undrained evolutions can be expressed in the general

[13]Such as sintered glass, for which the order of magnitude of $\varkappa$ is $10^{-12}/\text{m}^2$, see Plona T.J., Johnson D.L. (1980), 'Experimental study of two bulk compressional modes in water-saturated porous structures', *Ultrasonics Symporium, IEEE*, 868–872.

[14]For further details see Bourbié T., Coussy O., Zinszner B. (1987), *Acoustics of Porous Media*, Gulf Publishing Company/Editions Technip, Paris.

non-linear form:

$$\dot{\boldsymbol{\sigma}} = \mathbf{C}^u : \dot{\boldsymbol{\varepsilon}}; \quad \dot{\sigma}_{ij} = C^u_{ijkl} \dot{\varepsilon}_{kl} \tag{7.177}$$

where the fourth-order tensor $\mathbf{C}^u$ of components C^u_{ijkl} stands for the tangent undrained stiffness tensor which possibly depends on σ_{ij}. Owing to expression (1.26) for strain tensor $\boldsymbol{\varepsilon}$, strain rate $\dot{\boldsymbol{\varepsilon}}$ is expressed in the form:

$$\dot{\boldsymbol{\varepsilon}} = \frac{1}{2}(\nabla \dot{\boldsymbol{\xi}} + {}^t\nabla \dot{\boldsymbol{\xi}}); \quad \dot{\varepsilon}_{ij} = \frac{1}{2}\left(\frac{\partial \dot{\xi}_i}{\partial x_j} + \frac{\partial \dot{\xi}_j}{\partial x_i}\right) \tag{7.178}$$

The use of (7.152d) with $\mathbf{g} = \boldsymbol{\xi}$ and the application of the jump operator [[ ]] to (7.178) give:

$$c\,[[\dot{\varepsilon}]] = -\frac{1}{2}\left([[\ddot{\boldsymbol{\xi}}]] \otimes \mathbf{n} + \mathbf{n} \otimes [[\ddot{\boldsymbol{\xi}}]]\right) \tag{7.179a}$$

$$c\left[\left[\dot{\varepsilon}_{ij}\right]\right] = -\frac{1}{2}\left(\left[\left[\ddot{\xi}_i\right]\right] n_j + \left[\left[\ddot{\xi}_j\right]\right] n_i\right) \tag{7.179b}$$

Owing to the continuity and the symmetry $C^u_{ijkl} = C^u_{ijlk}$ of $\mathbf{C}^u$, (7.177) and (7.179) combine to give the following relations:

$$[[\dot{\boldsymbol{\sigma}}]] = -\frac{1}{c}\left(\mathbf{C}^u \cdot \mathbf{n}\right) \cdot \left[\left[\ddot{\boldsymbol{\xi}}\right]\right]; \quad \left[\left[\dot{\sigma}_{ij}\right]\right] = -\frac{1}{c} C^u_{ijkl} n_k \left[\left[\ddot{\xi}_l\right]\right] \tag{7.180}$$

Furthermore, applying (7.148) to $\partial \sigma_{ij}/\partial x_j$ and summing the resulting equations for $i = 1$ to 3, we also derive:

$$c\left[\left[\frac{\partial \sigma_{ij}}{\partial x_j}\right]\right] + \left[\left[\dot{\sigma}_{ij}\right]\right] n_j = 0; \quad c\,[[\nabla \cdot \boldsymbol{\sigma}]] + [[\dot{\boldsymbol{\sigma}}]] \cdot \mathbf{n} = 0 \tag{7.181}$$

Considering undrained evolutions such as $\boldsymbol{\xi}^f = \boldsymbol{\xi}$, the combination of (7.156) and (7.181) gives the dynamic relation:

$$\mathbf{n} \cdot [[\dot{\boldsymbol{\sigma}}]] - c\rho \left[\left[\ddot{\boldsymbol{\xi}}\right]\right] = 0 \tag{7.182}$$

Use of the latter in (7.180) finally yields:

$$\mathbf{A} \cdot \left[\left[\ddot{\boldsymbol{\xi}}\right]\right] = c^2 \left[\left[\ddot{\boldsymbol{\xi}}\right]\right]; \quad \mathbf{A} = \frac{1}{\rho}\mathbf{n} \cdot \mathbf{C} \cdot \mathbf{n}; \quad A_{jk} = \frac{1}{\rho} n_i C_{ijkl} n_l \tag{7.183}$$

Owing to property (7.183), the second-order tensor $\mathbf{A}$ is called the (undrained) acoustic tensor. Indeed, its eigenvalues and the related eigenvectors turn out to be the squares of the wavespeeds and the related eigenmovements.

Chapter 8

Poroplasticity

Poroplasticity is the ability of porous materials to undergo permanent strains and permanent changes in porosity and, as a consequence, permanent changes in fluid mass content. Still setting the viscosity effects aside (see Chapter 9), this chapter is devoted to the modelling of such behaviour, including hardening effects.

8.1 Poroplastic Behaviour

8.1.1 Plastic Strain and Plastic Porosity

Owing to the permanent strains and to the permanent changes in porosity, poroplastic evolutions are irreversible and, in contrast to poroelasticity, the strains ε_{ij} and the Lagrangian porosity ϕ do not suffice to characterize the current skeleton energy Ψ_s. Internal or hidden variables (see §3.4.2) must be added to capture the irreversible character of plasticity. These internal variables are the plastic strains ε_{ij}^p and the plastic porosity ϕ^p.

Since viscous effects related to the skeleton behaviour are not considered here, the physical time is actually involved only through heat and fluid mass transfers occurring between juxtaposed infinitesimal volumes. There is no intrinsic time attached to the skeleton plastic behaviour so that the plastic deformation occurs instantaneously in response to the stress and the fluid pressure increments whatever the actual time rate of the latter. The poroplastic evolution can be viewed as a succession of thermodynamic equilibrium states and depends only on the loading chronology. Accordingly, the constitutive equations linking the stress–fluid pressure history to the plastic strain–porosity history are appropriately formulated in incremental form.

Consider a porous material sample subjected to current stress σ_{ij} and current fluid pressure p. From this current state, let $d\sigma_{ij}$ and dp be an incremental loading in stress and fluid pressure and let $d\varepsilon_{ij}$ and $d\phi$ be the incremental strain and the incremental porosity they produce. The unloading process defined by the opposite increments $-d\sigma_{ij}$ and $-dp$ allow us to record the reversible or elastic increments $-d\varepsilon_{ij}^{el}$ and $-d\phi^{el}$. The

Poromechanics O. Coussy
 ISBN 0-470-84920-7

irreversible or plastic increments, $d\varepsilon_{ij}^p$ and $d\phi^p$, are defined through the relations:

$$d\varepsilon_{ij} = d\varepsilon_{ij}^{el} + d\varepsilon_{ij}^{p}; \quad d\phi = d\phi^{el} + d\phi^{p} \tag{8.1}$$

Plastic strain ε_{ij}^p and plastic (Lagrangian) porosity ϕ^p are defined as the integrals of the increments recorded along the loading path from an initial reference state in stress and fluid pressure to the current one. We write:

$$\varepsilon_{ij} = \varepsilon_{ij}^{el} + \varepsilon_{ij}^{p}; \quad \phi - \phi_0 = \phi^{el} + \phi^{p} \tag{8.2}$$

so that ϕ^p is eventually defined as the irreversible change of the porous volume per unit of initial volume $d\Omega_0$. The observable macroscopic volume plastic dilation ϵ^p undergone by the skeleton is due to both the plastic change in porosity and the volume plastic dilation ϵ_s^p undergone by the solid matrix. Applying (1.32), we obtain:

$$\epsilon^p = (1 - \phi_0)\epsilon_s^p + \phi^p \tag{8.3}$$

In soil and rock mechanics the plastic evolutions are caused by the irreversible relative sliding of the solid grains forming the matrix, so that the volume change of the matrix due uniquely to plasticity turns out to be negligible in the absence of any occluded porosity, resulting in $\epsilon_s^p = 0$ and entailing (see Fig. 8.1 below):

$$\phi^p = \epsilon^p \tag{8.4}$$

In order to capture the departure from matrix plastic incompressibility an heuristic assumption consists in postulating:[1]

$$\phi^p = \beta\epsilon^p \tag{8.5}$$

In contrast to the value $\beta = 1$ corresponding to a plastically incompressible matrix, that is (8.4), according to (8.3) the value $\beta = \phi_0$ corresponds to $\epsilon^p = \epsilon_s^p$, that is to a volumetric plastic strain of the skeleton due only to that of the solid matrix. It is then consistent to require β to satisfy inequalities $\phi_0 \leq \beta \leq 1$.

8.1.2 Poroplastic State Equations for the Skeleton

In the context of infinitesimal isothermal transformations and saturated porous materials, inequality (3.33) expressing the positiveness of the dissipation attached to the irreversible evolutions of the skeleton can be specialized in the form:

$$\sigma_{ij}\, d\varepsilon_{ij} + p\, d\phi - d\Psi_s \geq 0 \tag{8.6}$$

From the current state, when considering reversible or poroelastic evolutions, the values of internal plastic variables ε_{ij}^p and ϕ^p remain the same and there is no dissipation. For

[1] An even more general anisotropic assumption would have consisted in setting $\phi^p = \beta_{ij}\varepsilon_{ij}^p$.

such evolutions inequality (8.6) becomes an equality, leading to state equations:

$$\sigma_{ij} = \frac{\partial \Psi_s}{\partial \varepsilon_{ij}}; \quad p = \frac{\partial \Psi_s}{\partial \phi} \tag{8.7}$$

State equations (8.7) have been derived for elastic evolutions. They apply to any evolution as soon as free energy Ψ_s is continuously differentiable with regard to the whole set of state variables.

Free energy Ψ_s of the skeleton accounts for the energy which can be eventually recovered in mechanical form. Accordingly we let Ψ_s depend only on the reversible strain and the reversible change in porosity, that is:

$$\Psi_s = \Psi_s\left(\varepsilon_{ij} - \varepsilon_{ij}^p, \phi - \phi^p\right) \tag{8.8}$$

In (8.8) we did not consider the internal variables characterizing the possible hardening (or softening) state. This will be detailed in the specific sections of this chapter devoted to materials whose poroelasticity domain changes throughout plastic evolutions. Proceeding as in Chapter 4, from (8.7) and (8.8) we derive the state equations of linear poroelastoplasticity in the form:

$$\sigma_{ij} = C_{ijkl}\left(\varepsilon_{ij} - \varepsilon_{ij}^p\right) - b_{ij} p \tag{8.9a}$$

$$\phi - \phi^p - \phi_0 = b_{ij}\left(\varepsilon_{ij} - \varepsilon_{ij}^p\right) + \frac{p}{N} \tag{8.9b}$$

In the isotropic case they are specialized in the form:

$$\sigma = K(\epsilon - \epsilon^p) - bp \tag{8.10a}$$

$$s_{ij} = 2\mu\left(e_{ij} - e_{ij}^p\right) \tag{8.10b}$$

$$\phi - \phi^p - \phi_0 = b(\epsilon - \epsilon^p) + \frac{p}{N} \tag{8.10c}$$

8.1.3 Poroplastic State Equations for the Porous Material

Combining (4.59a) and (8.10c), we can get rid of the porosity variation $d\phi$ to the benefit of the variation in fluid mass content dm_f, resulting in:

$$\frac{dm_f}{\rho_f} = d\phi^p + b(d\epsilon - d\epsilon^p) + \frac{dp}{M} \tag{8.11}$$

where M is still defined by (4.61). Under the assumption of small perturbations (see §5.1.1), the integration of (8.11) gives:

$$p = M(-b(\epsilon - \epsilon^p) + v_f - \phi^p) \tag{8.12}$$

where v_f is the current change in fluid volume content per unit of initial volume (see (5.13)). For poroplastic materials v_f can be split into its reversible or elastic part and its irreversible or plastic part according to:

$$v_f = \frac{m_f - m_f^0}{\rho_f^0} = v_f^{el} + v_f^p \tag{8.13}$$

where we let:

$$v_f^{el} = \phi^{el} + \frac{\rho_f - \rho_f^0}{\rho_f^0}; \quad v_f^p = \phi^p \tag{8.14}$$

According to (8.14), the irreversible or plastic part v_f^p of v_f identifies with the plastic porosity ϕ^p so that the experimental determination of the former provides the means to determine the latter. The linear poroelastoplastic behaviour as captured by state equations (8.10)–(8.12) is experimentally illustrated for a limestone in Fig. 8.1.

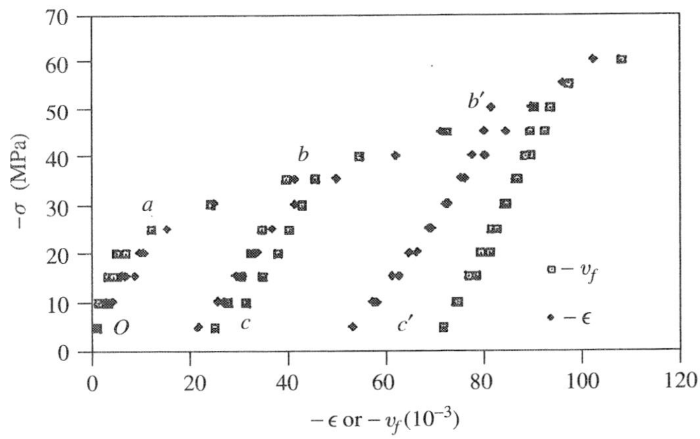

Figure 8.1: Experimental evidence of the poroplastic behaviour of a limestone. A limestone sample is subjected to loading–unloading cycles of confining pressure $-\sigma$ in drained conditions such that the fluid pressure remains equal to its initial value. Up to $-\sigma \simeq 20$ MPa the behaviour remains poroelastic since the same strain $-\epsilon$ and the same change in fluid volume content $-v_f$ are recorded along the loading and the unloading paths Oa and aO. Beyond this initial threshold the behaviour becomes poroplastic along the loading paths ab and bb' since irreversible or plastic strains and changes in fluid volume content are recorded along the unloading paths bc and $b'c'$. By letting $p = \phi^p = \epsilon^p = 0$ in (8.12), Biot's coefficient b is assessed by the ratio v_f/ϵ recorded along the poroelastic unloading paths aO, bc or $b'c'$ and is found here to be close to 0.9. Along the poroplastic loading paths ab and bb' the values of $-\epsilon$ and $-v_f$ are the sum of the poroelastic contributions $-\epsilon^{el}$ and $-\phi^{el}$ (with $\rho_f = \rho_f^0$ in (8.14)), and the poroplastic contributions $-\epsilon^p$ and $-\phi^p$. Since the values of $-\epsilon$ and $-v_f$ nearly coincide and the elastic contributions were found to be nearly the same ($b \simeq 0.9$), it can be concluded that $\epsilon^p = \phi^p$ and, consequently, that the solid matrix is plastically incompressible. In addition, the limestone exhibits a hardening behaviour since along the poroplastic loadings ab and bb' the elastic limit in confining pressure $-\sigma$ increases (by courtesy of F. Skozylas).

8.1.4 Domain of Poroelasticity and the Loading Function. Ideal and Hardening Poroplastic Material

Any loading is characterized by the stress components σ_{ij} and the fluid pressure p. Owing to symmetry $\sigma_{ij} = \sigma_{ji}$, the current loading (σ_{ij}, p) can be represented in the R^{6+1} loading space $\{\sigma_{ij} \times p\}$ relative to the six independent stress components σ_{ij} and to the fluid pressure p. In the loading space there exists an initial domain of poroelasticity including the zero loading point $(\sigma_{ij} = 0, p = 0)$ for a material devoid of any loading history. The poroelasticity domain C_E is such that the strain and the change in porosity remain reversible along any loading path starting from the origin and lying entirely within the domain (Fig. 8.2a, paths such as Oa). For an ideal poroplastic material, as sketched in Fig. 8.2a, the initial domain of poroelasticity is not altered by the plastic evolutions occurring along the loading paths such as ab. By contrast, for a hardening poroplastic material, as sketched in Fig. 8.2b, the initial domain of poroelasticity is altered by the plastic evolutions occurring along the loading paths such as ab. For an ideal poroplastic material the interior of poroelasticity domain C_E can be defined by means of the loading function $f(\sigma_{ij}, p)$ such as:

$$
\begin{aligned}
&f(\sigma_{ij}, p) < 0 \text{ for } (\sigma_{ij}, p) \text{ belonging to the interior of } C_E \\
&f(\sigma_{ij}, p) = 0 \text{ for } (\sigma_{ij}, p) \text{ belonging to the border of } C_E \\
&f(\sigma_{ij}, p) > 0 \text{ for } (\sigma_{ij}, p) \text{ belonging to the exterior of } C_E
\end{aligned}
\tag{8.15}
$$

The elasticity criterion is the condition $f(\sigma_{ij}, p) < 0$. The plasticity criterion is the condition $f(\sigma_{ij}, p) = 0$ while the yield surface is the border of C_E, that is the surface defined

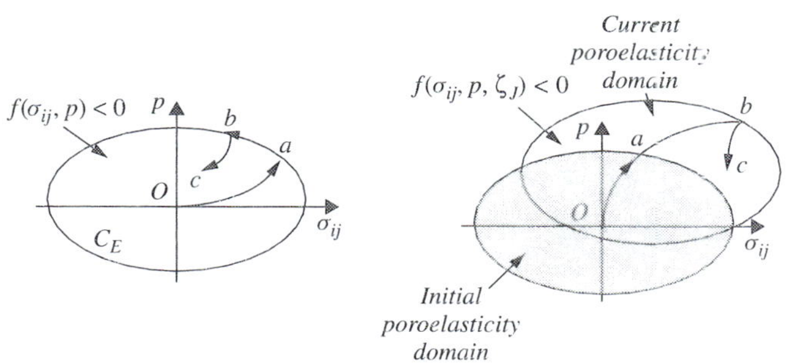

(a) Ideal poroplasticity (b) Hardening poroplasticity

Figure 8.2: (a) The initial domain of poroelasticity defined by $f(\sigma, p) < 0$ in the R^{6+1} loading space $\{\sigma \times p\}$ so that the strain and the change in porosity recorded along the loading paths such as Oa remain reversible. For an ideal poroplastic material, as sketched in (a), the initial domain of poroelasticity is not altered by plastic evolutions occurring along the loading paths such as ab. By contrast, for a hardening poroplastic material, as sketched in (b) and experimentally observed in Fig. 8.1, the initial domain of poroelasticity is altered by the plastic evolutions occurring along the loading paths such as ab. In order to account for the alteration the loading function must involve evolutionary hardening forces ζ_J so that the interior of the current domain of elasticity is defined by the condition $f(\sigma_{ij}, p, \zeta_J) < 0$.

by $f(\sigma_{ij}, p) = 0$. A plastically admissible loading (σ_{ij}, p) satisfies $f(\sigma_{ij}, p) \leq 0$. In the case of a hardening poroplastic material, the previous definitions apply to the current domain of elasticity, provided that the current loading function $f(\sigma_{ij}, p, \zeta_J)$ involves in addition hardening forces ζ_J accounting for the current hardening state.

The domain of elasticity of most plastic materials is convex and this property can be extended to porous materials whose poroplastic behaviour is due to that of the solid matrix. The fundamental property of a convex domain is that all points of a segment that joins any couple of points belonging to the border of the domain lie inside the domain. If loading function $f(\sigma_{ij}, p)$ is continuously differentiable with respect to both σ_{ij} and p the fundamental property is equivalent to the condition:

$$\forall(\sigma_{ij}, p) \text{ and } \forall(\tilde{\sigma}_{ij}, \tilde{p}) \neq (\sigma_{ij}, p) \text{ with } f(\sigma_{ij}, p) = f(\tilde{\sigma}_{ij}, \tilde{p}) = 0:$$

$$(\sigma_{ij} - \tilde{\sigma}_{ij})\frac{\partial f}{\partial \sigma_{ij}} + (p - \tilde{p})\frac{\partial f}{\partial p} \geq 0 \tag{8.16}$$

The plasticity criterion $f(\sigma_{ij}, p) = 0$ or $f(\sigma_{ij}, p, \zeta_J) = 0$ indicates when plastic evolutions may occur. The (plastic) flow rule specifies how they occur and its formulation is the objective of the two subsequent sections, separately addressing the ideal and the hardening poroplastic material.

8.2 Ideal Poroplasticity

8.2.1 The Flow Rule and the Plastic Work

Owing to the definition of the plasticity criterion, if the loading point (σ_{ij}, p) lies within the poroelasticity domain C_E, that is if $f(\sigma_{ij}, p) < 0$ (point c in Fig. 8.2a), the evolutions are poroelastic. If the loading point (σ_{ij}, p) is on the border of C_E but subsequently leaves it (point b in Fig. 8.2a), the evolutions are also poroelastic (elastic unloading). Since the value of loading function f becomes negative for a loading point (σ_{ij}, p) re-entering the poroelasticity domain C_E, the elastic unloading condition is $f = 0$ and $df < 0$, with:

$$df = \frac{\partial f}{\partial \sigma_{ij}} d\sigma_{ij} + \frac{\partial f}{\partial p} dp \tag{8.17}$$

Furthermore, if the loading point (σ_{ij}, p) is and subsequently remains on the border of C_E (point a in Fig. 8.2b) plasticity evolutions may occur. The corresponding loading now satisfies $f = df = 0$ and is said to be neutral with respect to the plasticity criterion.

Taking into account all the above remarks, the flow rule can now be more precisely expressed in the form:

$$d\varepsilon_{ij}^p = d\lambda\ h_{ij}^{\varepsilon}(\sigma_{ij}, p); \quad d\phi^p = d\lambda\ h_{\phi}(\sigma_{ij}, p) \tag{8.18}$$

In (8.18), $d\lambda$ is the so-called plastic multiplier that scales the intensity of plastic strain increments $(d\varepsilon_{ij}^p, d\phi^p)$ and satisfies the Kuhn–Tucker relations:

$$d\lambda \geq 0; \quad f \leq 0; \quad d\lambda \cdot f = 0; \quad d\lambda \cdot df = 0 \tag{8.19}$$

The couple $\left(h_{ij}^{\varepsilon}, h_{\phi}\right)$ defines the directions of the plastic strain increments $\left(d\varepsilon_{ij}^{p}, d\phi^{p}\right)$ in the loading space $\left\{\sigma_{ij} \times p\right\}$ and depends on the current loading (σ_{ij}, p).

For an ideal plastic material, the plastic multiplier $d\lambda$ remains undetermined by the material constitutive equations but does depend on the poroelastoplastic evolution of the whole structure formed by the porous material. In contrast, directions $\left(h_{ij}^{\varepsilon}, h_{\phi}\right)$ depend on the material constitutive equations and require a further analysis. Substitution of (8.8) into (8.6) and use of state equations (8.7) give:

$$\delta W^{p} = \sigma_{ij} d\varepsilon_{ij}^{p} + p d\phi^{p} \geq 0 \tag{8.20}$$

In the absence of any other state variable than $\varepsilon_{ij} - \varepsilon_{ij}^{p}$ and $\phi - \phi^{p}$ as arguments of Ψ_s, the dissipated energy related to the skeleton plastic evolution reduces to the infinitesimal plastic work δW^{p}, while the positiveness of the associated infinitesimal production of entropy is:

$$\delta S^{p} = \frac{\delta W^{p}}{T} \geq 0 \tag{8.21}$$

Owing to the (conventional) positiveness of plastic multiplier $d\lambda$, the positiveness (8.20) of the plastic work finally requires $\left(h_{ij}^{\varepsilon}, h_{\phi}\right)$ to satisfy:

$$\sigma_{ij} h_{ij}^{\varepsilon} + p h_{\phi} \geq 0 \tag{8.22}$$

8.2.2 Principle of Maximal Plastic Work and the Flow Rule. Standard and Non-standard Materials

In order to derive the flow rule the principle of maximal plastic work is now invoked.[2] The principle of maximal plastic work requires the actual plastic strain increment $\left(d\varepsilon_{ij}^{p}, d\phi^{p}\right)$ associated with the current loading (σ_{ij}, p) to maximize the plastic work over all plastically admissible loadings $(\tilde{\sigma}_{ij}, \tilde{p})$. It is written:

$$\begin{gathered} \forall(\tilde{\sigma}_{ij}, \tilde{p}), \quad f(\tilde{\sigma}_{ij}, \tilde{p}) \leq 0 : \\ (\sigma_{ij} - \tilde{\sigma}_{ij}) d\varepsilon_{ij}^{p} + (p - \tilde{p})\, d\phi^{p} \geq 0 \quad (\delta W^{p} \geq \delta \tilde{W}^{p}) \end{gathered} \tag{8.23}$$

In fact, according to (8.21) the principle of maximal plastic work relies on the more general principle of maximal production of entropy as often encountered in physics. Since the zero loading $(\tilde{\sigma}_{ij} = 0, \tilde{p} = 0)$ is plastically admissible, the principle of maximal plastic work ensures the fulfilment of condition (8.22).

In order to explore the consequences of the principle of maximal plastic work we rewrite (8.23) in the condensed form:

$$\forall \tilde{\mathbf{\Sigma}}, \ f\left(\tilde{\mathbf{\Sigma}}\right) \leq 0 : \ \left(\mathbf{\Sigma} - \tilde{\mathbf{\Sigma}}\right) \cdot d\mathbf{E}^{p} \geq 0 \tag{8.24}$$

[2]The formulation of the principle of maximal plastic work is originally due to R. Hill (see Hill R. (1950), *The Mathematical Theory of Plasticity*, Clarendon Press, Oxford).

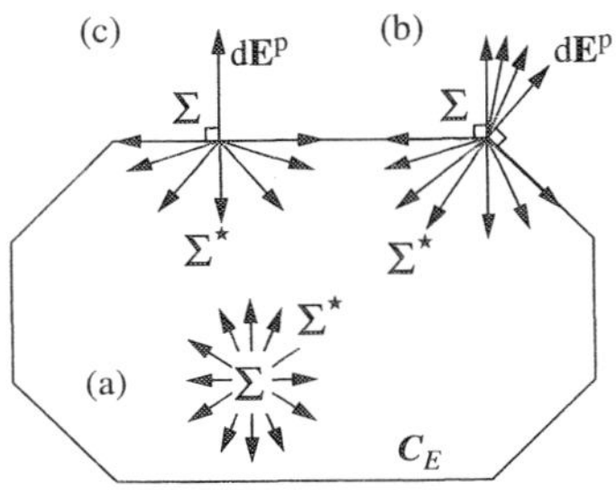

Figure 8.3: Principle of maximal plastic work and the normality of the flow rule. See the main text for an explanation of the different situations (a), (b) and (c) sketched in the figure.

where we denote $\mathbf{\Sigma} = (\sigma_{ij}, p)$ and $\mathbf{E}^p = (\varepsilon_{ij}^p, \phi^p)$. If $\mathbf{\Sigma}$ lies inside the poroelasticity domain C_E, vector $\mathbf{\Sigma} - \tilde{\mathbf{\Sigma}}$ can take any orientation so that, in order to fulfil condition (8.24), $d\mathbf{E}^p$ must be zero in conformity with the actual definition of the poroelasticity domain (case (a) of Fig. 8.3). Now let $\mathbf{\Sigma}$ be on the border of poroelasticity domain C_E. Since $\tilde{\mathbf{\Sigma}}$ cannot lie in the exterior of the domain of elasticity, vector $\tilde{\mathbf{\Sigma}} - \mathbf{\Sigma}$, for stress states $\tilde{\mathbf{\Sigma}}$ close to $\mathbf{\Sigma}$, can only point towards the interior of C_E lying in the immediate vicinity of $\mathbf{\Sigma}$. In order to fulfil (8.24), vector $d\mathbf{E}^p$ has to belong to the cone of outward normals to the border of C_E, resulting in the so-called normality of the flow rule (case (b) of Fig. 8.3). In any regular point of the border of C_E this cone reduces to the normal (case (c) of Fig. 8.3) and the normality of the flow rule is expressed in the form:

$$d\varepsilon_{ij}^p = d\lambda \frac{\partial f}{\partial \sigma_{ij}}; \quad d\phi^p = d\lambda \frac{\partial f}{\partial p}; \quad d\lambda \geq 0 \tag{8.25}$$

Substitution of (8.25) into (8.23) yields:

$$(\sigma_{ij} - \tilde{\sigma}_{ij}) \frac{\partial f}{\partial \sigma_{ij}} + (p - \tilde{p}) \frac{\partial f}{\partial p} \geq 0 \tag{8.26}$$

Comparing (8.16) and (8.26), we conclude that the principle of maximal plastic work also implies the convexity of the poroelasticity domain C_E in addition to the normality of the flow rule. Finally, let us consider the case of elastic unloading where $\mathbf{\Sigma}$, in addition to being on the border of poroelasticity domain C_E, subsequently leaves it so that:

$$\frac{\partial f}{\partial \sigma_{ij}} d\sigma_{ij} + \frac{\partial f}{\partial p} dp < 0 \tag{8.27}$$

Inequality (8.24) requires increment $d\mathbf{\Sigma} = (d\sigma_{ij}, dp)$ to satisfy inequality $d\mathbf{\Sigma} \cdot d\mathbf{E}^p = d\sigma_{ij} d\varepsilon_{ij}^p + dp d\phi^p \geq 0$. Combining the latter with (8.25) and (8.27) gives $d\lambda = 0$, that is $d\mathbf{E}^p = 0$ in agreement with the definition of elastic unloading.

Collecting the above results, we conclude that the principle of maximal plastic work again requires the plastic multiplier to satisfy the Kuhn–Tucker conditions (8.19), but also requires the yield surface to be convex and the directions of the plastic strain increments

$(d\varepsilon_{ij}^p, d\phi^p)$ to be normal to the latter.[3] Potential f is then said to be associated and the material is said to be standard. In the case of a non-associated potential, $g \neq f$, the flow rule of the related non-standard material can be written in the form:

$$d\varepsilon_{ij}^p = d\lambda \frac{\partial g}{\partial \sigma_{ij}}; \quad d\phi^p = d\lambda \frac{\partial g}{\partial p} \tag{8.28}$$

where $d\lambda$ is still required to satisfy (8.19). Owing to the positiveness condition (8.22) that is now not automatically satisfied, g is required to satisfy:

$$\sigma_{ij} \frac{\partial g}{\partial \sigma_{ij}} + p \frac{\partial g}{\partial p} \geq 0 \tag{8.29}$$

8.3 Hardening Poroplasticity

8.3.1 Hardening Variables and Trapped Energy

As ascertained for the limestone related to the experimental data reported in Fig. 8.1 and as sketched in Fig. 8.2b, the domain of elasticity of actual materials is generally altered by the occurrence of plastic evolutions. This alteration is accounted for by introducing hardening forces $\zeta_{J=1,N}$ whose strength changes during the plastic evolutions and modifies the current value of the loading function $f(\sigma_{ij}, p, \zeta_J)$.

To understand the energy side of hardening plasticity let us consider an ideal poroplastic material undergoing undrained plastic evolutions. In such undrained evolutions the poroplastic material apparently behaves as a plastic solid whose loading function $f(\sigma_{ij}, p)$ depends on hardening force $\zeta = p$. In the case of linear poroelasticity the free or elastic energy associated with this internal hardening force is $\frac{1}{2}pv_f^{el}$. In undrained evolutions the fluid mass content does not change. Substituting $v_f = 0$ into (8.13) and (8.14) we infer that $v_f^{el} = -\phi^p$. At the end of the loading–unloading process, the strain reduces to the plastic contribution. Putting $v_f = 0$ and $\epsilon = \epsilon^p$ into (8.12) we derive $p = Mv_f^{el} = -M\phi^p$. Accordingly, we conclude that a part U of the overall free energy stored by the undrained porous material during the loading process is not recovered at the end of the unloading process. This 'trapped' energy can be expressed in the form:

$$U = \frac{1}{2}M(\phi^p)^2 \tag{8.30}$$

U is the free energy associated with the elastic fluid volume change v_f^{el} counterbalancing the plastic change in porosity ϕ^p in order that the internal plastic evolution remains compatible with the saturation condition of the porous space under undrained conditions.

For the hardening poroplastic material the same phenomenon of energy trapping occurs on account of the kinematical compatibility conditions of the deformation at the microscopic scale of the grains forming the solid matrix. After a complete unloading process

[3] Indeed, the principle of maximal plastic work turns out to require the plastic dissipation mechanism to be normal in the sense defined in §3.4.2 (see in particular Coussy O. (1995), *Mechanics of Porous Continua*, John Wiley & Sons, Chichester).

the mesoscopic permanent volumetric dilation ϵ_s^p in (8.3) does not generally result from a homogeneous microscopic state of plasticization of the solid grains. Indeed, whatever the scale considered, only the total strain eventually derives from a displacement field so that an elastic contribution is required to ensure the kinematic compatibility of the total microscopic strain of the solid grains. The free energy represented by the elastic energy relative to this microscopic contribution is therefore trapped, even after complete unloading. In order to capture the current trapped energy U, we replace (8.8) by:

$$\Psi_s = W_s\left(\varepsilon_{ij} - \varepsilon_{ij}^p, \phi - \phi^p\right) + U(\chi_J) \tag{8.31}$$

In (8.31), hardening state variables χ_J are introduced in order to account for the current state of hardening resulting in the trapping of energy $U(\chi_J)$. For instance, in the case of an ideal poroplastic material subjected to undrained evolutions, (8.30) leads us to identify hardening variable χ with plastic porosity ϕ^p. Experimental evidence (e.g. the experiments reported in Fig. 8.1) shows that hardening effects do not significantly affect state equations (8.10). As anticipated by (8.31), this results in expressing overall free energy Ψ_s as the sum[4] of an energy $W_s\left(\varepsilon_{ij} - \varepsilon_{ij}^p, \phi - \phi^p\right)$ depending only on reversible variations and of the trapped energy $U(\chi_J)$ depending only on hardening state variables χ_J. The actual meaning of trapped energy U is finally obtained by expressing the positiveness of the dissipated energy. Substitution of (8.31) into (8.6) and use of state equations (8.7) give:

$$\delta W^p - dU \geq 0 \tag{8.32}$$

During an infinitesimal poroplastic evolution, the amount dU of free energy is actually trapped but, not being dissipated in heat form, it has to be removed from the infinitesimal plastic work δW^p. Accordingly, the trapped energy U can only be determined by the difference between the plastic work, as defined by (8.20), and the energy dissipated into heat. This imperatively requires calorimetric measurements.

8.3.2 Flow Rule for the Hardening Material. Hardening Modulus

The positiveness (8.32) of dissipated energy can be written explicitly:

$$\sigma_{ij} d\varepsilon_{ij}^p + p d\phi^p + \zeta_J d\chi_J \geq 0 \tag{8.33}$$

where the hardening force ζ_J is consistently recognized as the force associated with the hardening variable χ_J according to the state equation:

$$\zeta_J = -\frac{\partial U}{\partial \chi_J} \tag{8.34}$$

[4]This energy separation is irrelevant for the ideal poroplastic material subjected to undrained evolutions as previously analysed. Indeed 'hardening force' p appears explicitly in state equation (8.10a) whereas its expression derived from (8.10c) involves reversible variations of the state variables.

Variations $d\chi_J$ of hardening variables occur only during a (plastic) loading which produces plastic increments $\left(d\varepsilon_{ij}^p, d\phi^p\right)$. Accordingly, flow rule (8.18) of ideal poroplasticity is extended in the form:

$$d\varepsilon_{ij}^p = d\lambda\ h_{ij}^\varepsilon(\sigma_{ij}, p, \zeta_J) \tag{8.35a}$$

$$d\phi^p = d\lambda\ h_\phi(\sigma_{ij}, p, \zeta_J) \tag{8.35b}$$

$$d\chi_J = d\lambda\ h_J(\sigma_{ij}, p, \zeta_J) \tag{8.35c}$$

Plastic multiplier $d\lambda$ still obeys (8.19) provided that the arguments of the loading function f now include the hardening forces ζ_J. Owing to the positiveness of plastic multiplier $d\lambda$, the positiveness (8.33) of dissipated energy requires $\left(h_{ij}^\varepsilon, h_\phi, h_J\right)$ to satisfy:

$$\sigma_{ij}h_{ij}^\varepsilon + ph_\phi + \zeta_J h_J \geq 0 \tag{8.36}$$

The principle of maximal plastic work can again be invoked to specify the directions $\left(h_{ij}^\varepsilon, h_\phi\right)$ of the plastic increments $\left(d\varepsilon_{ij}^p, d\phi^p\right)$. This principle can possibly be extended to the hardening work $-dU = \zeta_J d\chi_J$ so that, in addition to (8.25), we can write:

$$d\chi_J = d\lambda \frac{\partial f}{\partial \zeta_J} \tag{8.37}$$

while the convexity of $f(\sigma_{ij}, p, \zeta_J)$ extends to hardening forces ζ_J so that (8.36) is automatically satisfied. Alternatively, in the case of a non-associated potential, $g \neq f$, the flow rule concerning the hardening variables is written in the form:

$$d\chi_J = d\lambda \frac{\partial g}{\partial \zeta_J} \tag{8.38}$$

In any case, as soon as a non-associated potential g is involved, condition (8.36) must be written accordingly.

Since f now includes hardening forces ζ_J, the consistency condition $df = 0$ of plastic loading must be reconsidered. To this end let $d_\zeta f$ be defined according to:

$$d_\zeta f = \frac{\partial f}{\partial \sigma_{ij}} d\sigma_{ij} + \frac{\partial f}{\partial p} dp \tag{8.39}$$

so that:

$$df = d_\zeta f + \frac{\partial f}{\partial \zeta_J} d\zeta_J \tag{8.40}$$

Use of (8.34) and (8.38), together with the consistency condition $df = 0$, allows us to identify the expression of the plastic multiplier $d\lambda$ in the form:

$$d\lambda = \frac{d_\zeta f}{H} \tag{8.41}$$

where H is the hardening modulus defined by:

$$H = -\frac{\partial f}{\partial \zeta_J}\frac{\partial \zeta_J}{\partial \chi_J}\frac{d\chi_J}{d\lambda} = \frac{\partial f}{\partial \zeta_J}\frac{\partial^2 U}{\partial \chi_J^2}\frac{\partial g}{\partial \zeta_J} \tag{8.42}$$

Plastic evolution involves positive values for the plastic multiplier $d\lambda$. Expression (8.41) of the latter leads us to distinguish actual hardening, where $H > 0$, from softening, where $H < 0$. Again using the incremental loading vector $d\boldsymbol{\Sigma} = (d\sigma_{ij}, dp)$, we first express $d_\zeta f$ in the condensed form:

$$d_\zeta f = \boldsymbol{\nu} \cdot d\boldsymbol{\Sigma} \tag{8.43}$$

where $\boldsymbol{\nu} = (\partial f/\partial \sigma_{ij}, \partial f/\partial p)$ stands for the outward normal to the yield surface $f = 0$ in loading space $\{\sigma_{ij} \times p\}$. For a hardening material such as $H > 0$, condition $d\lambda > 0$ implies that plastic evolutions actually do occur if and only if $d_\zeta f > 0$; that is, if and only if loading increment $d\boldsymbol{\Sigma}$ is oriented outwards with regard to the current poroelasticity domain (see Fig. 8.4a). In other words, the new loading $\boldsymbol{\Sigma} + d\boldsymbol{\Sigma}$ escapes from the current poroelasticity domain while carrying it along. By contrast, for a softening material such as $H < 0$ (negative hardening), condition $d\lambda > 0$ implies that plastic evolutions actually do occur if and only if $d_\zeta f < 0$; that is, if and only if loading increment $d\boldsymbol{\Sigma}$ is oriented inwards with regard to the current poroelasticity domain (see Fig. 8.4b). In other words, the new loading $\boldsymbol{\Sigma} + d\boldsymbol{\Sigma}$ enters the current poroelasticity domain while carrying it along. Accordingly, in the softening case plastic loading cannot be distinguished from elastic loading for an experiment carried out by imposing the incremental loading $d\boldsymbol{\Sigma}$. In both cases the new loading enters the current poroelasticity domain and the material response is not unique (see details in Fig. 8.4b). Only an experiment carried out by monitoring the strain increment $d\mathbf{E}$ produces the experimental evidence of softening. Finally, contrary to ideal plasticity, for both hardening and softening materials no plastic evolutions occur

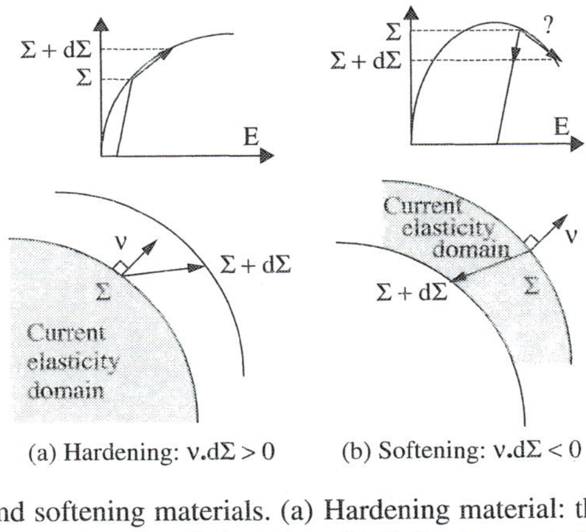

Figure 8.4: Hardening and softening materials. (a) Hardening material: the new loading point $\boldsymbol{\Sigma} + d\boldsymbol{\Sigma}$ escapes from the current domain of elasticity. (b) Softening material: the new loading point enters the current domain of elasticity so that plastic loading cannot be distinguished from elastic loading for experiments carried out by imposing the loading increment $d\boldsymbol{\Sigma}$.

in the case of a neutral incremental loading such as $d_\zeta f = 0$ so that the new loading $\Sigma + d\Sigma$ remains on the border of the current poroelasticity domain.

8.4 Usual Models of Poroplasticity

8.4.1 Poroplastic Effective Stress

In the case of matrix plastic incompressibility, substitution of (8.4) into (8.20) gives:

$$\delta W^p = \sigma'_{ij} d\varepsilon^p_{ij} \tag{8.44}$$

so that Terzaghi's effective stress defined by:

$$\sigma'_{ij} = \sigma_{ij} + p\,\delta_{ij} \tag{8.45}$$

is found to be the driving force of plastic strains. If the heuristic assumption (8.5) is adopted in order to explore the consequences of the departure from matrix plastic incompressibility, the driving force of the plastic strain can be written more generally as $\sigma'_{ij} = \sigma_{ij} + \beta p\,\delta_{ij}$. By contrast, the driving force of elastic strains is Biot's effective stress $\sigma''_{ij} = \sigma_{ij} + bp\,\delta_{ij}$ (see (4.26)). Recalling Biot's relation $b = 1 - K/K_s$, it is only in the case of elastic incompressibility ($K_s \to \infty, b = 1$) and plastic incompressibility ($\beta = 1$) of the matrix that both driving forces coincide. Furthermore, when the material also obeys the principle of maximal plastic work, a combination of (8.5) and (8.25) requires the loading function f to depend only upon σ'_{ij}. Accordingly, the plasticity criterion is expressed in the form $f(\sigma'_{ij}) = 0$, which can be confirmed by experiment as illustrated in Figs. 8.5[5] and 8.8. By extension, in the more general context of a non-associated flow rule both loading function f and non-associated potential g are generally expressed as a function only of σ'_{ij} so that non-associated flow rule (8.28) is written in the form:

$$d\varepsilon^p_{ij} = d\lambda \frac{\partial g}{\partial \sigma'_{ij}}; \quad d\phi^p = d\epsilon^p \tag{8.46}$$

8.4.2 Isotropic and Kinematic Hardening

Two main hardening modes are generally invoked to capture the transformations of the poroelasticity domain through plastic loadings:

- *Isotropic hardening*. In isotropic hardening the poroelasticity domain dilates (hardening) or contracts (softening) in an isotropic way around the origin of the effective loading space $\{\sigma'_{ij}\}$. A single hardening scalar force ζ is required to define the homothetical transformation of the initial reference hardening state into the current one (see Fig. 8.6a).

[5]See Kerbouche R., Shao J.-F., Skozylas F., Henry J.-P. (1995), 'On the poroplastic behavior of porous rocks', *European Journal of Mechanics, A/Solids*, **14**, (4), 3577–3587.

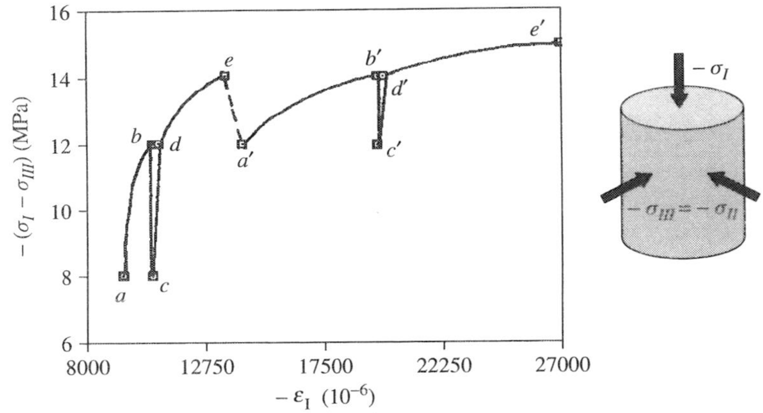

Figure 8.5: Checking the validity of the plasticity criterion in the form $f(\sigma'_{ij} = \sigma_{ij} + p\delta_{ij}) = 0$. Path ab corresponds to a plastic shear loading by increasing the axial pressure $-\sigma_I$ (from 28 MPa to 32 MPa), while holding constant both the confining pressure $-\sigma_{III} = -\sigma_{II}$ (= 20 MPa) and the pore pressure p (= 10 MPa). Path bc corresponds to the exact opposite elastic shear unloading. Path cd corresponds to a new elastic shear loading, but now by holding constant the axial pressure $-\sigma_I$ (= 28 MPa) and decreasing both the confining pressure $-\sigma_{III} = -\sigma_{II}$ (from 20 MPa to 16 MPa) and the pore pressure p (from 10 MPa to 16 MPa) in order to restore the same effective volumetric stress $-\frac{1}{3}\sigma'_{kk}$ (i.e. 42 MPa) as in the previous loading point b, but now achieved for distinct stresses and pore pressure. Subsequent loading path de corresponds to a new plastic loading, so that it can be concluded that the plasticity criterion is expressed as a function only of the effective stress σ'_{ij}. This can be confirmed by repeating the procedure (new loading path $a'b'c'd'e'$). The experimental data correspond to the same limestone as that in Fig. 8.1, which was reported to be plastically incompressible, that is $\phi^p = \epsilon^p$ (after Kerbouche *et al.* (1995), see footnote).

- *Kinematic hardening*. In kinematic hardening the poroelasticity domain moves in a rigid way in the effective loading space $\{\sigma'_{ij}\}$ (see Fig. 8.6b). A second-order tensor of components ζ_{ij} is required to define the translation transporting the initial reference yield surface to the current one (see Fig. 8.6b).

The two previous hardening models can be appropriately combined to yield an isotropic–kinematic hardening model.

8.4.3 The Usual Cohesive–Frictional Poroplastic Model

The Drucker–Prager criterion

For an isotropic material all the directions are equivalent so that the loading function f involves only the fluid pressure and the eigenvalues of the stress tensor, that is the principal stresses $\sigma_{J=I,II,III}$. The latter are expressed as a function of the three first invariants of stress tensor $\boldsymbol{\sigma}$. Let s be the deviator of $\boldsymbol{\sigma}$, reading:

$$\mathbf{s} = \boldsymbol{\sigma} - \sigma \mathbf{1}; \quad s_{ij} = \sigma_{ij} - \frac{1}{3}\sigma_{kk}\delta_{ij} \tag{8.47}$$

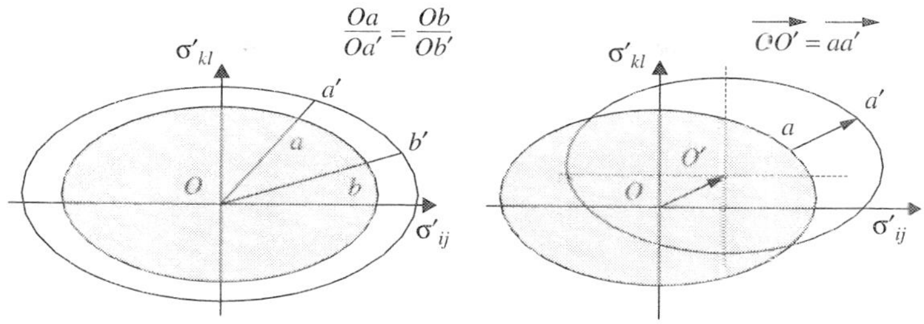

(a) Isotropic hardening (b) Kinematic hardening

Figure 8.6: Usual hardening models: (a) isotropic hardening; (b) kinematic hardening.

The second invariant of $\boldsymbol{\sigma}$ equals the (positive) second invariant τ of $\mathbf{s}$, that is:

$$\tau = \sqrt{\frac{1}{2}s_{ij}s_{ji}} = \sqrt{\frac{1}{6}\left[(\sigma_I - \sigma_{II})^2 + (\sigma_I - \sigma_{II})^2 + (\sigma_I - \sigma_{II})^2\right]} \tag{8.48}$$

The loading function of the usual isotropic poroplastic materials is generally well captured by involving only the first two stress invariants. In the case of ideal plasticity we write:

$$f = f(\sigma', \tau) \tag{8.49}$$

where $\sigma' = \sigma + p$, assuming that Terzaghi's effective stress concept is relevant (see §8.4.1).

Let $\mathbf{n}_{(\Delta)}$ be the vector orienting trisector (Δ) of the effective loading space $\{\sigma'_I \times \sigma'_{II} \times \sigma'_{III}\}$:

$$\mathbf{n}_{(\Delta)} = \frac{1}{\sqrt{3}}(\mathbf{e}_I + \mathbf{e}_{II} + \mathbf{e}_{III}) \tag{8.50}$$

where $\mathbf{e}_{J=I,II,III}$ stand for the principal stress directions. In the effective loading space $\{\sigma'_I \times \sigma'_{II} \times \sigma'_{III}\}$ where the origin is O, the projection Q of the loading point $P(\sigma'_I, \sigma'_{II}, \sigma'_{III})$ onto trisector (Δ) is such that $OQ = \sigma\sqrt{3}$, $PQ = \tau\sqrt{2}$. Consequently, as sketched in Fig. 8.7, the yield surface of isotropic materials defined by $f = 0$ is an axisymmetric surface around (Δ). Furthermore, noting that $\sigma' = \mathbf{n}_{(\Delta)} \cdot \boldsymbol{\sigma}' \cdot \mathbf{n}_{(\Delta)}$ and $(\boldsymbol{\sigma}' \cdot \mathbf{n}_{(\Delta)})^2 = \sigma'^2 + \frac{2}{3}\tau^2$, the plasticity criterion $f(\sigma', \tau) = 0$ turns out to be a relation which links the effective normal stress σ' and the shear stress $\tau\sqrt{\frac{2}{3}}$ exerted on the material facet oriented by $\mathbf{n}_{(\Delta)}$. Plastic yielding relates to the irreversible sliding between the solid material planes facing each other and the ones oriented by $\mathbf{n}_{(\Delta)}$. Owing to the physical interpretation of σ' and τ, the loading function is consistently sought in the form (see Fig. 8.8 for experimental evidence[6]):

$$f(\sigma', \tau) = \tau + \text{f}\sigma' - \text{c} \tag{8.51}$$

[6]The data reported in Fig. 8.8 are from Vincké O., Boutéca M., Piau J.-M., Fourmaintraux D. (1998), 'Study of the effective stress at failure', *Poromechanics, A tribute to M.A. Biot*, Proceedings of the First Biot Conference, ed. Thymus *et al.*, Balkema J.-F., Rotterdam.

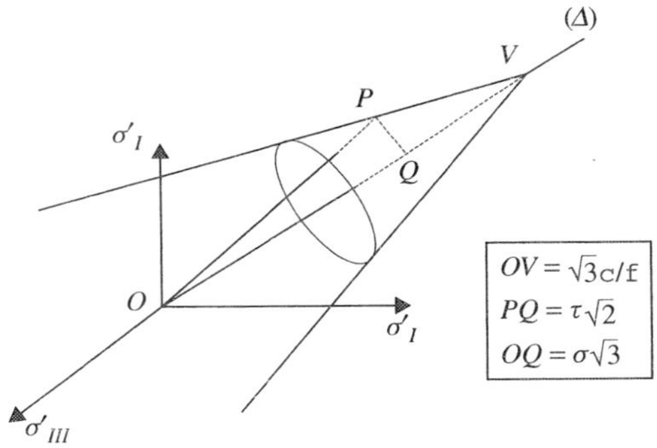

Figure 8.7: For isotropic poroplastic materials the yield surface defined by $f(\sigma', \tau) = 0$ is an axisymmetric surface around the trisector (Δ), specified as a cone for the Drucker–Prager criterion (8.51).

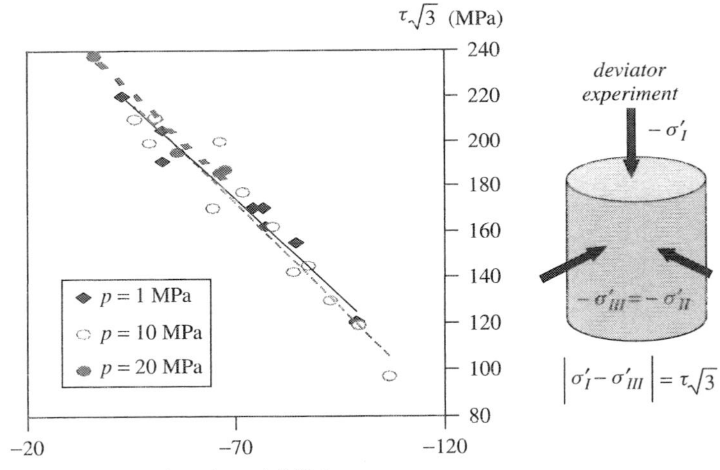

Figure 8.8: Investigation of the plasticity criterion of a limestone and checking the validity of Terzaghi's plastic effective stress (8.45) through a deviatoric experiment. This experiment consists of subjecting the sample to a shear force by progressively increasing the axial effective pressure $-\sigma'_I$ beyond a constant confining effective pressure $-\sigma'_{II} = -\sigma'_{III}$, so that $\sqrt{3}\tau = \left|\sigma'_I - \sigma'_{III}\right|$. The limestone is the one whose Biot's coefficient was equal to 0.63 in Fig. 4.1. It is instructive here to note that Terzaghi's plastic effective stress (8.45) does not coincide with Biot's elastic effective stress (4.26) (after Vincké *et al.* (1998) © Sweets & Zeitlinger, see footnote).

where $\mathtt{f}$ is a friction coefficient, while $\mathtt{c}$ is a cohesion property. The plasticity criterion $f(\sigma', \tau) = 0$ is the Drucker–Prager criterion, reducing to the von Mises criterion for a frictionless material such as $\mathtt{f} = 0$. In the effective loading space $\left\{\sigma'_I \times \sigma'_{II} \times \sigma'_{III}\right\}$ the related yield surface is an axisymmetric cone whose vertex V lies on (Δ) at a distance $\sqrt{3}\mathtt{c}/\mathtt{f}$ from origin O (see Fig. 8.7).

Flow rule

In §8.2.2 we saw that the flow rule could be derived from a potential g according to (8.46), the flow rule being either associated, $g = f$, or not, $g \neq f$. For isotropic materials, similar to f in (8.49) g is expressed in the form:

$$g = g(\sigma', \tau) \tag{8.52}$$

In order to derive the flow rule explicitly, using definitions (8.47) and (8.48) we first get the instructive relation:

$$\frac{\partial g(\sigma', \tau)}{\partial \sigma_{ij}} = \frac{1}{3}\frac{\partial g}{\partial \sigma'}\delta_{ij} + \frac{\partial g}{\partial \tau}\frac{s_{ij}}{2\tau} \tag{8.53}$$

With the help of (8.53) the flow rule (8.46) is specialized for isotropic materials in the form:

$$d\varepsilon_{ij}^{p} = d\lambda\left(\frac{1}{3}\frac{\partial g}{\partial \sigma'}\delta_{ij} + \frac{\partial g}{\partial \tau}\frac{s_{ij}}{2\tau}\right)\left(\Rightarrow d\phi^{p} = d\epsilon^{p} = d\lambda\frac{\partial g}{\partial \sigma'}\right) \tag{8.54}$$

The dilatancy factor quantifies the magnitude of the distortion compared with the volumetric strain and turns out to be useful to account for experimental results by way of potential g. Consider an experiment where the porous material sample is subjected to the hydrostatic effective loading $\sigma'_{ij} = \sigma'\delta_{ij}$, whereas a deviatoric loading is simultaneously superimposed and is such that only one component $s_{ij} = s_{12} = s_{21}$ is non-zero. The material isotropy implies that only the volumetric strain $\epsilon = \varepsilon_{kk}$ and the distortion $\gamma = 2\varepsilon_{12} = 2\varepsilon_{21}$ are non-zero. The plastic dilatancy factor we denote d is then defined by:

$$d\epsilon^{p} = \mathrm{d}\, d\gamma^{p} \tag{8.55}$$

where $\gamma^{p} = 2\left|\varepsilon_{12}^{p}\right| = 2\left|\varepsilon_{21}^{p}\right|$ is the (positive) plastic distortion accounting for the strength of the irreversible change undergone by the angle between material directions $\mathbf{e}_i$ and $\mathbf{e}_j$. Dilatancy factor d is a priori not constant and can be either positive or negative, the poroplastic material exhibiting either plastic dilation ($\mathrm{d} > 0$) or plastic contraction ($\mathrm{d} < 0$). As roughly sketched in Fig. 8.9 the macroscopic dilatancy results from the microscopic rugosity of the contact planes along which the relative irreversible sliding of the solid grains actually occurs. Definition (8.55) is extended to any evolution by letting:

$$d\gamma^{p} = \sqrt{\frac{1}{2}d\gamma_{ij}^{p}d\gamma_{ji}^{p}};\quad \gamma_{ij}^{p} = 2e_{ij}^{p};\quad e_{ij}^{p} = \varepsilon_{ij}^{p} - \frac{1}{3}\epsilon^{p}\delta_{ij} \tag{8.56}$$

so that (8.54) can be rewritten in the form:

$$de_{ij}^{p} = d\lambda\frac{\partial g}{\partial \tau}\frac{s_{ij}}{2\tau};\quad d\phi^{p} = d\epsilon^{p} = d\lambda\frac{\partial g}{\partial \sigma'} \tag{8.57}$$

yielding:

$$d\gamma^{p} = d\lambda\frac{\partial g}{\partial \tau} \tag{8.58}$$

When dilatancy factor d is constant with regard to σ' and τ, a combination of (8.55), (8.57) and (8.58) requires $g(\sigma', \tau)$ to be specialized as a function of $\tau + \mathrm{d}\sigma'$only. In addition, whatever the final choice of $g\,(.)$, the derivative $dg\,(.)/d\,(.)$ can be included in the expression of the plastic multiplier $d\lambda$, whereas g is defined irrespective of any constant with regard to σ' and τ. Function g is eventually chosen in the form:

$$g(\sigma', \tau) = \tau + \mathrm{d}\sigma' - \mathrm{c} \tag{8.59}$$

so that (8.57) reduces to:

$$d\varepsilon_{ij}^{p} = d\lambda \left(\frac{s_{ij}}{2\tau} + \frac{1}{3}\mathrm{d}\delta_{ij} \right) \tag{8.60}$$

yielding:

$$d\epsilon^{p} = d\lambda\, \mathrm{d}; \quad d\gamma^{p} = d\lambda \tag{8.61}$$

Use of (8.56) in (8.44) gives:

$$\delta W^{p} = \sigma' d\epsilon^{p} + \tau d\gamma^{p} \tag{8.62}$$

In the case of an ideal plastic material, the dissipated energy reduces to the plastic work. Substitution of (8.61) into (8.62) then leads us to express the positiveness of the latter in the form:

$$\delta W^{p} = \left[\frac{\mathrm{d}}{\mathrm{f}}\mathrm{c} + \left(1 - \frac{\mathrm{d}}{\mathrm{f}} \right) \tau \right] d\gamma^{p} \geq 0 \tag{8.63}$$

where we used the Drucker–Prager plasticity criterion, that is $\tau + \mathrm{f}\sigma' - \mathrm{c} = 0$. Since the plasticity criterion allows τ to take positive infinite values, whereas it has a lower bound of zero (see Fig. 8.9), the previous inequality requires the dilatancy factor to be positive and have an upper bound provided by the friction coefficient, irrespective of the value of the cohesion. Regardless of the positiveness of the dissipated energy, it is also usually required that no tensile stress is tolerated within the material in the absence of any cohesion. This means the poroelasticity domain has to lie entirely in the octant defined by $\sigma'_{J=I,II,II} \leq \mathrm{c}/\mathrm{f}$ and it eventually restricts possible values of the friction coefficient to $\mathrm{f} \leq \frac{\sqrt{3}}{2}$. Accordingly we can eventually write:

$$0 \leq \mathrm{d} \leq \mathrm{f} \leq \frac{\sqrt{3}}{2} \tag{8.64}$$

When the poroplastic material satisfies the principle of maximal plastic work, $g = f$ and $\mathrm{d} = \mathrm{f}$ so that (8.63) gives:

$$\delta W^{p} = \mathrm{c} d\gamma^{p} \tag{8.65}$$

and we recover that $\delta W^{p} \geq 0$ since the material satisfies the principle of maximal plastic work. More generally, irrespective of any specific poroplastic model it can be shown that the plastic work can be expressed as a function of the plastic strain increments only as soon as the material satisfies the principle of maximal plastic work.

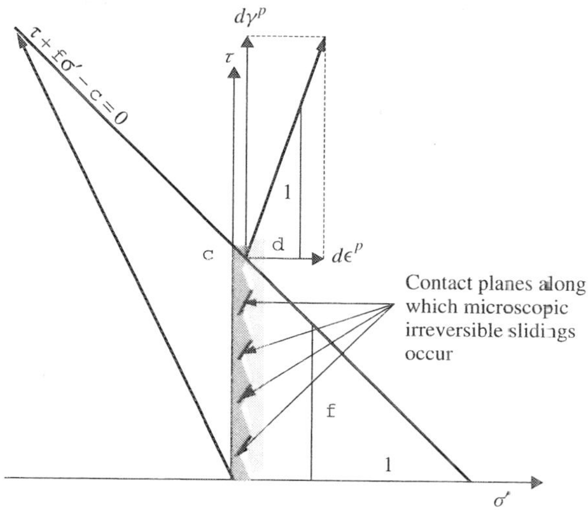

Figure 8.9: The macroscopic dilatancy results from the microscopic rugosity constituted by the contact planes along which the irreversible sliding of the solid grains occurs and orienting the direction of the plastic strain increment vector $\left(d\epsilon^p, d\gamma^p\right)$. The positiveness of the plastic work requires the product of the stress vector of components (σ', τ) with the plastic strain increment vector $\left(d\epsilon^p, d\gamma^p\right)$ to be positive. Accordingly, the dilatancy factor $\mathtt{d}$ must be positive and have an upper bound provided by the friction coefficient $\mathtt{f}$.

Hardening laws

A Drucker–Prager material exhibiting isotropic hardening is obtained by extending (8.51) and (8.59) in the form:

$$f(\sigma', \tau, \zeta) = \tau + \mathtt{f}\sigma' - \mathtt{c} + \zeta; \quad g(\sigma', \tau, \zeta) = \tau + \mathtt{d}\sigma' - \mathtt{c} + \zeta \tag{8.66}$$

When adopting (8.66), the flow rule (8.60) related to the plastic strain still holds, whereas the flow rule (8.38) related to the hardening variable χ and the expression (8.41) for the plastic multiplier are specialized in the form:

$$d\chi = d\lambda = \frac{d\tau + \mathtt{f}d\sigma'}{H} \tag{8.67}$$

Comparing (8.61) and (8.67) eventually identifies the hardening variable χ with the plastic distortion γ^p. The positiveness of dissipated energy $\Phi_1 dt$, now:

$$\Phi_1 dt = \sigma' d\epsilon^p + \tau d\gamma^p + \zeta d\gamma^p \geq 0 \tag{8.68}$$

The plasticity criterion $f = 0$, where f is given by (8.66), provides ζ as a function of τ and σ'. Substitution of the latter into (8.68) leads to identify $\Phi_1 dt$ with the plastic work (8.63) of the underlying ideal poroplastic material. As a consequence, its positiveness requires the same condition, that is $0 \leq \mathrm{d} \leq \mathrm{f}$. Use of (8.66) allows us to specialize the general expression (8.42) of the hardening modulus H in the form:

$$H(\gamma^p) = \frac{\partial^2 U}{\partial \gamma^{p2}} \tag{8.69}$$

where $U(\gamma^p)$ is the trapped energy as a function of the hardening variable $\chi = \gamma^p$. In practice it is eventually H instead of U that is experimentally determined through the measured incremental history (8.67) of γ^p.

A Drucker–Prager material exhibiting kinematic hardening is obtained by extending (8.51) and (8.59) in the form:

$$f\left(\sigma', s_{ij}, \zeta_{ij}\right) = \sqrt{\frac{1}{2}(s_{ij} + \omega_{ij})(s_{ij} + \omega_{ij})} + \mathrm{f}(\sigma' + \zeta) - \mathrm{c} \tag{8.70a}$$

$$g\left(\sigma', s_{ij}, \zeta_{ij}\right) = \sqrt{\frac{1}{2}(s_{ij} + \omega_{ij})(s_{ij} + \omega_{ij})} + \mathrm{d}(\sigma' + \zeta)\sigma' - \mathrm{c} \tag{8.70b}$$

where we let:

$$\zeta = \frac{1}{3}\zeta_{kk}; \quad \omega_{ij} = \zeta_{ij} - \zeta\delta_{ij} \tag{8.71}$$

so that the current yield surface $f\left(\sigma', s_{ij}, \zeta_{ij}\right) = 0$ is obtained by transporting the reference yield surface $f\left(\sigma', s_{ij}, 0\right) = 0$ through the translation of strength ζ. Based on expressions (8.70) for potentials f and g, the flow rule and the plastic multiplier can be derived from (8.28) and (8.41). Since σ'_{ij} and ζ_{ij} are involved in the same algebraic way in flow rules (8.28) and (8.38), and also in expressions (8.70) for f and g, the hardening variables χ_{ij} associated with ζ_{ij} are identified with ε^p_{ij}, whereas the hardening variable χ associated with ζ is identified with $\varepsilon^p = \phi^p$. Moreover, the dissipated energy $\Phi_1 dt$ is again identified with the plastic work (8.63) of the underlying ideal poroplastic material.

8.4.4 The Cam–Clay Model

The Cam–Clay criterion

The usual model as detailed in the previous section covers a large class of cohesive–frictional materials but it turns out to be irrelevant for capturing the plastic behaviour of soils such as clays. Indeed, soils cannot sustain tensile stress, nor infinite confining pressure, whereas the observed dilatancy can be either positive (dilation) or negative (contraction), depending on the ratio between the shear τ and the confining effective pressure $-\sigma'$. The mechanical behaviour of soils and rocks is largely explored through the deviatoric experiment consisting of subjecting the sample to an increasing axial pressure $-\sigma_I$ beyond the confining pressure $-\sigma_{II} = -\sigma_{III}$ so that $\tau\sqrt{3} = |\sigma_I - \sigma_{III}|$ (see Fig. 8.8). As is common in soil and rock mechanics, it is convenient to note:

$$p' = -\sigma'; \quad q = \tau\sqrt{3} \tag{8.72}$$

in order to consider positive values, that is $p' \geq 0$, and to investigate the plastic loading function in the form $f(p', q)$.

In investigating the behaviour of clays, in particular the clay of the River Cam, the Cambridge School[7] designed the so-called 'Cam–Clay' model originally intended to account for the plastic behaviour of saturated clays. This model, now widely used and extended to other materials and to unsaturated situations,[8] is particularly attractive since it covers the whole range of behaviours expected for plastic materials: ideal, hardening and softening plasticity. Adopting the notation in (8.72), the loading function finally retained to account for the yield surface of clays is expressed in the form:

$$2f(p', q, p_{co}) = \left(p' - \frac{1}{2}p_{co}\right)^2 + \frac{q^2}{M^2} - \frac{1}{4}p_{co}^2 \tag{8.73}$$

where M is a material constant, while p_{co} is the effective consolidation pressure. Consolidation pressure p_{co} is the upper bound of the admissible current effective pressure p' and turns out to be the maximum effective pressure to which the material has been subjected during the past plastic loadings. Indeed, the loading surface is an ellipse centred at $p' = \frac{1}{2}p_{co}$ and having $\frac{1}{2}p_{co}$ and $\frac{1}{2}M\,p_{co}$ as half-axes respectively along the p' direction and the q direction. As hardening force p_{co} varies, the ellipse transforms homothetically and the hardening/softening mode is isotropic (see Fig. 8.10).

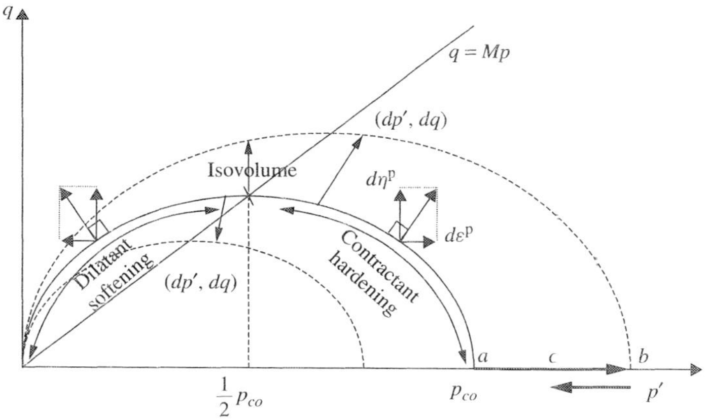

Figure 8.10: Poroplastic Cam–Clay model. The loading surface is an ellipse. The material exhibits plastic contraction for $p' > \frac{1}{2}p_{co}$ and plastic dilatancy for $p' < \frac{1}{2}p_{co}$, whereas for $p' = \frac{1}{2}p_{co}$ the plastic evolution occurs at constant volume. For $p' \neq \frac{1}{2}p_{co}$ the material exhibits isotropic hardening. By contrast, for $p' = \frac{1}{2}p_{co}$ and therefore for $q' = \frac{1}{2}M\,p_{co}$, plastic flow occurs indefinitely at constant loading (ideal plastic material) so that the line defined by $q = Mp'$ is commonly said to be the locus of critical states.

[7]A founding paper is Roscoe K.H., Schofield A.N., Wroth C.P. (1958), 'On the yielding of soils', *Geotechnique*, **8**, 22–53.

[8]See Alonzo E.E., Gens A., Josa A. (1990), 'A constitutive model for partially saturated soils', *Geotechnique*, **40**, 405–430.

Flow rule

In addition to (8.72), it is convenient to let:

$$\varepsilon^p = -\epsilon^p; \quad \eta^p = \frac{1}{\sqrt{3}}\gamma^p \tag{8.74}$$

so that ε^p is the contraction and η^p is equal to $\frac{2}{3}\,|\varepsilon_I - \varepsilon_{III}|$ in the deviatoric experiment. Revisiting (8.62) we can alternatively write:

$$\delta W^p = p' d\varepsilon^p + q d\eta^p \tag{8.75}$$

The Cam–Clay model assumes the validity of the principle of maximal plastic work so that the potential is associated and the flow rule can be expressed in the form:

$$d\varepsilon^p = d\lambda \frac{\partial f}{\partial p'} = d\lambda \left(p' - \frac{1}{2}p_{co}\right); \quad d\eta^p = d\lambda \frac{\partial f}{\partial q} = d\lambda \frac{q}{M^2} \tag{8.76}$$

where:

$$d\lambda = \frac{1}{H}\left[\left(p' - \frac{1}{2}p_{co}\right)dp' + \frac{q\,dq}{M^2}\right] \tag{8.77}$$

According to (8.76), the material undergoes plastic dilation ($d\varepsilon^p < 0$) for effective pressures lower than half the consolidation pressure ($p' < \frac{1}{2}p_{co}$), whereas it undergoes plastic contraction ($d\varepsilon^p > 0$) in the inverse case ($p' > \frac{1}{2}p_{co}$). For $p' > \frac{1}{2}p_{co}$ the plastic evolution occurs without any change in volumetric plastic strain ($d\varepsilon^p = 0$) (see Fig. 8.10).

Hardening law. Critical state

The solid matrix of clays is formed of subparticles or platelets. The hardening state of clays relates to the strength of the interactions between the latter and therefore to their irreversible densification degree. Since the subparticles do not undergo any volume change, their current densification degree is well captured by the current plastic contraction ε^p which is identified as the appropriate hardening variable χ of the argument of the trapped energy $U = U(\chi = \varepsilon^p)$. Since the trapped energy must increases with both ε^p and the current consolidation pressure p_{co}, it is consistent to write:

$$p_{co} = \frac{\partial U}{\partial \varepsilon^p} \tag{8.78}$$

According to (8.34), it is $\zeta = -p_{co}$ that is eventually identified as the hardening force associated with plastic contraction $\chi = \varepsilon^p$. The condition (8.33) of positiveness of the dissipated energy can be written:

$$f = 0: \quad p' d\varepsilon^p + q d\eta^p - p_{co} d\varepsilon^p \geq 0 \tag{8.79}$$

which is actually satisfied over the whole range of admissible effective pressure p' since it can be shown that:

$$f = 0: \quad p'd\varepsilon^p + q d\eta^p - p_{co} d\varepsilon^p = d\lambda \times \frac{1}{2} p_{co}(p_{co} - p') \tag{8.80}$$

Substitution of $\zeta = -p_{co}$ and $\chi = \varepsilon^p$ into (8.42) and use of (8.73) allow us to express the hardening modulus in the form:

$$H = \frac{1}{2} p' \left(p' - \frac{1}{2} p_{co} \right) \frac{dp_{co}}{d\varepsilon^p} \tag{8.81}$$

The determination of H requires knowledge of the state equation linking ε^p to p_{co}. The latter can be obtained from an experiment performed at $q = 0$ and where the total void ratio e is plotted against the effective pressure p'. The resulting experimental data such as those reported in Fig. 8.11 are well captured according to the law:

$$p_{co} = p_{co}^0 \exp[-(\lambda - \kappa)\, e^p]; \quad \lambda > \kappa \tag{8.82}$$

Besides, use of (1.34) provides the relation linking the plastic contraction ε^p to the plastic void ratio e^p:

$$e^p = -(1 + e_0)\varepsilon^p \tag{8.83}$$

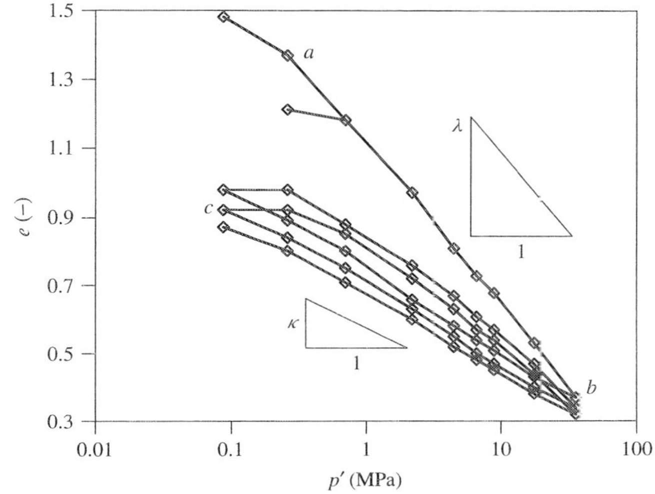

Figure 8.11: Poroplastic behaviour of an artificial compacted clay in a experiment where $q = 0$ and where the total void ratio e is plotted against the effective pressure p'. The clay behaves elastically up to some effective pressure threshold representing the maximum effective pressure to which the material has been subjected in all the past plastic loadings (point a in Figs. 8.10–8.11). Along a poroplastic loading path $de = -\lambda dp'/p'$ (path ab in Figs. 8.10–8.11), while $de - de^p = -\kappa dp'/p'$ along an elastic unloading path (path bc in Figs. 8.10–8.11). This results in $de^p = -(\lambda - \kappa)\, dp_{co}/p_{co}$ since $p' = p_{co}$ along the plastic loading path followed at $q = 0$.

Equations (8.73), (8.78), (8.81)–(8.83) combine to give the following expressions of the trapped energy U:

$$U(\varepsilon^p) = \frac{p_{co}^0}{(\lambda-\kappa)(1+e_0)} \exp\left[(\lambda-\kappa)(1+e_0)\varepsilon^p\right] \tag{8.84}$$

and the hardening modulus H:

$$H = \frac{1}{2}(\lambda-\kappa)(1+e_0)p_{co}p'\left(p' - \frac{1}{2}p_{co}\right) \tag{8.85}$$

When $p' > \frac{1}{2}p_{co}$, the plastic contraction ($d\varepsilon^p > 0$) causes hardening ($H > 0$). Conversely, when $p' < \frac{1}{2}p_{co}$, the plastic dilation ($d\varepsilon^p < 0$) causes softening ($H < 0$). For $p' = \frac{1}{2}p_{co}$ and therefore for $q = \frac{1}{2}Mp_{co}$ plastic flow occurs indefinitely at constant loading ($H = 0$, ideal plasticity). For this reason line $q = Mp'$ is commonly said to be the locus of critical states and $\frac{1}{2}p_{co}$ the critical effective consolidation pressure (see Fig. 8.10).

8.5 Advanced Analysis

8.5.1 Uniqueness of Solution

Analogously to (5.149), the state equations of linear poroplasticity (8.9) can be written in the form:

$$\sigma_{ij} = \frac{\partial W}{\partial \varepsilon_{ij}^{el}}; \quad p = \frac{\partial W}{\partial v_f^{el}} \tag{8.86}$$

where $W\left(\varepsilon_{ij}^{el}, v_f^{el}\right)$ is the reduced potential defined by (5.150). Let $W^*(\sigma_{ij}, p)$ be the Legendre–Fenchel transform of $W\left(\varepsilon_{ij}^{el}, v_f^{el}\right)$:[9]

$$W^*(\sigma_{ij}, p) = \sigma_{ij}\varepsilon_{ij}^{el} + pv_f^{el} - W\left(\varepsilon_{ij}^{el}, v_f^{el}\right) \tag{8.87}$$

so that (8.86) can be inverted according to:

$$\varepsilon_{ij}^{el} = \frac{\partial W^*}{\partial \sigma_{ij}}; \quad v_f^{el} = \frac{\partial W^*}{\partial p} \tag{8.88}$$

The linearity of state equations (8.9) linking (σ_{ij}, p) to $\left(\varepsilon_{ij}^{el}, v_f^{el}\right)$ gives the useful relation:

$$W^*(\sigma_{ij}, p) = W\left(\varepsilon_{ij}^{el}, v_f^{el}\right) = \frac{1}{2}\left(\boldsymbol{\sigma} : \boldsymbol{\varepsilon}^{el} + p\, v_f^{el}\right) \tag{8.89}$$

[9]The general explicit expression of W^* is:

$$W^*(\sigma_{ij}, p) = \frac{1}{2}\sigma_{ij}C_{ijkl}^{-1}\sigma_{kl} + \sigma_{ij}C_{ijkl}^{-1}b_{kl}\,p + \frac{1}{2M}\left(1 + Mb_{ij}C_{ijkl}^{-1}b_{kl}\right)p^2$$

According to stability conditions (5.153), energy $W^*(\sigma_{ij}, p)$ is always positive as soon as σ_{ij} and p are non-zero. In addition, use of (8.86) and (8.89) leads to:

$$\boldsymbol{\sigma} : \dot{\boldsymbol{\varepsilon}}^{el} + p\,\dot{v}_f^{el} = \frac{dW^*}{dt} \tag{8.90}$$

Now consider two possible solutions (σ_{ij}, p) and $\left(\sigma'_{ij}, p'\right)$ for the poroplastic problem under consideration. Inequality (5.157) holds irrespective of the specific skeleton constitutive equations so that the combination of the latter with (8.90) gives:

$$\begin{aligned} &\frac{d}{dt}\int_\Omega W^*\left(\sigma'_{ij} - \sigma_{ij}, p' - p\right)d\Omega \\ &\quad \leq -\int_\Omega \left[\left(\sigma'_{ij} - \sigma_{ij}\right)\left(\dot{\varepsilon}^{p'}_{ij} - \dot{\varepsilon}^{p}_{ij}\right) + \left(p' - p\right)\left(\dot{v}^{p'}_f - \dot{v}^{p}_f\right)\right]d\Omega \end{aligned} \tag{8.91}$$

For an ideal poroplastic material both possible solutions (σ_{ij}, p) and $\left(\sigma'_{ij}, p'\right)$ are required to lie within the invariant domain of poroelasticity C_E. If in addition the material obeys the principle of maximal plastic work, applying inequality (8.23) successively to both possible solutions furnishes:

$$\left(\sigma'_{ij} - \sigma_{ij}\right)d\varepsilon^{p'}_{ij} + (p' - p)dv^{p'} \geq 0; \quad \left(\sigma_{ij} - \sigma'_{ij}\right)d\varepsilon^{p}_{ij} + (p - p')dv^{p} \geq 0 \tag{8.92}$$

so that the right hand side of inequality (8.91) turns out to be negative, yielding:

$$\frac{d}{dt}\int_\Omega W^*\left(\sigma'_{ij} - \sigma_{ij}, p' - p\right)d\Omega \leq 0 \tag{8.93}$$

Energy $W^*\left(\sigma'_{ij} - \sigma_{ij}, p' - p\right)$ is always positive as soon as $\boldsymbol{\sigma}' \neq \boldsymbol{\sigma}$ and $p' \neq p$ and can be used to express the distance between two solutions. Accordingly inequality (8.93) expresses that the distance with regard to energy between two possible solutions can only decrease. Consequently, since both of them are required to meet the same initial conditions, the distance equal to zero at $t = 0$ remains so – the solution is unique.

The uniqueness of the solution with regard to the stress and the fluid pressure does not a priori ensure the uniqueness with regard to the skeleton velocity nor to the rate in fluid volume content. Substitution of (8.13)–(8.14) and of the second term in (8.88) into diffusion equation (5.8) gives:

$$\dot{\phi}^p = -\frac{d}{dt}\left(\frac{\partial W^*}{\partial p}\right) - k\nabla^2 p \tag{8.94}$$

Substitution of flow rule (8.25) into (8.94) allows us to express the rate $d\lambda/dt$ in the form:

$$\frac{d\lambda}{dt} = \frac{1}{\partial f/\partial p}\left[-\frac{d}{dt}\left(\frac{\partial W^*}{\partial p}\right) - k\nabla^2 p\right] \tag{8.95}$$

The above expression ensures the uniqueness of the plastic multiplier history and, consequently, the uniqueness of the rates of both the strain and the fluid volume content. It is worth noting that the uniqueness of the plastic multiplier history is achieved by the regularization afforded by the diffusion equation and therefore by the viscous effects associated with the fluid flow. Indeed, the uniqueness of the plastic multiplier history could not have been proven for a monophasic ideal plastic material. By contrast, if the principle of maximal plastic work is extended to the hardening work with no possibility of softening, the uniqueness of the solution extends to the hardening variables and is achieved irrespective of the regularization effects associated with the viscous flow of the fluid.

8.5.2 Limit Analysis

The kinematical theorem

Limit analysis aims at determining the critical loading provoking the collapse of structures without having to determine the behaviour prior to failure. In the framework of ideal plasticity this determination can be carried out by analysing the maximum rate of dissipated energy that the structure can oppose in relation to the work rate of external forces in the likely modes of plastic collapse.

Under the hypothesis of small quasistatic perturbations, when considering only gravity forces, namely $\mathbf{f} = \mathbf{g}$, momentum equation (2.19) simplifies to the form:

$$\nabla \cdot \boldsymbol{\sigma}' - \nabla p + \left(\rho_s(1-\phi_0) + \rho_f \phi_0\right)\mathbf{g} = 0 \tag{8.96}$$

where $\boldsymbol{\sigma}' = \boldsymbol{\sigma} + p\mathbf{1}$ stands for Terzaghi's effective stress (8.45). Use of Darcy's law (3.39) allows us to rewrite (8.96) in the form

$$\nabla \cdot \boldsymbol{\sigma}' + \mathbf{f}' = 0 \tag{8.97}$$

where we note:

$$\mathbf{f}' = \frac{1}{k}\mathcal{V} + \rho'\mathbf{g}; \quad \rho' = (1-\phi_0)(\rho_s - \rho_f) \tag{8.98}$$

Provided that Terzaghi's effective stress concept applies to the constitutive porous material, (8.97) can be roughly interpreted as an effective momentum equation related to the skeleton. Effective volumetric forces $\mathbf{f}'$ result from viscous drag forces through the term $\frac{1}{k}\mathcal{V}$ and from gravity forces through the term $\rho'\mathbf{g}$, where the effective density ρ' is considered in order to remove the Archimedes vertical lifting force. We multiply (8.97) by any possible candidate $\mathbf{V}$ describing the plastic collapse and we integrate over the volume Ω occupied by the structure. Using the divergence theorem in addition, the procedure gives the following balance of work rates:

$$\mathcal{P}'_{def} = \mathcal{P}' \tag{8.99}$$

where we note:

$$\mathcal{P}'_{def} = \int_\Omega \sigma'_{ij} d_{ij} d\Omega; \quad \mathcal{P}' = \int_{\partial\Omega} \mathbf{T}' \cdot \mathbf{V} da + \int_\Omega \mathbf{f}' \cdot \mathbf{V} d\Omega; \tag{8.100}$$

In (8.100), d_{ij} is the rate of deformation:

$$d_{ij} = \frac{1}{2}\left(\frac{\partial V_i}{\partial x_j} + \frac{\partial V_j}{\partial x_i}\right) \tag{8.101}$$

while $\mathbf{T}' = \boldsymbol{\sigma}' \cdot \mathbf{n}$ is the effective stress vector, $\mathbf{n}$ being the outward unit normal to the border $\partial\Omega$. Balance equation (8.99) states that the effective strain work rate is equal to the effective work rate of the external forces whatever the velocity field $\mathbf{V}$ considered.

Assuming the validity of the effective stress concept, let $f(\sigma'_{ij})$ be the loading function of the ideal poroplastic material constituting the structure. The collapse of the structure will unavoidably occur if one velocity field does exist such that the associated work rate of the external forces exceeds the rate of dissipated energy that the structure can offer in the same field, so that (8.99) definitively cannot be fulfilled. This statement can be expressed in the form of the kinematical theorem of limit analysis:

$$\exists \mathbf{V} : \ \Phi(\mathbf{V}) = \sup_{f(\sigma'_{ij}) \leq 0} \mathcal{P}'_{def} < \mathcal{P}' \Longrightarrow \text{Collapse} \tag{8.102}$$

According to the principle of maximal plastic work (see §8.2.2 and Fig. 8.3), whatever the velocity field $\mathbf{V}$ under consideration the maximum value of $\sigma'_{ij} d_{ij}$, and consequently of $\mathcal{P}'_{def}$, is achieved for the effective stress σ'_{ij} located on the yield surface $f(\sigma'_{ij}) = 0$ where d_{ij} is normal. The infinitesimal strain work rate $\sigma'_{ij} d_{ij}$ is eventually identified with the plastic work rate. For instance, for the usual cohesive–frictional model studied in §8.4.3, (8.65) gives:

$$\Phi(\mathbf{V}) = \int_{\Omega} c\dot{\gamma}\, d\Omega; \quad \dot{\gamma} = \sqrt{\frac{1}{2}\dot{\gamma}_{ij}\dot{\gamma}_{ji}} \tag{8.103}$$

where:

$$\dot{\gamma} = \sqrt{\frac{1}{2}\dot{\gamma}_{ij}\dot{\gamma}_{ji}}; \quad \dot{\gamma}_{ij} = 2\left(d_{ij} - \frac{1}{3} d_{kk}\delta_{ij}\right) \tag{8.104}$$

Limit analysis and the stability of slopes

In order to illustrate (8.102) let us consider a soil layer of uniform thickness h in the $\mathbf{e}_1$ direction resting on a impervious rigid bedrock inclined at angle α to the horizontal. The layer is subjected both to the vertical gravity body force $g\mathbf{e}_z$ and to a steady fluid flow parallel to the direction of the layer of strength $\mathcal{V}_1$ (see Fig. 8.12). The layer consists of an ideal poroplastic material obeying the usual cohesive–frictional model studied in §8.4.3 so that (8.102) and (8.103) apply.

In order to determine the critical flow provoking the collapse of the layer, we consider the following possible candidate of mode of plastic collapse:

$$\mathbf{V} = V\frac{x_2}{h}\mathbf{e}_1; \quad \gamma_{12} = \gamma_{21} = \frac{V}{h}; \quad (\text{other } \gamma_{ij} = 0) \tag{8.105}$$

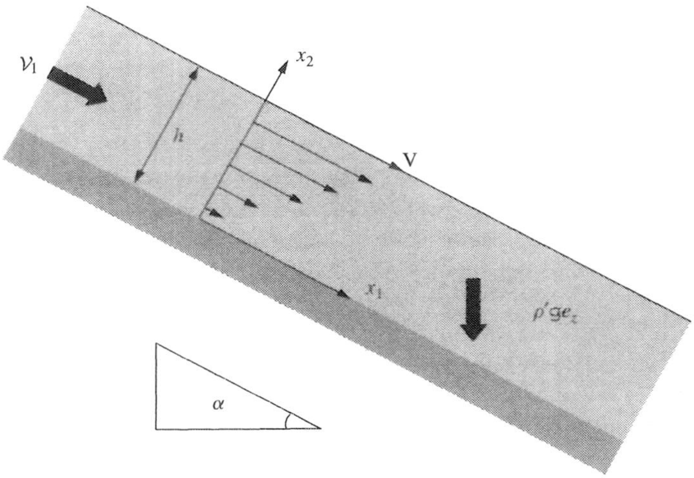

Figure 8.12: The kinematical theorem of limit analysis applied to the analysis of the poroplastic collapse of a soil layer subjected to vertical gravity forces and a steady state fluid flow.

Using (8.103), the maximal rate $\Phi(\mathbf{V})$ of dissipated energy that the layer can provide per length unit in the $\mathbf{e}_1$ direction to resist the possible collapse mode (8.105) is:

$$\Phi(\mathbf{V}) = \mathrm{c}V \tag{8.106}$$

which is to be compared with the external work rate $\mathcal{P}'$ provided in the velocity field $\mathbf{V}$ by the effective force $\mathbf{f}' = \frac{1}{k}\mathcal{V}_1\mathbf{e}_1 + \rho' \mathrm{g}\mathbf{e}_Z$, that is:

$$\mathcal{P}'_{def} = \left(\frac{1}{k}\mathcal{V}_1 + \rho' \mathrm{g} \sin\alpha\right)\frac{Vh}{2} \tag{8.107}$$

Applying the kinematical theorem (8.102) of limit analysis, the critical strength $\mathcal{V}_1 = \mathcal{V}_{cr}$ provoking the collapse of the layer has an upper bound according to:

$$\frac{\mathcal{V}_{cr}}{k\rho' \mathrm{g} \sin\alpha} < \frac{2\mathrm{c}}{\rho' \mathrm{g} h \sin\alpha} - 1 \tag{8.108}$$

8.5.3 Thermal and Chemical Hardening

Thermal hardening

As illustrated in Fig. 8.13 for a clay,[10] poroplastic geomaterials are often observed to exhibit thermal hardening, such that the stress threshold causing the plastic yielding of

[10]The experimental data are from Sultan N., Delage P., Cui Y.J. (2002), 'Temperature effects on the volume change behaviour of Boom clay', *Engineering Geology* **64**, (2–3), 135–145.

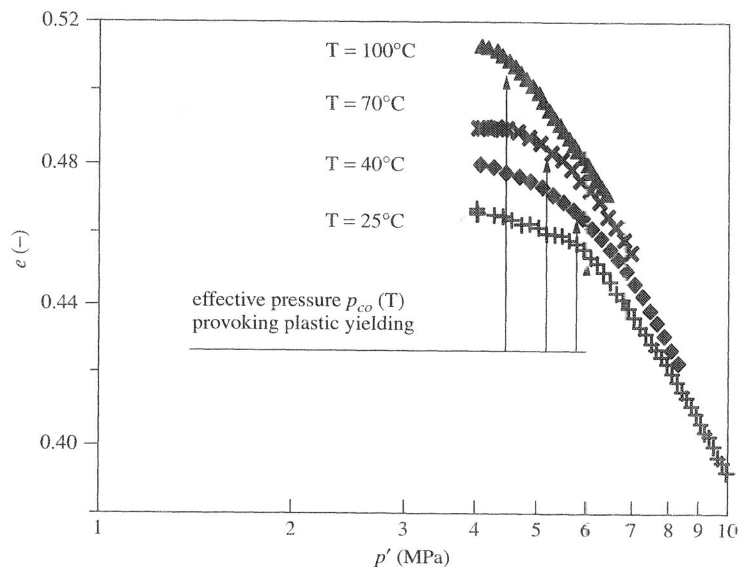

Figure 8.13: Experimental evidence of thermal hardening. The effective (consolidation) pressure causing irreversible plastic changes in the void ratio decreases with temperature for clay samples previously subjected to $T = 20°C$ (from Sultan *et al.* (2002), see footnote 10).

the material depends on temperature. In order to include this thermoplastic coupling, (8.31) is extended in the form:

$$\Psi_s = W_s\big(\varepsilon_{ij} - \varepsilon_{ij}^p, \phi - \phi^p, T\big) + U(\chi_J, T) \tag{8.109}$$

Accordingly, the state equations are now written:

$$\sigma_{ij} = \frac{\partial W_s}{\partial \varepsilon_{ij}}; \quad p = \frac{\partial W_s}{\partial \phi}; \quad S_s = -\frac{\partial W_s}{\partial T} - \frac{\partial U}{\partial T}; \quad \zeta_J = -\frac{\partial U}{\partial \chi_J} \tag{8.110}$$

so that (8.33) expressing the dissipation positiveness remains unchanged while flow rules (8.25) or (8.28), (8.37) or (8.38) still apply. Nevertheless, since hardening forces ζ_J now depend on both temperature and hardening state variables χ_J, the expression for the plastic multiplier $d\lambda$ must be reconsidered. To this end let $d_\chi f$ be defined by:

$$d_\chi f = \frac{\partial f}{\partial \sigma_{ij}} d\sigma_{ij} + \frac{\partial f}{\partial p} dp + \frac{\partial f}{\partial \zeta} \frac{\partial \zeta_J}{\partial T} dT \tag{8.111}$$

so that:

$$df = d_\chi f + \frac{\partial f}{\partial \zeta_J} \frac{\partial \zeta_J}{\partial \chi_J} d\chi_J \tag{8.112}$$

Use of (8.38) and of the last of state equations (8.110), together with consistency condition $df = 0$, allows us to identify the expression of the plastic multiplier $d\lambda$ in the form:

$$d\lambda = \frac{d_\chi f}{H} \tag{8.113}$$

where H is the hardening modulus whose expression is still given by (8.42).

For instance, the Cam–Clay model presented in §8.4.4 can be extended in order to account for thermal hardening. The experimental results reported in Fig. 8.13 are well captured according to:

$$p_{co} = p_{co}^0 (T) \exp\left[-(\lambda - \kappa)\, e^p\right]; \quad \lambda > \kappa \tag{8.114}$$

where coefficients λ and κ have the same definition as in Fig. 8.11. The remaining equations related to the Cam–Clay model are unchanged provided that, instead of (8.77) and according to (8.113), the plastic multiplier is expressed in the form:

$$d\lambda = \frac{1}{H}\left[\left(p' - \frac{1}{2}p_{co}\right) dp' + \frac{q dq}{M^2} - \frac{1}{2} p' \frac{p_{co}}{p_{co}^0} \frac{dp_{co}^0}{dT} dT\right] \tag{8.115}$$

Chemical softening/hardening

The dissolution process affects the poroelastic properties of a porous material (see for instance Table 4.1), but also reduces the strength properties (see Table 8.1[11]), leading to chemical softening when strength properties are addressed in the context of poroelasticicty. In order to include this chemoplastic coupling, (8.31) is extended in the form:

$$\Psi_s = W_s\left(\varepsilon_{ij} - \varepsilon_{ij}^p, \phi - \phi_{ch} - \phi^p, \phi_{ch}\right) + U(\chi_J, \phi_{ch}) \tag{8.116}$$

where the expression of W_s is given by (4.197) so that the state equations can be derived accordingly, leading us to replace ε_{ij} and ϕ in (3.167) and (4.194) by $\varepsilon_{ij} - \varepsilon_{ij}^p$ and $\phi - \phi^p$, respectively. Inequality (8.33) expressing the positiveness of the plastic contribution to the overall dissipation turns out to be unchanged so that flow rules (8.25) or (8.28), (8.37) or (8.38) still apply. Nevertheless, since hardening forces ζ_J now depend both on ϕ_{ch} and on hardening state variables χ_J, the expression for the plastic multiplier $d\lambda$ must be reconsidered. Proceeding as in the previous section, the expression for the plastic multiplier $d\lambda$ can be written in the form (8.113) provided that (8.111) is replaced by:

$$d_\chi f = \frac{\partial f}{\partial \sigma_{ij}} d\sigma_{ij} + \frac{\partial f}{\partial p} dp + \frac{\partial f}{\partial \zeta}\frac{\partial \zeta_J}{\partial \phi_{ch}} d\phi_{ch} \tag{8.117}$$

As reported in Table 8.1, for materials obeying the Drucker criterion (8.51) the cohesion is generally more affected by the dissolution process than the friction coefficient is. Such a

[11] For the experimental data reported in Table 8.1 see Heukamp F.H., Ulm F.-J., Germaine J.T. (2002), 'Residual design strength of cement-based materials for nuclear waste storage systems', *Nuclear Engineering Design*, **211**, (1), 51–60.

Table 8.1: Reduction of cohesion and friction strength properties related to Drucker criterion (8.51) of cement-based materials due to leaching (from Heukamp *et al.* (2001), see footnote 11).

	Cohesion	Friction coefficient
Unleached cement paste	$c = 17.1$ MPa	$f = 0.82$
Leached cement paste	$c = 1.1$ MPa	$f = 0.56$
Unleached mortar	$c = 9.8$ MPa	$f = 1.1$
Leached mortar	$c = 0.96$ MPa	$f = 0.81$

reduction of cohesion c for a material exhibiting isotropic hardening such as that captured by (8.66) can be accounted for by letting:

$$U(\chi = \gamma^p, \phi_{ch}) = -\gamma^p c z_{ch}(\phi_{ch}) + z_{mk}(\phi_{ch}) U_0(\gamma^p) \quad (z'_{mk}(\phi_{ch}) < 0) \tag{8.118}$$

so that:

$$\zeta = -\frac{\partial U}{\partial \gamma^p} = c z_{ch}(\phi_{ch}) - z_{mk}(\phi_{ch}) \frac{\partial U_0}{\partial \gamma^p} \tag{8.119}$$

The first term in (8.119) accounts for the chemical softening that would occur in the absence any loading. Indeed, substitution of (8.119) into (8.66) results in the current cohesion $c(1 - z_{ch}(\phi_{ch}))$ irrespective of the term associated with U_0. The latter term relates to the usual mechanical hardening associated with plastic deformation, which is now affected by the dissolution process through the factor $z_{mk}(\phi_{ch})$. The latter accounts for the dissolution effect in such a way that energy $U_0(\gamma^p)$ which would be trapped during a plastic deformation in the absence of dissolution, that is for $z_{mk}(\phi_{ch} = 0) = 1$, is reduced by the factor $z_{mk}(\phi_{ch})$. As a consequence, the same factor $z_{mk}(\phi_{ch})$ affects the hardening modulus. Indeed, substitution of (8.118) into (8.69) gives:

$$H = z_{mk}(\phi_{ch}) H_0; \quad H_0 = \frac{\partial^2 U_0}{\partial \gamma^{p2}} \tag{8.120}$$

The previous modelling was relative to the chemical softening following to a dissolution process. It similarly applies to the chemical hardening following to a chemical reaction that strengthens the material. Indeed, denoting ξ as the reaction extent (see §3.6.3 and §4.4.4), (8.118) is appropriately replaced by:[12]

$$U(\chi = \gamma^p, \xi) = \gamma^p c z_{ch}(\xi) + z_{mk}(\xi) U_0(\gamma^p) \quad (z'_{mk}(\xi) > 0) \tag{8.121}$$

[12]For chemical hardening induced by the setting of concrete the theory is fully developed in Ulm F.-J., Coussy O. (1998), 'Couplings in early-age concrete: from material modeling to structural design', *Journal of Solids and Structures*, **35**, (31–32), 4295–4311.

8.5.4 Localization of Deformation

As illustrated in Fig. 8.14[13] a band of intense deformation and negligible thickness is often observed in triaxial tests performed on geomaterials. The search for the conditions of formation of such a band can be carried out through a two-step analysis: the first step consists in deriving the incremental constitutive equations of plasticity while the second step analyses the conditions required for the formation of discontinuities related to such a narrow band.

Incremental constitutive equations of poroplasticity

The state equations of poroplasticity (8.9) can be written in the general incremental form:

$$d\sigma_{ij} = C_{ijkl}\left(d\varepsilon_{kl} - d\varepsilon_{kl}^{p}\right) - b_{ij}dp \tag{8.122a}$$

$$d\phi - d\phi^{p} = b_{ij}\left(d\varepsilon_{ij} - d\varepsilon_{ij}^{p}\right) + \frac{dp}{N} \tag{8.122b}$$

Use of (8.122) in (8.39) gives:

$$d_{\zeta} f = \frac{\partial f}{\partial \sigma_{ij}} C_{ijkl}\left(d\varepsilon_{kl} - d\varepsilon_{kl}^{p}\right) + \left(\frac{\partial f}{\partial p} - b_{ij}\frac{\partial f}{\partial \sigma_{ij}}\right) dp \tag{8.123}$$

Figure 8.14: Formation of a band of intense deformation in an axisymmetric triaxial test performed on a Hostunsand (from Desrues (1988), see footnote).

---

[13]The illustration is by courtesy of Desrues J. (1984), 'La localisation de la déformation dans les milieux granulaires', *Thèse de Doctorat d'Etat*, Université de Grenoble.

Substitution of flow rule (8.28) into (8.123) and of the resulting expression in (8.41) allows us to express the plastic multiplier in the alternative form:

$$d\lambda = \frac{\frac{\partial f}{\partial \sigma_{ij}} C_{ijkl} d\varepsilon_{kl} + \left(\frac{\partial f}{\partial p} - b_{ij} \frac{\partial f}{\partial \sigma_{ij}} \right) dp}{H + \frac{\partial f}{\partial \sigma_{ij}} C_{ijkl} \frac{\partial g}{\partial \sigma_{kl}}} \tag{8.124}$$

In turn, substitution of (8.124) into flow rule (8.28) provides expressions for $d\varepsilon_{kl}^{p}$ and $d\phi^{p}$ that combine with incremental state equations (8.122) to give the following incremental constitutive equations of poroplasticity:

$$d\sigma_{ij} = C_{ijkl}^{T} d\varepsilon_{kl} - b_{ij}^{T} dp \tag{8.125a}$$

$$d\phi = \beta_{ij}^{T} d\varepsilon_{ij} + \frac{dp}{N^{T}} \tag{8.125b}$$

where tangent properties C_{ijkl}^{T}, b_{ij}^{T}, β_{ij}^{T} and N^{T} are expressed in the form:

$$C_{ijkl}^{T} = C_{ijkl} - \frac{C_{ijmn} \frac{\partial g}{\partial \sigma_{mn}} \frac{\partial f}{\partial \sigma_{pq}} C_{pqkl}}{H + \frac{\partial f}{\partial \sigma_{mn}} C_{mnpq} \frac{\partial g}{\partial \sigma_{pq}}} \tag{8.126a}$$

$$b_{ij}^{T} = b_{ij} + \frac{\frac{\partial f}{\partial p} - b_{mn} \frac{\partial f}{\partial \sigma_{mn}}}{H + \frac{\partial f}{\partial \sigma_{mn}} C_{mnpq} \frac{\partial g}{\partial \sigma_{pq}}} C_{ijkl} \frac{\partial g}{\partial \sigma_{kl}} \tag{8.126b}$$

$$\beta_{ij}^{T} = b_{ij} + \frac{\frac{\partial g}{\partial p} - b_{mn} \frac{\partial g}{\partial \sigma_{mn}}}{H + \frac{\partial f}{\partial \sigma_{mn}} C_{mnpq} \frac{\partial g}{\partial \sigma_{pq}}} C_{ijkl} \frac{\partial f}{\partial \sigma_{kl}} \tag{8.126c}$$

$$\frac{1}{N^{T}} = \frac{1}{N} + \frac{\left(\frac{\partial f}{\partial p} - b_{mn} \frac{\partial f}{\partial \sigma_{mn}} \right) \left(\frac{\partial g}{\partial p} - b_{mn} \frac{\partial g}{\partial \sigma_{mn}} \right)}{H + \frac{\partial f}{\partial \sigma_{mn}} C_{mnpq} \frac{\partial g}{\partial \sigma_{pq}}} \tag{8.126d}$$

so that the standard poroelastic symmetries $C_{ijkl}^{T} = C_{klij}^{T}$ and $b_{ij}^{T} = \beta_{ij}^{T}$ are recovered only for an associated flow rule such as $f = g$. In the case of an isotropic material, C_{ijkl} and b_{ij} are specialized in the form:

$$C_{ijkl} = \left(K - \frac{2}{3}\mu \right) \delta_{ij}\delta_{kl} + \mu(\delta_{ik}\delta_{jl} + \delta_{il}\delta_{jk}); \quad b_{ij} = b\delta_{ij} \tag{8.127}$$

Combining (4.58) and (8.125b) provides the incremental constitutive equation relative to the porous material:

$$\frac{dm_f}{\rho_f} = \beta_{ij}^{T} d\varepsilon_{ij} + \frac{dp}{M^{T}}; \quad \frac{1}{M^{T}} = \frac{1}{N^{T}} + \frac{\phi}{K_f} \tag{8.128}$$

so that:

$$d\sigma_{ij} = C^{uT}_{ijkl} d\varepsilon_{kl} - b^T_{ij} M^T \frac{dm_f}{\rho_f}; \quad C^{uT}_{ijkl} = C^T_{ijkl} + b^T_{ij}\beta^T_{kl} M^T \tag{8.129}$$

Adopting the usual isotropic cohesion–frictional model developed in §8.4.3, (8.51) and (8.59) give:

$$\frac{\partial f}{\partial \sigma_{ij}} = \frac{s_{ij}}{2\tau} + \frac{1}{3}\mathtt{f}\delta_{ij}; \quad \frac{\partial f}{\partial p} = \mathtt{f} \tag{8.130a}$$

$$\frac{\partial g}{\partial \sigma_{ij}} = \frac{s_{ij}}{2\tau} + \frac{1}{3}\mathtt{d}\delta_{ij}; \quad \frac{\partial g}{\partial p} = \mathtt{d} \tag{8.130b}$$

so that:

$$C_{ijkl}\frac{\partial f}{\partial \sigma_{kl}} = \frac{\partial f}{\partial \sigma_{kl}} C_{klij} = \mathtt{f}K\delta_{ij} + \mu\frac{s_{ij}}{\tau}; \quad b_{ij}\frac{\partial f}{\partial \sigma_{ij}} = b\mathtt{f} \tag{8.131a}$$

$$C_{ijkl}\frac{\partial g}{\partial \sigma_{kl}} = \frac{\partial g}{\partial \sigma_{kl}} C_{klij} = \mathtt{d}K\delta_{ij} + \mu\frac{s_{ij}}{\tau}; \quad b_{ij}\frac{\partial g}{\partial \sigma_{ij}} = b\mathtt{d} \tag{8.131b}$$

Use of (8.127) and (8.131) in (8.126) finally produces:

$$C^T_{ijkl} = \left(K - \frac{2}{3}\mu\right)\delta_{ij}\delta_{kl} + \mu(\delta_{ik}\delta_{jl} + \delta_{il}\delta_{jk}) - \frac{\left(\mathtt{d}K\delta_{ij} + \mu\frac{s_{ij}}{\tau}\right)\left(\mathtt{f}K\delta_{kl} + \mu\frac{s_{kl}}{\tau}\right)}{H + \mu + \mathtt{f}\mathtt{d}K} \tag{8.132a}$$

$$b^T_{ij} = b\delta_{ij} + \frac{\mathtt{f}\,(1-b)}{H + \mu + \mathtt{f}\mathtt{d}K}\left(\mathtt{d}K\delta_{ij} + \mu\frac{s_{ij}}{\tau}\right) \tag{8.132b}$$

$$\beta^T_{ij} = b\delta_{ij} + \frac{\mathtt{d}\,(1-b)}{H + \mu + \mathtt{f}\mathtt{d}K}\left(\mathtt{f}K\delta_{ij} + \mu\frac{s_{ij}}{\tau}\right) \tag{8.132c}$$

$$\frac{1}{N^T} = \frac{1}{N} + \frac{\mathtt{f}\mathtt{d}\,(1-b)^2}{H + \mu + \mathtt{f}\mathtt{d}K} \tag{8.132d}$$

Critical hardening modulus

No scalar remains undetermined in the formulation of the constitutive equations of plastic hardening materials so that discontinuities of velocity cannot occur within such materials. Therefore the discontinuities attached to the localization of deformation does not concern the skeleton velocity, $\mathbf{V} = \dot{\boldsymbol{\xi}}$, but its gradient. Adopting the notation of §7.4.3 and (7.178), we let:

$$[[\nabla\mathbf{V}]] = \boldsymbol{v} \otimes \mathbf{n}; \quad [[\dot{\boldsymbol{\varepsilon}}]] = \frac{1}{2}(\boldsymbol{v} \otimes \mathbf{n} + \mathbf{n} \otimes \boldsymbol{v}) \tag{8.133}$$

where $\mathbf{n}$ stands for the unit vector normal to the plane of localization while $\boldsymbol{v}$ is the polarization vector quantifying the direction and the strength of the discontinuity.

During the process of localization of the deformation, owing to diffusion effects associated with Darcy's law the fluid cannot sustain a discontinuity relative to the pressure gradient ∇p. Indeed, substitution of (8.128) into (5.8) and application of the jump operator [[ ]] to the resulting equation lead to:

$$\beta_{ij}^T \dot{\varepsilon}_{ij} + \frac{1}{M^T} \dot{p} = k \nabla^2 p \tag{8.134}$$

If the strength of the discontinuity relative to the pressure gradient ∇p were to differ both from zero and from infinity, reading $[[\nabla p]] \neq 0$ and $[[\nabla p]] \neq \infty$, $\nabla^2 p$ would be infinite. Since $[[\dot{\boldsymbol{\varepsilon}}]]$ remains finite, diffusion equation (8.134) would require $\dot{p}$ to be infinite too. This is not allowed by the kinematical compatibility condition (7.149), which requires $[[\dot{p}]]$ to be proportional to $[[\nabla p]] \cdot \mathbf{n}$ and therefore to remain finite as $[[\nabla p]]$ is. From the analysis of (8.134) we eventually conclude:[14]

$$[[\dot{p}]] = 0 \tag{8.135}$$

whose substitution into (8.125a) leads to:

$$[[\dot{\sigma}_{ij}]] = C_{ijkl}^T [[\dot{\varepsilon}_{ij}]] \tag{8.136}$$

In quasistatics, (7.182) gives:

$$\mathbf{n}.\,[[\dot{\boldsymbol{\sigma}}]] = 0 \tag{8.137}$$

so that (8.133), (8.136) and (8.137) require the polarization vector $\boldsymbol{v}$ to satisfy:

$$(\mathbf{n} \cdot \mathbf{C}^T \cdot \mathbf{n}) \cdot \boldsymbol{v} = 0 \tag{8.138}$$

The localization of deformation becomes possible when the previous equation admits a non-zero solution for $\boldsymbol{v}$, yielding:

$$\det(\mathbf{n} \cdot \mathbf{C}^T \cdot \mathbf{n}) = 0 \tag{8.139}$$

According to (7.183), $\mathbf{n} \cdot \mathbf{C}^T \cdot \mathbf{n}$ is proportional to the acoustic tensor related to the sole skeleton. Condition (8.139) shows that the localization of deformation can occur when one of the acceleration wavespeeds relative to the skeleton becomes zero. Indeed, as soon as one wavespeed becomes zero, any localization of the strain energy cannot be followed by its spontaneous redistribution within the material through wave transport.

Substitution of expression (8.132a) for $\mathbf{C}^T$ related to the usual cohesion–frictional model into the localization condition (8.139) produces:

$$\frac{H}{\mu} = \frac{\left(\mathtt{d}K + \mu \frac{s_{11}}{\tau}\right)\left(\mathtt{f}K + \mu \frac{s_{11}}{\tau}\right)}{\mu \left(K + \frac{4}{3}\mu\right)} + \left(\frac{s_{12}^2}{\tau} + \frac{s_{13}^2}{\tau}\right) - \frac{\mathtt{d}\mathtt{f}K + \mu}{\mu} \tag{8.140}$$

[14] Analogously, owing to thermal diffusion effects, the material cannot sustain a discontinuity in temperature gradient ∇T and a similar analysis would lead to $[[\dot{T}]] = 0$.

where we introduced the coordinate reference system such as $\mathbf{e}_1 = \mathbf{n}$. The hardening modulus H generally decreases with the hardening process so that the search for the early localization turns out to the search for the greatest value H_{cr} of the hardening modulus H given by (8.140) over all the possible stress states. Detailed calculations show that the greatest value is achieved when the direction of intermediary principal stress σ_{II} ($\sigma_I \geq \sigma_{II} \geq \sigma_{III}$) belongs to the localization plane and when the strength s_{II} of its deviatoric part is given by:

$$\frac{s_{II}}{\tau} = -\frac{1}{3}\left(\mathtt{d} + \mathtt{f}\right) \tag{8.141}$$

The associated critical modulus is then shown to be in the form:[15]

$$\frac{H_{cr}}{\mu} = \frac{1}{9}\left(\mathtt{d} - \mathtt{f}\right)^2 \frac{1+\nu}{1-\nu} \tag{8.142}$$

where ν is the drained Poisson coefficient. The above expression is consistent with the fact that localization of the deformation will never occur for a hardening material satisfying the principle of maximal plastic work, that is for $\mathtt{d} = \mathtt{f}$.

[15]The determination of the critical hardening modulus associated with the localization of deformation was originally derived in the founding paper Rudnicki J.W., Rice J.R. (1975), 'Conditions of the localization of deformation in pressure-sensitive materials', *Journal of Mechanics and Physics of Solids*, **23**, 371–394.

Chapter 9

Poroviscoelasticity

As long as viscous effects are not considered the skeleton behaviour does not depend on the loading rates and any evolution of the system can be considered as a sequence of equilibrium states. The skeleton response occurs in totality and simultaneously to an infinitesimal loading, while there is no further evolution under constant loading. By contrast, when viscous phenomena are involved, the response is partially delayed. The modelling of such a hereditary behaviour is the aim of this chapter.

9.1 Poroviscoelastic Behaviour

9.1.1 Viscous Strain and Viscous Porosity

Consider a sample of porous material subjected to the current stress σ_{ij} and the current pore pressure p and perform an instantaneous unloading process restoring a zero stress and a zero pore pressure. Only the reversible or elastic parts ε_{ij}^{el} and ϕ^{el} of the strain and the porosity, respectively, are immediately recovered. The viscoelastic parts ε_{ij}^{v} and ϕ^{v} are defined through the relations:

$$\varepsilon_{ij} = \varepsilon_{ij}^{el} + \varepsilon_{ij}^{v}; \qquad \phi - \phi_0 = \phi^{el} + \phi^{p} \tag{9.1}$$

The viscous volumetric dilation ϵ^{v} undergone by the skeleton is due to the viscous change in porosity and to the viscous volumetric dilation ϵ_s^{v} undergone by the solid matrix. Applying (1.32), we obtain:

$$\epsilon^{v} = (1 - \phi_0)\, \epsilon_s^{v} + \phi^{v} \tag{9.2}$$

In soil and rock mechanics the viscoelastic evolutions are caused by the viscous relative sliding occurring between the solid grains forming the matrix and generally due to the adsorbed water, so that the volume change of the matrix due uniquely to viscosity turns out to be negligible in the absence of any occluded porosity, resulting in $\epsilon_s^{v} = 0$ and entailing:

$$\phi^{v} = \epsilon^{v} \tag{9.3}$$

Poromechanics O. Coussy
 ISBN 0-470-84920-7

In order to capture the departure from matrix viscous incompressibility an heuristic assumption consists in setting:

$$\phi^v = \beta\epsilon^v \tag{9.4}$$

By contrast to the value $\beta = 1$ corresponding to a viscous incompressible matrix, that is (9.3), according to (9.2) the value $\beta = \phi_0$ corresponds to $\epsilon^v = \epsilon_s^v$; that is, to a volumetric viscous strain of the skeleton due only to that of the solid matrix. It is then consistent to require β to satisfy inequalities $\phi_0 \leq \beta \leq 1$.

9.1.2 Poroviscoelastic State Equations for the Skeleton

In the context of infinitesimal isothermal transformations and saturated porous materials, inequality (3.33) expressing the positiveness of the dissipation attached to the irreversible evolutions of the skeleton is specialized in the form:

$$\sigma\, d\epsilon + s_{ij}\, de_{ij} + p\, d\phi - d\Psi_s \geq 0 \tag{9.5}$$

From the current state, when considering reversible or instantaneous poroelastic evolutions, the values of internal viscous variables ε_{ij}^v and ϕ^v remain the same and there is no dissipation. For such evolutions inequality (9.5) becomes an equality, leading to the state equations:

$$\sigma = \frac{\partial\Psi_s}{\partial\epsilon}; \quad s_{ij} = \frac{\partial\Psi_s}{\partial e_{ij}}; \quad p = \frac{\partial\Psi_s}{\partial\phi} \tag{9.6}$$

State equations (9.6) have been derived for poroelastic evolutions. They apply to any evolution as soon as free energy Ψ_s is continuously differentiable with regard to the whole set of state variables.

For the same physical reasons as the ones developed in hardening plasticity (see §8.3.1) we infer the existence of a trapped energy, that is an energy that is not recovered but not dissipated in the instantaneous elastic unloading process. Anticipating the validity of assumption (9.4), the trapped energy has to depend on the viscous strain only. We finally write:

$$\Psi_s = W_s\left(\epsilon - \epsilon^v, e_{ij} - e_{ij}^v, \phi - \beta\epsilon^v\right) + U\left(\epsilon^v, e_{ij}^v\right) \tag{9.7}$$

Restricting consideration to linear isotropic porous materials, we adopt for W_s and U the following expressions:

$$\begin{aligned} W_s\left(\epsilon - \epsilon^v, e_{ij} - e_{ij}^v, \phi - \beta\epsilon^v\right) &= \frac{1}{2}K_0\left(\epsilon - \epsilon^v\right)^2 + \mu_0\left(e_{ij} - e_{ij}^v\right)^2 \\ &\quad + \frac{1}{2}N_0\left[b(\epsilon - \epsilon^v) - \phi - \beta\epsilon^v - \phi_0\right]^2 \end{aligned} \tag{9.8a}$$

$$U\left(\epsilon^v, e_{ij}^v\right) = \frac{1}{2}\mathcal{K}\epsilon^{v2} + \mathcal{G}e_{ij}^{v2} \tag{9.8b}$$

Substitution of the latter into (9.6) gives the state equations in the form:

$$\sigma = K_0(\epsilon - \epsilon^v) - b_0 p \tag{9.9a}$$

$$s_{ij} = 2\mu_0\left(e_{ij} - e^v_{ij}\right) \tag{9.9b}$$

$$p = N_0\left[-b_0(\epsilon - \epsilon^v) + \phi - \beta\epsilon^v - \phi_0\right] \tag{9.9c}$$

Returning finally to the porous material and introducing the change in fluid volume content v_f (see (5.13)), the latter relation is replaced by:

$$v_f = \frac{m_f - m_f^0}{\rho_f^0} = b_0\epsilon + (\beta - b_0)\,\epsilon^v + \frac{p}{M_0}; \quad \frac{1}{M_0} = \frac{1}{N_0} + \frac{\phi_0}{K_f} \tag{9.10}$$

9.1.3 Complementary Evolution Laws

Substitution of (9.7)–(9.9) into (9.5) gives:

$$\left(\sigma + \beta p - \mathcal{K}\epsilon^v\right)\dot{\epsilon}^v + \left(s_{ij} - 2\mathcal{G}e^v_{ij}\right)\dot{e}^v_{ij} \geq 0 \tag{9.11}$$

Assuming that the viscous dissipative mechanism is normal (see §3.4.2) we introduce the convex dissipation potential $\mathcal{D}\left(\dot{\epsilon}^v, \dot{e}^v_{ij}\right)$ such that:

$$\sigma + \beta p - \mathcal{K}\epsilon^v = \frac{\partial\mathcal{D}}{\partial\dot{\epsilon}^v}; \qquad s_{ij} - 2\mathcal{G}e^v_{ij} = \frac{\partial\mathcal{D}}{\partial\dot{e}^v_{ij}} \tag{9.12}$$

In the isotropic case $\mathcal{D}$ is expressed in the form:

$$\mathcal{D}\left(\dot{\epsilon}^v, \dot{e}^v_{ij}\right) = \frac{1}{2}\varsigma\dot{\epsilon}^{v2} + \eta\dot{e}^{v2}_{ij}; \qquad \varsigma \geq 0, \quad \eta \geq 0 \tag{9.13}$$

where the positiveness of ς and η guarantees the convexity of $\mathcal{D}$. Substitution of (9.13) into (9.12) furnishes the explicit form of the complementary evolution laws:

$$\sigma + \beta p = \mathcal{K}\epsilon^v + \varsigma\dot{\epsilon}^v; \qquad s_{ij} = 2\mathcal{G}e^v_{ij} + 2\eta\dot{e}^v_{ij} \tag{9.14}$$

so that ς and η are identified respectively as the volumetric viscous coefficient and the shear viscous coefficient.

9.2 Functional Approach to Linear Poroviscoelasticity

9.2.1 Creep Test. Instantaneous and Relaxed Properties. The Trapped Energy

The creep test consists in subjecting the porous material sample to a time step in stress and pore pressure and in recording the resulting delayed strain history. The loading history is expressed in the mathematical form:

$$\sigma_{ij}(t) = \Delta\sigma_{ij} H(t); \quad p(t) = \Delta p H(t) \tag{9.15}$$

where $H(t)$ is the Heaviside step function, $H(t<0)=0$, $H(t>0)=1$. In response to the loading history (9.15) the response in viscous strain cannot be discontinuous. Indeed, a discontinuous viscous strain would require the viscous strain rate to be infinite, which is not allowed on account of the finite character of the stress and of the fluid pressure imposed by the complementary evolution laws (9.14). As a consequence, the instantaneous response ϵ_{0^+}, $e_{ij}^{0^+}$ and ϕ_{0^+} in strain and porosity to the loading (9.15) is poroelastic, resulting in $\epsilon_{0^+}^v = e_{ij}^{v0^+} = 0$ and corresponding to point a_0 in Fig. 9.1. According to (9.9), the instantaneous response satisfies:

$$\Delta\sigma = K_0\epsilon_{0^+} - b_0 p \tag{9.16a}$$

$$\Delta s_{ij} = 2\mu_0 \, e_{ij}^{0^+} \tag{9.16b}$$

$$\Delta p = N_0\left[-b_0\epsilon_{0^+} + \phi_{0^+} - \phi_0\right] \tag{9.16c}$$

so that K_0, b_0, μ_0 and N_0 are interpreted as the instantaneous poroelastic properties of the skeleton.

As time flows, the porous material creeps: at constant stress and fluid pressure, it undergoes a delayed strain history corresponding to path a_0a_∞ in Fig. 9.1. The long time response obtained as time goes to infinity and corresponding to point a_∞ in Fig. 9.1 is achieved at zero rates of the viscous strain. Letting $\dot{\epsilon}^v = \dot{e}_{ij}^v = 0$ in (9.14), the resulting equation combines with (9.9) to give the long time response ϵ_∞, e_{ij}^∞ and ϕ_∞ corresponding to point a_∞ in Fig. 9.1:

$$\Delta\sigma = K_\infty\epsilon_\infty - b_\infty p \tag{9.17a}$$

$$\Delta s_{ij} = 2\mu_\infty \, e_{ij}^\infty \tag{9.17b}$$

$$\Delta p = N_\infty\left[-b_\infty\epsilon_\infty + \phi_\infty - \phi_0\right] \tag{9.17c}$$

where the delayed or relaxed poroelastic properties K_∞, b_∞, μ_∞ and N_∞ are given by:

$$\frac{1}{K_\infty} = \frac{1}{K_0} + \frac{1}{\mathcal{K}}; \quad \frac{1}{\mu_\infty} = \frac{1}{\mu_0} + \frac{1}{\mathcal{G}} \tag{9.18a}$$

$$\frac{b_\infty}{K_\infty} = \frac{b_0}{K_0} + \frac{\beta}{\mathcal{K}}; \quad \frac{1}{N_\infty} = \frac{1}{N_0} + \frac{(\beta - b_\infty)(\beta - b_0)}{\mathcal{K}} \tag{9.18b}$$

Returning to the energy side, from (9.7)–(9.9) and (9.14)–(9.18) we derive:

$$\Psi_s(t \to \infty) = \Psi_s\left(t = 0^+\right) + U\left(\epsilon_\infty^v, e_{ij}^{v\infty}\right) \tag{9.19}$$

Once the long time response is established, we consider an instantaneous unloading process that restores the initial zero stress and fluid pressure state. The viscous contribution to the strain cannot undergo any instantaneous change so that only the elastic part of the strain is recovered in the unloading process. As a consequence, the strain recorded at the end of the instantaneous unloading process (path $a_\infty \varepsilon_\infty^v$ in Fig. 9.1) is the viscous strain ε_{ij}^v corresponding to point ε_v^∞ in Fig. 9.1. In addition, the porous material instantaneously releases only the part $\Psi_s(t = 0^+)$ (area of triangle $Oa_0\varepsilon_0$ or $\varepsilon_\infty^v a_\infty \varepsilon_\infty$ in Fig. 9.1) of the asymptotic free energy $\Psi_s(t \to \infty)$ (area $Oa_0a_\infty\varepsilon_\infty$ in Fig. 9.1), whereas the part $U\left(\epsilon_\infty^v, e_{ij}^{v\infty}\right)$ remains trapped (aera of parallelogram $\varepsilon_0 a_0 a_\infty \varepsilon_\infty$ or $Oa_0a_\infty\varepsilon_\infty^v$ in Fig. 9.1). Eventually, although now being unloaded, the poroviscoelastic material still deforms until the zero strain state is reached (path $\varepsilon_\infty^v O$ in Fig. 9.1). In the meantime the whole trapped energy $U\left(\epsilon_\infty^v, e_{ij}^{v\infty}\right)$ is eventually dissipated in heat form through viscous effects, in opposition to the trapped energy of hardening plasticity The energy picture so sketched for the creep test, and also the definition of the instantaneous and relaxed properties, are quite general. Only the actual time history of the delayed straining effects, that is the rate at which the related point moves along path a_0a_∞, is specific to the poroviscoelastic material under consideration. This time history can be more generally addressed through the functional approach developed in the next section.

9.2.2 Creep and Relaxation Functions

For loading history (9.15), time integration of (9.14) gives:

$$\epsilon^v(t) = (\Delta\sigma + \beta\Delta p)\frac{1}{\mathcal{K}}\left[1 - \exp\left(-\frac{t}{\tau_c}\right)\right] \tag{9.20a}$$

$$e_{ij}^v(t) = \frac{\Delta s_{ij}}{2\mathcal{G}}\left[1 - \exp\left(-\frac{t}{\vartheta_c}\right)\right] \tag{9.20b}$$

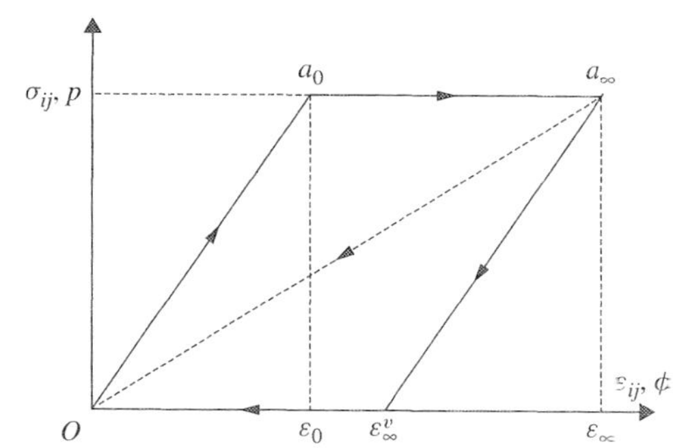

Figure 9.1: Creep test and trapped energy. After being instantaneously deformed (path Oa_0) the poroviscoelastic material creeps, undergoing a delayed deformation history at subsequently constant loading (path a_0a_∞). In a later unloading process the porous material instantaneously releases only that part of the asymptotic free energy corresponding to the instantaneous elastic strain (triangle $Oa_0\varepsilon_0$ or $\varepsilon_\infty^v a_\infty\varepsilon_\infty$), so that the part corresponding to the viscoelastic strain remains trapped (parallelogram $\varepsilon_0a_0a_\infty\varepsilon_\infty$ or $Oa_0a_\infty\varepsilon_\infty^v$). The latter is eventually dissipated in heat form during the creep process, restoring a zero deformation (path $\varepsilon_\infty^v O$). Path $a_\infty O$ corresponds to an unloading process performed infinitely slowly so that no dissipation occurs. The energy picture so sketched is analogous to the one relative to the consolidation process (see Fig. 5.4).

where τ_c and ϑ_c are the creep characteristic times defined by:

$$\tau_c = \frac{\varsigma}{\mathcal{K}}; \quad \vartheta_c = \frac{\eta}{\mathcal{G}} \tag{9.21}$$

Substitution of (9.20) into (9.9a)–(9.9b) provides the strain history relative to the creep test in the form:

$$\epsilon\,(t) = K^{-1}\,(t)\,\Delta\sigma - B\,(t)\,\Delta p \tag{9.22a}$$

$$e_{ij}\,(t) = \frac{1}{2}\mu^{-1}\,(t)\;\Delta s_{ij} \tag{9.22b}$$

where $K^{-1}\,(t)$, $B\,(t)$ and $\mu^{-1}\,(t)$ stand for the creep functions:

$$K^{-1}\,(t) = \left[\frac{1}{K_\infty} + \left(\frac{1}{K_\infty} - \frac{1}{K_0}\right)\exp\left(-\frac{t}{\tau_c}\right)\right] H\,(t) \tag{9.23a}$$

$$B\,(t) = \left[\frac{b_\infty}{K_\infty} + \left(\frac{b_\infty}{K_\infty} - \frac{b_0}{K_0}\right)\exp\left(-\frac{t}{\tau_c}\right)\right] H\,(t) \tag{9.23b}$$

$$\mu^{-1}\,(t) = \left[\frac{1}{\mu_\infty} + \left(\frac{1}{\mu_\infty} - \frac{1}{\mu_0}\right)\exp\left(-\frac{t}{\vartheta_c}\right)\right] H\,(t) \tag{9.23c}$$

Since the behaviour is linear, the strain history resulting from any history in stress $\sigma_{ij}\,(t)$ and fluid pressure $p\,(t)$ is derived by summing the infinitesimal creep responses to the successive infinitesimal loadings $d\sigma_{ij}\,(t)$ and $dp(t)$. We write:

$$\epsilon = K^{-1} \odot \sigma + B \odot p \tag{9.24a}$$

$$e_{ij} = \frac{1}{2}\mu^{-1} \odot s_{ij} \tag{9.24b}$$

In (9.24), $f \odot g$ stands for the Stieltjes convolution product of functions $f\,(t)$ and $g\,(t)$. Restricting consideration here to functions of zero value for negative time and considering possible discontinuities of function g at time t_i, including the one occurring at the origin of time caused by the possible discontinuity in loading, the Stieltjes convolution product $f \odot g$ can be expressed in the form:

$$f \odot g = \int_0^t f(t-u)dg(u) + \sum_i f(t-t_i)\left[g\left(t_i^+\right) - g\left(t_i^-\right)\right] \tag{9.25}$$

The convolution product has the same properties as the ordinary product: commutativity, associativity and distributivity with regard to addition, and the existence of an inverse f^{-1} satisfying $f^{-1} \odot g = H$, where the Heaviside function H plays the role of the neutral element. Accordingly, we anticipated in (9.24) that K^{-1} and μ^{-1} were the inverses with respect to the convolution product of the relaxation functions K and μ, that is:

$$K\,(t) = \left[K_\infty + (K_\infty - K_0)\exp\left(-\frac{t}{\tau_r}\right)\right] H\,(t) \tag{9.26a}$$

$$\mu\,(t) = \left[\mu_\infty + (\mu_\infty - \mu_0)\exp\left(-\frac{t}{\vartheta_r}\right)\right] H\,(t) \tag{9.26b}$$

where τ_r and ϑ_r are the relaxation characteristic times defined by:

$$\tau_r = \frac{K_\infty}{K_0}\tau_c; \quad \vartheta_r = \frac{\mu_\infty}{\mu_0}\vartheta_c \tag{9.27}$$

Some further calculations also give $B = K^{-1} \odot b$ where $b(t)$ is the function:

$$b(t) = \left[b_\infty + (b_\infty - b_0)\exp\left(-\frac{t}{\tau_r}\right)\right] H(t) \tag{9.28}$$

so that (9.24) can be rewritten in the form:

$$\sigma = K \odot \epsilon - b \odot p; \quad s_{ij} = 2\mu \odot e_{ij} \tag{9.29}$$

It is then worthwhile to note that Biot's poroelastic effective stress $\sigma'' = \sigma + bp$ is extended in the poroviscoelastic form $\sigma'' = \sigma + b \odot p$.

In order to derive the missing equation related to the change in porosity, rather than further exploring the previous creep test, we consider a mixed relaxation–creep test. The latter consists in subjecting the porous material sample to time steps in volumetric strain and in pore pressure:

$$\epsilon(t) = \Delta\epsilon H(t); \quad p(t) = \Delta p H(t) \tag{9.30}$$

and in recording the resulting delayed stress history. Substitution of (9.30) into the first equation of (9.29) gives the related stress history:

$$\sigma(t) = K(t)\,\Delta\epsilon - b(t)\,\Delta p \tag{9.31}$$

Combining the latter with (9.9a) and (9.30) we obtain the following expression for ϵ^v:

$$\epsilon^v(t) = \left(1 - \frac{K(t)}{K_0}\right)\Delta\epsilon + \frac{b(t) - b_0}{K_0}\Delta p \tag{9.32}$$

Substitution of (9.30) and (9.32) into (9.9c) and use of (9.18) and (9.26a) give the porosity history in the form:

$$\phi - \phi_0 = b(t)\,\Delta\epsilon + N^{-1}(t)\,\Delta p \tag{9.33}$$

where we note:

$$N^{-1}(t) = \left[\frac{1}{N_\infty} + \left(\frac{1}{N_\infty} - \frac{1}{N_0}\right)\exp\left(-\frac{t}{\tau_r}\right)\right] H(t) \tag{9.34a}$$

$$N(t) = \left[N_\infty + (N_\infty - N_0)\exp\left(-\frac{N_0 t}{N_\infty \tau_r}\right)\right] H(t) \tag{9.34b}$$

Surprisingly, the function $N^{-1}(t)$ involves the characteristic relaxation time τ_r, although $N^{-1}(t)$ appears to play the role of a creep function since it governs the history in porosity in response to a loading in pore pressure. However, the creep is only apparent since in the meantime the skeleton has to relax in response to the imposed volumetric strain defined in (9.30). Finally, invoking the linearity, from (9.33) we derive the missing constitutive equation in the form:

$$\phi - \phi_0 = b \odot \epsilon + N^{-1} \odot p \tag{9.35}$$

Returning to the porous material and introducing the change in fluid volume content v_f (see (5.13)), the latter relation has to be replaced by:

$$v_f = b \odot \epsilon + M^{-1} \odot p; \quad M^{-1}(t) = N^{-1}(t) + \frac{\phi_0}{K_f} H(t) \tag{9.36}$$

Up to now the poroviscoelastic behaviour has been captured through a single viscous strain tensor ε_{ij}^{v}. More general constitutive equations of poroviscoelasticity can be obtained by increasing the number of independent internal variables in accordance with the number of characteristic times required to account for the actual creep and relaxation functions. Indeed, still invoking the linearity, constitutive equations (9.29) and (9.35) can be extended to any linear poroviscoelastic material whatever the experimental creep and relaxation functions are. More generally, invoking the underlying superposition principle, the functional approach gives the constitutive equations of linear poroviscoelasticity in the form:

$$\sigma_{ij} = C_{ijkl} \odot \varepsilon_{kl} - b_{ij} \odot p; \quad \phi - \phi_0 = b_{ij} \odot \varepsilon_{ij} + N^{-1} \odot p \tag{9.37}$$

Nevertheless, the functional approach based on the sole superposition principle fails to capture the usual symmetry properties, that is the symmetry relations $C_{ijkl} = C_{klij}$ and the one implying that function b or functions b_{ij} have to be the same in (9.29) and (9.35), or in both equations (9.37). These symmetry properties are based on the existence of potentials Ψ_s and $\mathcal{D}$ whose general determination, whatever the creep and relaxation functions are in functional constitutive equations (9.29) and (9.35), is left to §9.4.

9.2.3 Poroviscoelastic Properties and Constituent Properties

Since the convolution product has the same properties as the normal product, the poroelastic compatibility relations (4.35) can be extended in the form:

$$b = 1 - K \odot K_s^{-1}; \quad N^{-1} = (b - \phi_0) \odot K_s^{-1} \tag{9.38}$$

where 1 stands for $H(t)$ and K_s^{-1} for the creep function of the solid matrix. For instance, the following expression:

$$K_s^{-1}(t) = \left[\frac{1}{K_{s\infty}} + \left(\frac{1}{K_{s\infty}} - \frac{1}{K_{s0}}\right) \exp\left(-\frac{t}{\tau_{sc}}\right)\right] H(t) \tag{9.39}$$

where:

$$\tau_{sc} = \frac{\varsigma_s}{\mathcal{K}_s} = \varsigma_s \left(\frac{1}{K_{s\infty}} - \frac{1}{K_{s0}}\right) \tag{9.40}$$

leads us to identify the poroviscoelastic material as the one we analyzed extensively in the previous sections with the set of relations:

$$b_{0 \text{ or } \infty} = 1 - \frac{K_{0 \text{ or } \infty}}{K_{s0 \text{ or } \infty}}; \quad \frac{1}{N_{0 \text{ or } \infty}} = \frac{b_{0 \text{ or } \infty} - \phi_0}{K_{s0 \text{ or } \infty}}; \quad \frac{\varsigma}{\mathcal{K}} = \frac{\varsigma_s}{\mathcal{K}_s} \tag{9.41}$$

In addition, (9.18) and the above relations combine to deliver the particular relations:

$$\beta = \phi_0 = 1 - \frac{\mathcal{K}}{\mathcal{K}_s}; \quad \varsigma = (1 - \phi_0)\, \varsigma_s \tag{9.42}$$

The latter relations indicate that relations (9.38) are based on the assumption of an homogeneous matrix leading us to identify $\epsilon^v = \epsilon_s^v$ as recovered by the combination of (9.2), (9.4) and (9.42).

9.3 Primary and Secondary Consolidation

In this section we extend to the poroviscoelastic material the analysis of the consolidation of a poroelastic layer we performed in §5.2.2. Indeed the actual consolidation of a clay layer results from the combination of two processes. The primary consolidation of the layer is the process we analysed in §5.2.2. It is mainly due to the progressive transfer of the loading from the saturating fluid to the solid matrix on account of the diffusion of the fluid pressure. By contrast, the secondary consolidation process involves the delayed creep of the layer due to the relative viscous micro-slidings of the platelets forming the solid matrix of the clay. These viscous slidings result from the lubrication induced at the contact between the platelets by the water film electrically bound to the platelets and not contributing to the fluid flow. The two consolidation processes are generally well separated in time, the secondary consolidation occurring much later than the primary one.

To examine this two-step consolidation process, the poroviscoelastic behaviour of the clay can be roughly captured through the model we developed in §9.1. A justification for such a choice is given in §9.4.2 (see comments following (9.75)). Since the platelets do not undergo any volume change, both the elastic and the viscous volumetric strains are due only to the change in porosity. In addition, the fluid flow is assumed to be incompressible. These incompressibility assumptions result in:

$$b_0 = \beta = 1; \quad \frac{1}{M_0} = 0 \tag{9.43}$$

so that (9.10) reduces to the volume conservation condition:

$$v_f = \frac{m_f - m_f^0}{\rho_f^0} = \epsilon \tag{9.44}$$

Substitution of the latter into (5.8) gives the 1D diffusion equation:

$$\frac{\partial \epsilon}{\partial t} = k \frac{\partial^2 p}{\partial z^2} \tag{9.45}$$

Beyond the instantaneous application of the vertical constant loading $\sigma_{zz} = -\varpi$ at the upper surface $z = 0$, the vertical equilibrium requires $\sigma_{zz} = -\varpi$, while the displacement in the soil layer reduces to the vertical one, namely $\boldsymbol{\xi} = \xi\,(z, t)\,\mathbf{e}_z$. Use of (9.43) in state equations (9.9) and in the complementary evolution laws (9.14) gives:

$$\varepsilon_{zz} = \frac{\partial \xi}{\partial z} = \epsilon = \epsilon^v + \frac{p - \varpi}{\mathrm{E}_0}; \quad \mathrm{E}_0 = K_0 + \frac{4}{3}\mu_0 \tag{9.46}$$

and:

$$p - \varpi = \mathcal{E}\epsilon^v + \varrho\dot{\epsilon}^v; \quad \mathcal{E} = \mathcal{K} + \frac{4}{3}\mathcal{G}; \quad \varrho = \varsigma + \frac{4}{3}\eta \tag{9.47}$$

Owing to the undrained character of the instantaneous response (see §5.1.2) and to the assumed incompressibility of the constituents, the overall volumetric strain ϵ cannot undergo any time discontinuity. From the analysis carried out in §9.2.1 the viscous strain ϵ^v also cannot undergo any time discontinuity. Accordingly, (9.46) and (9.47) give the instantaneous response of the soil layer, that is:

$$p\left(z, t = 0^+\right) = \varpi; \quad \frac{\partial \xi}{\partial z}\left(z, t = 0^+\right) = 0 \tag{9.48}$$

Getting rid of ϵ^v between (9.46) and (9.47), we obtain the following differential equation:

$$\epsilon + \tau_v \frac{\partial \epsilon}{\partial t} = \frac{p - \varpi}{\mathrm{E}_\infty} + \frac{\tau_v}{\mathrm{E}_0}\frac{\partial p}{\partial t} \tag{9.49}$$

where we note:[1]

$$\tau_v = \frac{\varrho}{\mathcal{E}}; \quad \frac{1}{\mathrm{E}_\infty} = \frac{1}{\mathrm{E}_0} + \frac{1}{\mathcal{E}} \tag{9.50}$$

As in §5.2.2 we consider a soil layer of thickness h subjected to the hydraulic boundary conditions:

$$z = 0:\ p = 0; \quad z = h:\ \frac{\partial p}{\partial z} = 0 \tag{9.51}$$

The problem at hand eventually results in solving (9.45) and (9.49) with initial and boundary conditions (9.48) and (9.51).

The condition of time scale separation of the primary consolidation and the secondary consolidation is expressed in the form:

$$\tau = k\frac{\mathrm{E}_0}{h^2}; \quad \vartheta = \frac{\tau}{\tau_v} \ll 1 \tag{9.52}$$

where τ is the characteristic time associated with the primary consolidation process and involving the instantaneous oedometric modulus E_0. In order to explore the primary consolidation we let:

$$\bar{t} = \frac{t}{\tau}; \quad \overline{p} = \frac{p}{\varpi} \tag{9.53}$$

[1]It is worthwhile to note that the relaxed oedometric modulus $1/\mathrm{E}_\infty$ differs from $1/(K_\infty + \frac{4}{3}\mu_\infty)$.

Substitution of (9.53) into (9.49) gives:

$$\epsilon + \frac{1}{\vartheta}\frac{\partial \epsilon}{\partial \bar{t}} = \frac{\varpi}{\mathrm{E}_\infty}\left(\bar{p} - 1\right) + \frac{\varpi}{\mathrm{E}_0}\frac{1}{\vartheta}\frac{\partial \bar{p}}{\partial \bar{t}} \tag{9.54}$$

Use of the time scale separation condition (9.52) leads us to neglect the first term with respect to the second one in both members of (9.54). Accordingly, time integration and the use of initial conditions (9.48) give:

$$\epsilon\left(z, \bar{t}\right) = \frac{\varpi}{\mathrm{E}_0}\left(\bar{p}\left(z, \bar{t}\right) - 1\right) \tag{9.55}$$

Substituting the latter into (9.45), we recover the diffusion equation (5.76) governing the poroelastic consolidation to be solved with the same boundary conditions (9.51), so that poroelastic solutions (5.87)–(5.88) apply to the primary consolidation under the time scale separation condition (9.52). From (5.79), where we put $K_u \to \infty$ and $K = K_0$, we derive $s_0 = 0$ and $s_\infty = \varpi h/\mathrm{E}_0$. Substituting these expressions into (5.88), the settlement $s\left(\bar{t}\right) = \xi(z = 0, \bar{t})$ related to the primary consolidation can be expressed in the form:

$$s\left(\bar{t}\right) = \frac{\varpi h}{\mathrm{E}_0}\left[1 - \sum_{n=0}^{n=\infty} \frac{8}{\pi^2\left(2n+1\right)^2} \times \exp\left(-\frac{\left(2n+1\right)^2\pi^2}{4}\bar{t}\right)\right] \tag{9.56}$$

In order to explore the secondary consolidation we now let:

$$\tilde{t} = \frac{t}{\tau_v}; \quad \bar{z} = \frac{z}{h} \tag{9.57}$$

Substitution of (9.57) into (9.45) gives:

$$\frac{\partial \epsilon}{\partial \tilde{t}} = \frac{\varpi}{\mathrm{E}_0}\frac{1}{\vartheta}\frac{\partial^2 \bar{p}}{\partial \bar{z}^2} \tag{9.58}$$

Use of the time scale separation condition (9.52) leads us to neglect the term on the left hand side of (9.58) so that space integration and the use of boundary conditions (9.48) give $\bar{p} = 0$. Accordingly, (9.49) and (9.55) give:

$$\epsilon\left(z, \tilde{t} \to 0\right) = -\frac{\varpi}{\mathrm{E}_0}; \quad \epsilon + \frac{\partial \epsilon}{\partial \tilde{t}} = -\frac{\varpi}{\mathrm{E}_\infty} \tag{9.59}$$

whose solution is:

$$\epsilon\left(z, \tilde{t}\right) = -\frac{\varpi}{\mathrm{E}_\infty} + \left(\frac{\varpi}{\mathrm{E}_\infty} - \frac{\varpi}{\mathrm{E}_0}\right)\exp\left(-\tilde{t}\right) \tag{9.60}$$

Substituting (9.60) into (9.46) and integrating over the layer thickness, and taking into account the boundary condition $\xi(z = h, \tilde{t}) = 0$, we derive the settlement $s\left(\tilde{t}\right) = \xi(z = 0, \tilde{t})$ related to the secondary consolidation in the form:

$$s\left(\tilde{t}\right) = \frac{\varpi h}{\mathrm{E}_\infty} + \left(\frac{\varpi h}{\mathrm{E}_0} - \frac{\varpi h}{\mathrm{E}_\infty}\right)\exp\left(-\tilde{t}\right) \tag{9.61}$$

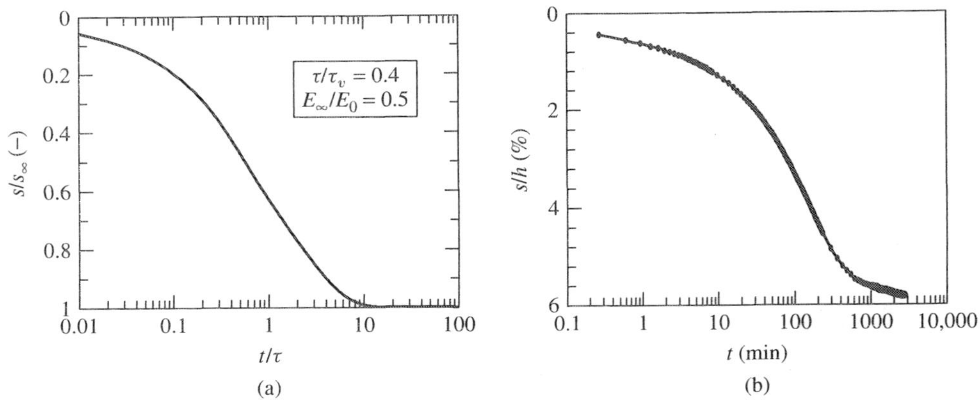

Figure 9.2: The two-step consolidation process: (a) the consolidation curve as predicted by (9.62) using the condition (9.52) of time scale separation of the primary and secondary consolidation processes; (b) the actual consolidation curve obtained for a consolidation overpressure of $\varpi =$ 2MPa for an artificially compacted clay (by courtesy of Y.-J. Cui).

The mixed solution accounting for both the primary and the secondary conditions is obtained by adding solutions (9.56) and (9.61) and removing the shared limit $\varpi h/\mathrm{E}_0$ as $\bar{t}$ and $\tilde{t}$ go respectively to infinity and to zero. We finally obtain:

$$s(t) = \frac{\varpi h}{\mathrm{E}_\infty}\left[1 - \exp\left(-\frac{t}{\tau_v}\right)\right] + \frac{\varpi h}{\mathrm{E}_0}\left[\exp\left(-\frac{t}{\tau_v}\right) - \sum_{n=0}^{n=\infty}\frac{8}{\pi^2(2n+1)^2}\exp\left(-\frac{(2n+1)^2\pi^2}{4}\frac{t}{\tau}\right)\right] \tag{9.62}$$

The predicted consolidation curve (9.62) is illustrated in Fig. 9.2a. Its shape compares well with the shape of the experimental consolidation curve found for an artificially compacted clay and reported in Fig. 9.2b.

9.4 Advanced Analysis

9.4.1 Poroviscoplasticity

Poroviscoelasticity imposes no limitation on stress and pore pressure. On the other hand, poroplasticity does not account for any viscous effects so that the deformation occurs in totality and simultaneously with the loading variation. Poroviscoplasticity aims at accounting for both the delayed character of the deformation and the existence of a current stress threshold beyond which irreversible strains are recorded. For instance, the delayed character of the deformation can be due to the viscous contacts existing between the solid grains composing the matrix and can be possibly enhanced by a thin film of absorbed water. On the other hand, the irreversible slidings occur only when the forces between the grains exceed some current threshold.

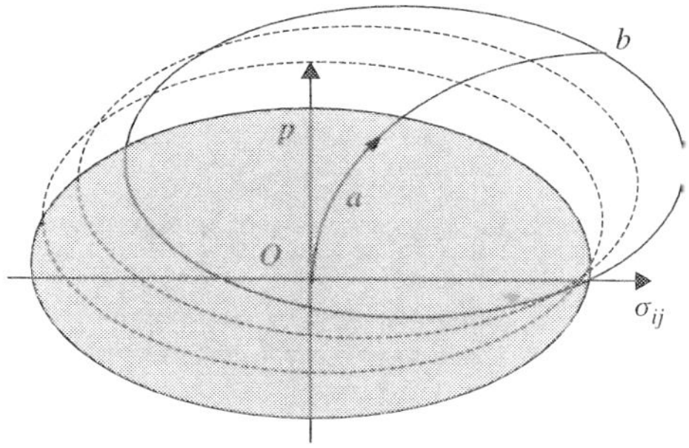

Figure 9.3: Poroviscoplastic behaviour. As long as the loading point remains within the current domain of elasticity the deformation is elastic (point a). In contrast to poroplasticity, the loading point may instantaneously leave the current elastic domain (point b). Owing to the combined action of viscous and hardening effects the current elastic domain can evolve under constant loading (successive domains of elasticity–dashed lines). Creep ends when the current loading point is on the boundary of the current elasticity domain (point b).

More precisely, poroviscoplasticity aims at accounting for the following behaviour: as long as the loading in stress and pore pressure is below some current threshold, the porous material deforms instantaneously and elastically (point a in Fig. 9.3). When the loading point leaves the current elastic domain, in addition to the elastic strains irreversible strains occur in a delayed manner with possible hardening effects resulting in a change in the current elastic domain. In contrast with poroplasticity, the loading point may instantaneously leave the current elasticity domain to be situated at its exterior (point b in Fig. 9.3). Creep under constant loading (viscous effect) can only occur beyond a loading threshold (plastic effect). Owing to the combined action of viscous and hardening effects, creep may also concern the current elastic domain so that the latter, in contrast to poroplasticity, still evolves under constant loading. Creep ends when the current loading point is on the boundary of the current elasticity domain. The positiveness of the dissipation (8.33) extends to poroviscoplasticity in the form:

$$\sigma_{ij}\dot{\varepsilon}_{ij}^{vp} + p\dot{\phi}^{vp} + \zeta_J \dot{\chi}_J \geq 0 \tag{9.63}$$

where the term $\zeta_J \dot{\chi}_J$ does not have to be considered in the case of ideal poroviscoplasticity. We restrict ourselves to the case of an incompressible viscoplastic matrix, resulting in $\phi^{vp} = \epsilon^{vp}$ so that (9.63) can be rewritten in the form:

$$\sigma'_{ij}\dot{\varepsilon}_{ij}^{vp} + \zeta_J \dot{\chi}_J \geq 0 \tag{9.64}$$

where $\sigma'_{ij} = \sigma_{ij} + p\delta_{ij}$ is Terzaghi's effective stress. According to the general approach to complementary laws developed in §3.4.2, the viscoplastic flow rule can be formulated with the help of a convex dissipation potential $\mathcal{D}^*(\sigma'_{ij})$ such that:

$$\dot{\varepsilon}_{ij}^{vp} = \frac{\partial \mathcal{D}^*}{\partial \sigma'_{ij}} \tag{9.65}$$

In order to account for the existence of a current domain of elasticity and for the main features of the poroviscoplastic behaviour, the usual models of viscoporoplasticity consist in adopting the following expression for $\mathcal{D}^*$:

$$\mathcal{D}^*\left(\sigma'_{ij}\right) = \frac{1}{2\eta}\left\langle f\left(\sigma'_{ij}, \zeta_J\right)\right\rangle^2 \tag{9.66}$$

where η is a viscosity coefficient and where $\langle f \rangle = \frac{1}{2}\left(|f| + f\right)$ stands for the positive part of the loading function f. A more general expression for $\mathcal{D}^*$ consists in replacing the exponent 2 in (9.66) by the exponent $n+1$ where n is not necessarily an integer, but has to be positive in order for the potential $\mathcal{D}^*$ to remain convex. Substitution of (9.66) into (9.65) gives:

$$\dot{\varepsilon}_{ij}^{vp} = \frac{1}{\eta}\langle f \rangle \frac{\partial f}{\partial \sigma'_{ij}} \tag{9.67}$$

so that $f \leq 0$ defines the current domain of elasticity in the effective stress space. By analogy with plasticity, a non-associated flow rule consists in adopting a potential $h \neq f$ such that:

$$\dot{\varepsilon}_{ij}^{vp} = \frac{1}{\eta}\langle f \rangle \frac{\partial h}{\partial \sigma'_{ij}} \tag{9.68}$$

If the viscosity η tends to zero in (9.67) or (9.68), $\langle f \rangle$ has to tend to zero too and $\langle f \rangle / \eta$ becomes undetermined so that ideal plastic behaviour is recovered. On the contrary, if the function f cannot take negative values the behaviour is viscoelastic. For instance, the poroviscoelastic domain developed in §9.1 is derived from the potential:

$$\mathcal{D}^*\left(\sigma'_{ij}\right) = \frac{1}{2\varsigma}\left(\sigma' - \mathcal{K}\epsilon^v\right)^2 + \frac{1}{4\eta}\left(s_{ij} - 2\mathcal{G}e_{ij}^v\right)^2 \tag{9.69}$$

Finally, by analogy with plasticity, the flow rule related to the hardening variables is:

$$\dot{\chi}_J = \frac{1}{\eta}\langle f \rangle \frac{\partial f}{\partial \zeta_J} \quad \text{or} \quad \dot{\chi}_J = \frac{1}{\eta}\langle f \rangle \frac{\partial h}{\partial \zeta_J} \tag{9.70}$$

9.4.2 Functional Approach to the Thermodynamics of Poroviscoelasticity

Based on (9.36) and (9.37) the functional approach to poroviscoelasticity consists in expressing the constitutive equations of the porous material in the form:

$$\sigma_{ij} = C_{ijkl} \odot \varepsilon_{kl} - b_{ij} \odot M \odot v_f; \quad p = -M \odot b_{ij} \odot \varepsilon_{ij} + M \odot v_f \tag{9.71}$$

When the porous material behaviour obeys the general model extending the isotropic one we analysed in §9.1 and §9.2.2, the constitutive equations (9.71) are equivalent to:

$$\sigma_{ij} = \frac{\partial W}{\partial \varepsilon_{ij}}; \quad p = \frac{\partial W}{\partial v_f} \tag{9.72a}$$

$$\sigma_{ij} + \beta_{ij} p = C_{ijkl}\varepsilon_{ij}^{v} + \eta_{ijkl}\dot{\varepsilon}_{kl}^{v} \tag{9.72b}$$

where, similar to (5.150), W is the elastic energy per unit of initial volume of the closed system formed from the skeleton and the fluid mass initially contained in volume $d\Omega_0$. It is expressed in the form:

$$\begin{aligned} W = &\frac{1}{2}\left(\varepsilon_{ij} - \varepsilon_{ij}^{v}\right)C_{ijkl}\left(\varepsilon_{kl} - \varepsilon_{kl}^{v}\right) \\ &+ \frac{1}{2}M\left(b_{ij}\left(\varepsilon_{ij} - \varepsilon_{ij}^{v}\right) - \left(v_f - \beta_{ij}\varepsilon_{ij}^{v}\right)\right)^2 + \frac{1}{2}\varepsilon_{ij}^{v}\mathcal{C}_{ijkl}\varepsilon_{kl}^{v} \end{aligned} \tag{9.73}$$

In addition, similar to (9.11) the positiveness of the viscous dissipation Φ can be written:

$$\Phi = \left(\sigma_{ij} + \beta_{ij} p - \mathcal{C}_{ijkl}\varepsilon_{ij}^{v}\right)\dot{\varepsilon}_{kl}^{v} = \eta_{ijkl}\dot{\varepsilon}_{ij}^{v}\dot{\varepsilon}_{kl}^{v} \geq 0 \tag{9.74}$$

This section aims to express the elastic energy W and the dissipation Φ associated with such a formulation by means of just the relaxation functions C_{ijkl}. b_{ij} and M involved in (9.71).

As a preliminary problem let us consider the constitutive equation:

$$\sigma = K \odot \epsilon \tag{9.75}$$

where K is given by (9.26a). For instance, the undrained evolutions of a material exhibiting a double porous network are governed by (9.75). Indeed, the general solution of (4.131) is expressed in the form (9.75) provided that K_0 and K_∞ in (9.26a) are identified respectively with K_U (see (4.129)) and K_u (see (4.126)), while the characteristic creep time τ_c in (9.27) is identified with time τ_η (see (4.132)). To some extent this remark justifies a posteriori the choice of the poroviscoelastic model retained in §9.3 to account for the secondary consolidation process due to the relative viscous micro-slidings of the platelets forming the solid matrix of the clay, provided that the latter are eventually caused by fluid microdiffusion effects as in squirt flow.

Adopting the notation of §9.1 and §9.2.2, the constitutive equation (9.75) is equivalent to the combined equations:

$$\sigma = K_0(\epsilon - \epsilon^{v}); \quad \sigma = \mathcal{K}\epsilon^{v} + \eta\dot{\epsilon}^{v} \tag{9.76}$$

so that:

$$w = \frac{1}{2}K_0(\epsilon - \epsilon^{v})^2 + \frac{1}{2}\mathcal{K}\epsilon^{v2} : \quad \sigma = \frac{\partial w}{\partial \epsilon} \tag{9.77}$$

For such a viscous model the dissipated energy rate is:

$$\varphi = (\sigma - \mathcal{K}\epsilon^{v})\dot{\epsilon}^{v} = \eta\dot{\epsilon}^{v2} \geq 0 \tag{9.78}$$

In order to express w and φ as a function solely of K, use of the identity:

$$2w\,(t) = \left[K_0(\epsilon\,(2t-u) - \epsilon^v\,(2t-u))\left(\epsilon\,(u) - \epsilon^v\,(u)\right)\right]_{u=0}^{u=t} + \left[\mathcal{K}\epsilon^v\,(2t-u)\,\epsilon^v\,(u)\right]_{u=0}^{u=t} \tag{9.79}$$

first provides:

$$\begin{aligned}2w\,(t) = &\int_0^t K_0\left(\epsilon\,(2t-u) - \epsilon^v\,(2t-u)\right)\left(\dot\epsilon\,(u) - \dot\epsilon^v\,(u)\right)du \\ &- \int_0^t K_0\left(\epsilon\,(u) - \epsilon^v\,(u)\right)\left(\dot\epsilon\,(2t-u) - \dot\epsilon^v\,(2t-u)\right)du \\ &+ \int_0^t \mathcal{K}\left[\epsilon^v\,(2t-u)\,\dot\epsilon^v\,(u) - \epsilon^v\,(u)\,\dot\epsilon^v\,(2t-u)\right]du \\ &+ \int_0^t \eta\left[\dot\epsilon^v\,(2t-u)\,\dot\epsilon^v\,(u) - \dot\epsilon^v\,(u)\,\dot\epsilon^v\,(2t-u)\right]du\end{aligned} \tag{9.80}$$

The latter and (9.76) combine to give the following relation:

$$w\,(t) = \frac{1}{2}\left(\int_0^t - \int_t^{2t}\right)\sigma\,(2t-u)\,\dot\epsilon\,(u)\,du \tag{9.81}$$

Use of (9.75) together with definition (9.25) of the convolution product allows us to rewrite the previous equation in the form:

$$\begin{aligned}w\,(t) = &\frac{1}{2}\int_0^t \dot\epsilon\,(u)\,du \int_0^{2t-u} K\,(2t-u-v)\,\dot\epsilon\,(v)\,dv \\ &- \frac{1}{2}\int_t^{2t} \dot\epsilon\,(u)\,du \int_0^{2t-u} K\,(2t-u-v)\,\dot\epsilon\,(v)\,dv\end{aligned} \tag{9.82}$$

Permutation of the integration order in the second integral gives:

$$\begin{aligned}&\int_t^{2t} \dot\epsilon\,(u)\,du \int_0^{2t-u} K\,(2t-u-v)\,\dot\epsilon\,(v)\,dv \\ &= \int_0^t \dot\epsilon\,(v)\,dv \int_0^{2t-v} K\,(2t-v-u)\,\dot\epsilon\,(u)\,du\end{aligned} \tag{9.83}$$

Exchanging the role of u and v on the right hand side of the previous equation and substituting the resulting equation into (9.82) give:

$$w\,(t) = \frac{1}{2}\int_0^t \int_0^t K\,(2t-u-v)\,\dot\epsilon\,(u)\,\dot\epsilon\,(v)\,du\,dv \tag{9.84}$$

A similar procedure applied to expression (9.78) for the viscous dissipation φ gives:

$$\varphi(t) = -\int_0^t \int_0^t \dot{K}(2t-u-v)\,\dot{\epsilon}(u)\,\dot{\epsilon}(v)\,du\,dv \tag{9.85}$$

Functional formulae (9.84)–(9.85) are known as the Staverman and Schwarzl formulae.[2] Returning to the porous material, the Staverman and Schwarzl formulae can be extended according to:

$$\begin{aligned} W(t) = {} & \frac{1}{2}\int_0^t \int_0^t C_{ijkl}(2t-u-v)\,\dot{\varepsilon}_{ij}(u)\,\dot{\varepsilon}_{kl}(v)\,du\,dv \\ & - \int_0^t \int_0^t (M \odot b_{ij})(2t-u-v)\,\dot{\varepsilon}_{ij}(u)\,\dot{v}_f(v)\,du\,dv \\ & + \frac{1}{2}\int_0^t \int_0^t M(2t-u-v)\,\dot{v}_f(u)\,\dot{v}_f(v)\,du\,dv \end{aligned} \tag{9.86}$$

and:

$$\begin{aligned} \Phi(t) = {} & -\int_0^t \int_0^t \dot{C}_{ijkl}(2t-u-v)\,\dot{\varepsilon}_{ij}(u)\,\dot{\varepsilon}_{kl}(v)\,du\,dv \\ & 2\int_0^t \int_0^t \dot{\overline{(M \odot b_{ij})}}(2t-u-v)\,\dot{\varepsilon}_{ij}(u)\,\dot{v}_f(v)\,du\,dv \\ & - \int_0^t \int_0^t \dot{M}(2t-u-v)\,\dot{v}_f(u)\,\dot{v}_f(v)\,du\,dv \end{aligned} \tag{9.87}$$

Apparently the determination of the energy history between times 0 and t requires knowledge of the relaxation functions between times 0 and $2t$. Nevertheless, owing to their analytical character the relaxation functions are the sum of their expansions as Taylor series so that knowledge on any interval of time suffices for their determination over the whole range of positive time. From a more physical standpoint, owing to the non-ageing character of the behaviour depicted by (9.71), from the very first relaxation experiment undergone in the past the material acquires knowledge of how to relax in the future.

[2]Staverman A.J., Schwarzl F. (1952), 'Thermodynamics of viscoelastic behaviour', *Proceedings of the Koninklijke Nederlandse Akademie van Wettenshappen*, Series **B**, Vol. 55.

Appendix A

Differential Operators

A.1 Orthonormal Cartesian Coordinates

A point M has coordinates x_i $(i = 1, 2, 3)$, respectively x, y, z, in a vector base $\mathbf{e}_i$ $(i = 1, 2, 3)$, respectively $(\mathbf{e}_x, \mathbf{e}_y, \mathbf{e}_z)$. The position vector **OM** is:

$$\mathbf{OM} = x_i \mathbf{e}_i = x\mathbf{e}_x + y\mathbf{e}_y + z\mathbf{e}_z$$

Scalar Field $a(\mathbf{x}, t)$

$$\nabla a \ (= \operatorname{grad} a) = \frac{\partial a}{\partial x_i}\mathbf{e}_i$$

Vector Field $\mathbf{b}(\mathbf{x}, t)$

$$\mathbf{b}(\mathbf{x}, t) = b_i(\mathbf{x}, t)\mathbf{e}_i$$

$$\nabla \cdot \mathbf{b} \ (= \operatorname{div} \mathbf{b}) = \frac{\partial b_i}{\partial x_i}$$

$$\nabla \times \mathbf{b} \ (= \operatorname{curl} \mathbf{b}) = \begin{pmatrix} \dfrac{\partial}{\partial x} \\ \dfrac{\partial}{\partial y} \\ \dfrac{\partial}{\partial z} \end{pmatrix} \times \begin{pmatrix} b_x \\ b_y \\ b_z \end{pmatrix}$$

$$= \left(\frac{\partial b_z}{\partial x} - \frac{\partial b_y}{\partial z}\right)\mathbf{e}_x + \left(\frac{\partial b_x}{\partial z} - \frac{\partial b_z}{\partial x}\right)\mathbf{e}_y + \left(\frac{\partial b_y}{\partial x} - \frac{\partial b_x}{\partial y}\right)\mathbf{e}_z$$

Poromechanics O. Coussy
 ISBN 0-470-84920-7

$$\nabla \mathbf{b} \ (= \operatorname{grad} \mathbf{b}) = \frac{\partial b_i}{\partial x_j} \mathbf{e}_i \otimes \mathbf{e}_j = \begin{pmatrix} \frac{\partial b_x}{\partial x} & \frac{\partial b_x}{\partial y} & \frac{\partial b_x}{\partial z} \\ \frac{\partial b_y}{\partial x} & \frac{\partial b_y}{\partial y} & \frac{\partial b_y}{\partial z} \\ \frac{\partial b_z}{\partial x} & \frac{\partial b_z}{\partial y} & \frac{\partial b_z}{\partial z} \end{pmatrix}$$

where $\otimes$ stands for the tensorial product.

Second-order Tensor Field c(x, t)

$$\mathbf{c}(\mathbf{x}, t) = c_{ij} \mathbf{e}_i \otimes \mathbf{e}_j$$

$$\nabla \cdot \mathbf{c} \ (= \operatorname{div} \mathbf{c}) = \frac{\partial c_{ij}}{\partial x_j} \mathbf{e}_i$$

A.2 Cylindrical Coordinates

A point M is defined by the cylindrical coordinates r, θ, z in a local orthonormal basis $(\mathbf{e}_r, \mathbf{e}_\theta, \mathbf{e}_z)$. This is shown in Fig. A.1a. Within a Cartesian coordinate system the position vector **OM** is:

$$\mathbf{OM} = r \cos\theta \mathbf{e}_x + r \sin\theta \mathbf{e}_y + z \mathbf{e}_z$$

Derivation of **OM** with regard to the cylindrical coordinates gives:

$$\frac{\partial \mathbf{OM}}{\partial r} = \cos\theta \mathbf{e}_x + \sin\theta \mathbf{e}_y = \mathbf{e}_r$$

$$\frac{\partial \mathbf{OM}}{\partial \theta} = r(-\sin\theta \mathbf{e}_x + \cos\theta \mathbf{e}_y) = r \mathbf{e}_\theta$$

$$\frac{\partial \mathbf{OM}}{\partial z} = \mathbf{e}_z$$

Thus:

$$\frac{\partial \mathbf{e}_r}{\partial \theta} = \mathbf{e}_\theta; \qquad \frac{\partial \mathbf{e}_\theta}{\partial r} = -\mathbf{e}_r$$

These derivatives of the base vectors need to be considered in the differential operators.

Scalar Field $a(r, \theta, z, t)$

$$\nabla a \ (= \operatorname{grad} a) = \frac{\partial a}{\partial r} \mathbf{e}_r + \frac{1}{r} \frac{\partial a}{\partial \theta} \mathbf{e}_\theta + \frac{\partial a}{\partial z} \mathbf{e}_z$$

Vector Field b(r, θ, z, t)

$$\mathbf{b}(r,\theta,z,t) = b_r(r,\theta,z,t)\mathbf{e}_r + b_\theta(r,\theta,z,t)\mathbf{e}_\theta + b_z(r,\theta,z,t)\mathbf{e}_z$$

$$\nabla\cdot\mathbf{b}\ (=\operatorname{div}\mathbf{b}) = \frac{1}{r}\frac{\partial}{\partial r}(rb_r) + \frac{1}{r}\frac{\partial b_\theta}{\partial\theta} + \frac{\partial b_z}{\partial z}$$

$$\nabla\times\mathbf{b}\ (=\operatorname{curl}\mathbf{b}) = \left(\frac{1}{r}\frac{\partial b_z}{\partial\theta} - \frac{\partial b_\theta}{\partial z}\right)\mathbf{e}_r + \left(\frac{1}{r}\frac{\partial b_r}{\partial z} - \frac{\partial b_z}{\partial r}\right)\mathbf{e}_\theta$$

$$+\left(\frac{1}{r}\frac{\partial(rb_\theta)}{\partial r} - \frac{1}{r}\frac{\partial b_r}{\partial\theta}\right)\mathbf{e}_z$$

$$\nabla\mathbf{b}\ (=\operatorname{grad}\mathbf{b}) = \begin{pmatrix} \dfrac{\partial b_r}{\partial r} & \dfrac{1}{r}\left(\dfrac{\partial b_r}{\partial\theta} - b_\theta\right) & \dfrac{\partial b_r}{\partial z} \\ \dfrac{\partial b_\theta}{\partial r} & \dfrac{1}{r}\left(\dfrac{\partial b_\theta}{\partial\theta} + b_r\right) & \dfrac{\partial b_\theta}{\partial z} \\ \dfrac{\partial b_z}{\partial r} & \dfrac{1}{r}\dfrac{\partial b_z}{\partial\theta} & \dfrac{\partial b_z}{\partial z} \end{pmatrix}$$

We verify that $\nabla\cdot\mathbf{b}\ (=\operatorname{div}\mathbf{b}) = (\nabla\mathbf{b})_{ii}\ \ (=\operatorname{tr}(\operatorname{grad}\mathbf{b}))$.

Second-order Symmetric Tensor Field c(r, θ, z, t)

$$\mathbf{c}(\mathbf{x},t) = c_{ij}(r,\theta,z,t)\mathbf{e}_i\otimes\mathbf{e}_j$$

$$\nabla\cdot\mathbf{c}\ (\operatorname{div}\mathbf{c}) = \left(\frac{\partial c_{rr}}{\partial r} + \frac{1}{r}\frac{\partial c_{r\theta}}{\partial\theta} + \frac{\partial c_{rz}}{\partial z} + \frac{c_{rr} - c_{\theta\theta}}{r}\right)\mathbf{e}_r$$

$$+\left(\frac{\partial c_{\theta r}}{\partial r} + \frac{1}{r}\frac{\partial c_{\theta\theta}}{\partial\theta} + \frac{\partial c_{\theta z}}{\partial z} + 2\frac{c_{r\theta}}{r}\right)\mathbf{e}_\theta$$

$$+\left(\frac{\partial c_{zr}}{\partial r} + \frac{1}{r}\frac{\partial c_{z\theta}}{\partial\theta} + \frac{\partial c_{zz}}{\partial z} + \frac{c_{zr}}{r}\right)\mathbf{e}_z$$

A.3 Spherical Coordinates

A point M is defined by the spherical coordinates r, θ, φ in a local orthonormal basis $(\mathbf{e}_r, \mathbf{e}_\theta, \mathbf{e}_\varphi)$. This is shown in Fig. A.1b. Within a Cartesian coordinate system the position vector **OM** is:

$$\mathbf{OM} = r\sin\theta\cos\varphi\mathbf{e}_x + r\sin\theta\sin\varphi\mathbf{e}_y + r\cos\theta\mathbf{e}_z$$

Derivation with respect to spherical coordinates:

$$\frac{\partial\mathbf{OM}}{\partial r} = \sin\theta\cos\varphi\mathbf{e}_x + \sin\theta\sin\varphi\mathbf{e}_y + \cos\theta\mathbf{e}_z = \mathbf{e}_r$$

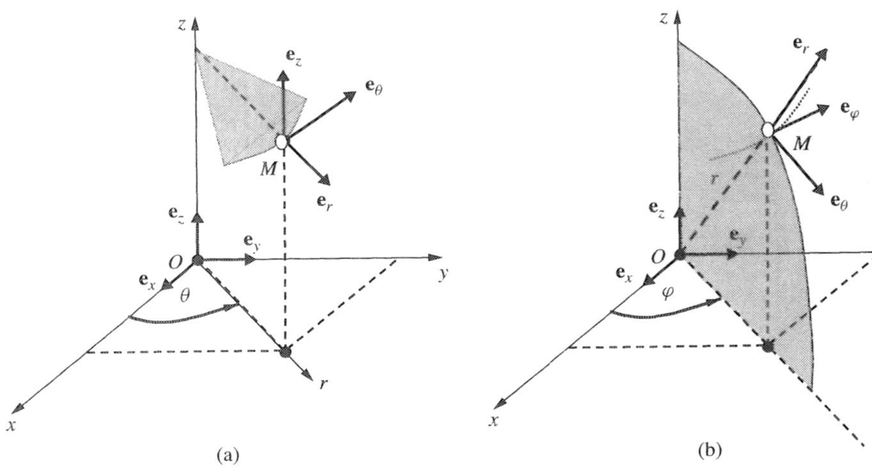

Figure A.1: Coordinate systems: (a) cylindrical coordinates; (b) spherical coordinates. (from Coussy 2004, reprinted by permission of Pearson Education, Inc.).

$$\frac{\partial \mathbf{OM}}{\partial \theta} = r(\cos\theta\cos\varphi\mathbf{e}_x + \cos\theta\sin\varphi\mathbf{e}_y - \sin\theta\mathbf{e}_z) = r\mathbf{e}_\theta$$

$$\frac{\partial \mathbf{OM}}{\partial \varphi} = r\sin\theta(-\sin\varphi\mathbf{e}_x + \cos\varphi\mathbf{e}_y) = r\sin\theta\mathbf{e}_\varphi$$

Thus:

$$\begin{array}{lll} \mathbf{e}_{r,r} = 0 & \mathbf{e}_{r,\theta} = \mathbf{e}_\theta & \mathbf{e}_{r,\varphi} = \sin\theta\mathbf{e}_\varphi \\ \mathbf{e}_{\theta,r} = 0 & \mathbf{e}_{\theta,\theta} = -\mathbf{e}_r & \mathbf{e}_{\theta,\varphi} = \cos\theta\mathbf{e}_\varphi \\ \mathbf{e}_{\varphi,r} = 0 & \mathbf{e}_{\varphi,r} = 0 & \mathbf{e}_{\varphi,\varphi} = -\sin\theta\mathbf{e}_r - \cos\theta\mathbf{e}_\theta \end{array}$$

where $\mathbf{e}_{\alpha,\beta} = \partial\mathbf{e}_\alpha/\partial\beta$.

Scalar Field $a(r, \theta, \varphi, t)$

$$\nabla a \ (= \operatorname{grad} a) = \frac{\partial a}{\partial r}\mathbf{e}_r + \frac{1}{r}\frac{\partial a}{\partial \theta}\mathbf{e}_\theta + \frac{1}{r\sin\theta}\frac{\partial a}{\partial \varphi}\mathbf{e}_\varphi$$

Vector Field $\mathbf{b}(r, \theta, \varphi, t)$

$$\mathbf{b}(r,\theta,\varphi,t) = b_r(r,\theta,\varphi,t)\mathbf{e}_r + b_\theta(r,\theta,\varphi,t)\mathbf{e}_\theta + b_\varphi(r,\theta,\varphi,t)\mathbf{e}_z$$

$$\nabla\cdot\mathbf{b} \ (= \operatorname{div}\mathbf{b}) = \frac{1}{r^2}\frac{\partial}{\partial r}(r^2 b_r) + \frac{1}{r\sin\theta}\frac{\partial}{\partial\theta}(\sin\theta\, b_\theta) + \frac{1}{r\sin\theta}\frac{\partial b_\varphi}{\partial\varphi}$$

$$\nabla \times \mathbf{b}\ (= \operatorname{curl} \mathbf{b}) = \frac{1}{r \sin\theta}\left(\frac{\partial}{\partial\theta}(\sin\theta\, b_\varphi) - \frac{\partial b_\theta}{\partial\varphi}\right)\mathbf{e}_r + \frac{1}{r}\left(\frac{1}{\sin\theta}\frac{\partial b_r}{\partial\varphi} - \frac{\partial}{\partial r}(r b_\varphi)\right)\mathbf{e}_\theta$$

$$+ \frac{1}{r}\left(\frac{\partial (r b_\theta)}{\partial r} - \frac{\partial b_r}{\partial\theta}\right)\mathbf{e}_z$$

$$\nabla \mathbf{b}\ (= \operatorname{grad} \mathbf{b}) = \begin{pmatrix} \dfrac{\partial b_r}{\partial r} & \dfrac{1}{r}\left(\dfrac{\partial b_r}{\partial\theta} - b_\theta\right) & \dfrac{1}{r}\left(\sin\theta\dfrac{\partial b_r}{\partial\varphi} - b_\varphi\right) \\ \dfrac{\partial b_\theta}{\partial r} & \dfrac{1}{r}\left(\dfrac{\partial b_\theta}{\partial\theta} + b_r\right) & \dfrac{1}{r}\left(\dfrac{1}{\sin\theta}\dfrac{\partial b_\theta}{\partial\varphi} - \cot\theta\, b_\varphi\right) \\ \dfrac{\partial b_\varphi}{\partial r} & \dfrac{1}{r}\dfrac{\partial b_\varphi}{\partial\theta} & \dfrac{1}{r}\left(\dfrac{1}{\sin\theta}\dfrac{\partial b_\varphi}{\partial\varphi} - \cot\theta\, b_\theta + b_r\right) \end{pmatrix}$$

Second-order Symmetric Tensor Field $\mathbf{c}(r, \theta, \varphi, t)$

$$\mathbf{c}(\mathbf{x}, t) = c_{ij}(r, \theta, \varphi, t)\mathbf{e}_i \otimes \mathbf{e}_j$$

$$\nabla \cdot \mathbf{c}\ (= \operatorname{div} \mathbf{c}) = \left(\frac{\partial c_{rr}}{\partial r} + \frac{1}{r}\frac{\partial c_{r\theta}}{\partial\theta} + \frac{1}{r\sin\theta}\frac{\partial c_{r\varphi}}{\partial\varphi} + \frac{1}{r}(2c_{rr} - c_{\theta\theta} - c_{\varphi\varphi} + c_{r\theta}\cot\theta)\right)\mathbf{e}_r$$

$$+ \left(\frac{\partial c_{\theta r}}{\partial r} + \frac{1}{r}\frac{\partial c_{\theta\theta}}{\partial\theta} + \frac{1}{r\sin\theta}\frac{\partial c_{\theta\varphi}}{\partial\varphi} + \frac{1}{r}[(c_{\theta\theta} - c_{\varphi\varphi})\cot\theta + 3c_{r\theta}]\right)\mathbf{e}_\theta$$

$$+ \left(\frac{\partial c_{zr}}{\partial r} + \frac{1}{r}\frac{\partial c_{\varphi\theta}}{\partial\theta} + \frac{1}{r\sin\theta}\frac{\partial c_{\varphi\varphi}}{\partial\varphi} + \frac{1}{r}(3c_{r\varphi} + 2\cot\theta\, c_{\theta\varphi})\right)\mathbf{e}_\varphi$$

Bibliography

Selected Books

1. Atkins P.W. (1990), *Physical Chemistry*, Fourth Edition, Oxford University Press.
2. Barenblatt G.I. (1966), *Scaling, Self-Similarity, and Intermediate Asymptotics*, Cambridge Texts in Applied Mathematics, Cambridge University Press.
3. Bear J. (1988), *Dynamics of Fluids in Porous Media*, Dover, New York (reprint of 1972 edition published by Elsevier, New York).
4. Bear J., Bachmat Y. (1990), *Introduction to Modeling of Transport in Porous Media*, Kluwer, Amsterdam.
5. Bourbié T., Coussy O., Zinszner B. (1987), *Acoustics of Porous Media*, Editions, Technip, Paris.
6. Carman P.C. (1956), *Flow of gases through porous media*, Butterworth Scientific, London.
7. Carmichäel R.S. (1982), *CRC Handbook of Physical Properties of Rocks*, Vol. II, CRC Press, Boca Raton, FL.
8. Carslaw A.R., Jaeger C.J. (1960), *Conduction of Heat in Solids*, Clarendon Press, Oxford.
9. Černy R., Rovnanikova P. (2002), *Transport Processes in Concrete*, Spon Press, London.
10. Chen W.F., Han D.J. (1988), *Plasticity for Structural Engineers*, Springer-Verlag, New York.
11. Crank J. (1975), *The Mathematics of Diffusion*, Second Edition, Oxford University Press.
12. Coussy O. (1995), *Mechanics of Porous Continua*, John Wiley & Sons, Chichester.
13. Coussy O., Ulm J.-F. (2004), *Mechanics and Durability of Solids, Vol. II, Durability Mechanics*, Prentice Hall, Upper Saddle Liver, NJ.
14. de Boer R. (2000), *Theory of Porous Media, Highlights in Historical Development and Current State*, Verlag/Hersteller, Springer, Berlin.
15. de Groot S.R., Mazur P. (1983), *Non-equilibrium Thermodynamics*, Dover, New York.

Poromechanics O. Coussy
 ISBN 0-470-84920-7

16. de Marsily J. (1986), *Quantitative Hydrogeology. Groundwater Hydrology for Engineers*, Academic Press, New York.
17. Detournay E., Cheng A.H.-D. (1993), 'Fundamentals of Poroelasticity', *Comprehensive Rock Engineering*, Vol. II, ed. Hudson J., Pergamon Press, Oxford and New York.
18. Dormieux L., Bourgeois E. (2002), *Introduction à la micromécanique des milieux poreux*, Presses de l'Ecole Nationale des Ponts et Chaussées, Paris.
19. Dullien F.A.L. (1979), *Porous media: fluid transport and pore structure*, Academic Press, New York.
20. Fredlund D.G., Rahardjo H. (1993), *Soil Mechanics for Unsaturated Soils*, John Wiley & Sons.
21. Halphen B., Salençon J. (1987), *Elasto-plasticité*, Presse de l'Ecole Nationale des Ponts et Chaussées, Paris.
22. *Handbook of mathematical functions* (1964), ed. Abramowitz M. and Stegun I.A. Dover, New York.
23. Hill R. (1950), *The Mathematical Theory of Plasticity*, Clarendon Press, Oxford.
24. Hunter J. (2001), *Foundations of Colloid Science*, Second Edition, Oxford University Press.
25. Israelachvili J. (1991), *Intermolecular and surface forces*, Second Edition, Academic Press.
26. Kowalski S.J. (2003), *Thermomechanics of Drying Processes*, Lecture Notes in Applied and Computational Mechanics, 8, Springer-Verlag, Berlin.
27. Lemaitre J., Chaboche J.-L. (1998), *Mechanics of Solid Materials*, Cambridge University Press.
28. Lewis R.W., Schrefler B.A. (1998), *The Finite Element Method in the Static and Dynamic Deformation and Consolidation of Porous Media*, Second edition, John Wiley & Sons.
29. Logan D.J. (2000), *Transport Modeling in Hydrogeochemical Systems*, Interdisciplinary Applied Mathematics, 15, Springer-Verlag, Berlin.
30. Mandel J. (1966), *Cours de Mécanique des Milieux Continus*, Gauthiers Villars, Paris.
31. *Mécanique des Sols Non Saturés* (2002), ed. Coussy O. and Fleureau J.-M., Hermès, Paris.
32. *Mechanics of Porous Media* (1995), ed. Charlez P. Lecture Notes of the Mechanics of Porous Media Summer School, Balkema, Rotterdam.
33. *Poromechanics, A tribute to M.A. Biot* (1998), Proceedings of the First Biot Conference, ed. J.-F. Thymus *et al.*, Balkema, Rofferdam.
34. *Poromechanics II* (2002), Proceedings of the Second Biot Conference, ed. Auriault J.-L., Balkema Rofferdam.
35. Russel W.B., Saville D.A., Schowalter W.R. (1989), *Colloidal Dispersions*, Cambridge University Press.
36. Salençon J. (1983), *Calcul à la Rupture et Analyse Limite*, Presse de l'Ecole Nationale des Ponts et Chaussées, Paris.

37. Salençon J. (2001), *Handbook of Continuum Mechanics: General concepts, Thermoelasticity*, Springer-Verlag, Berlin.
38. Sih G.C., Michopoulos J.G., Chou S. (1986), *Hygrothermoelasticity*, Martinus Nijhoff, Dordrecht.
39. Ulm J.-F., Coussy O. (2003), *Mechanics and Durability of Solids, Vol. I, Solid Mechanics*, Prentice Hall, Upper Saddle Liver, NJ.
40. Wang H.F. (2000), *Theory of Linear Poroelasticity*, Princeton series in geophysics, Princeton University Press.

Selected Papers

41. Abousleiman Y., Cheng A.H.-D., Cui L., Detournay E., Roegiers J.-C. (1996), 'Mandel's problem revisited', *Géotechnique*, **46**, 187–195.
42. Adenot F. (1992), 'Durabilité du Béton: Caractérisation et Modélisation des Processus Physiques et Chimiques de Dégradation du Ciment', *PhD Thesis*, Université d'Orléans.
43. Alonzo E.E., Delage P. (1995), 'Unsaturated Soils', *Proceedings of the First International Conference on Unsaturated Soils (Unsat 95)*, Paris, Vol. 3, Balkema, Rotterdam.
44. Alonzo E.E., Gens A., Josa A. (1990), 'A constitutive model for partially saturated soils', *Geotechnique*, **40**, 405–430.
45. Baroghel-Bouny V., Mainguy M., Lassabatère T., Coussy O. (1999), 'Characterization and identification of equilibrium and transfer moisture properties for ordinary and high performance cementitious materials', *Cement and Concrete Research*, **29**, 1225–1238.
46. Bažant Z.P. (1972), 'Thermodynamics of hindered adsorption and its implications for hardened cement paste and concrete', *Cement and Concrete Research*, **2**, (1), 1–16.
47. Bemer E., Boutéca M., Vincké O., Hoteit N., Ozanam O. (2001), 'Poromechanics: from linear poroelasticity to non-linear poroelasticity and poroviscoelasticity', *Oil & Gaz Science and Technology, Rev. IFP*, **56**, (6), 531–544.
48. Berryman J.G. (1980), 'Confirmation of Biot's theory', *Applied Physics Letters*, **37**, 382–384.
49. Bigas J.-P. (1994), 'Diffusion des ions chlorures dans les mortiers', *PhD Thesis*, Génie civil - INSA Toulouse.
50. Biot M.A. (1941), 'General theory of three dimensional consolidation', *Journal of Applied Physics*, **12**, 155–164.
51. Biot M.A. (1956), 'The theory of propagation of elastic waves in a fluid-saturated porous solid, I lower frequency range', Journal of the Acoustical Society of America, **28**, 168–178.
52. Biot M.A. (1962), 'Mechanics of deformation and acoustic propagation in porous media', *Journal of Applied Physics*, **27**, 459–467.

53. Biot M.A. (1972), 'Theory of finite deformation of porous solids', *Indiana University Mathematical Journal*, **24**, 579–620.
54. Biot M.A. (1977), 'Variational Lagrangian-thermodynamics of non isothermal finite strain. Mechanics of porous solid and thermomolecular diffusion', *International Journal of Solids and Structures*, **13**, 579–597.
55. Biot M.A., Willis D.G. (1957), 'The elastic coefficients of theory of consolidation', *Journal of Applied Mechanics*, **24**, 594–601.
56. Bishop A.W., Blight G.E.E. (1963), 'Some aspects of effective stress in saturated and partly saturated soils', *Geotechnique*, **2**, 177–197.
57. Bourgeois E. (1997), 'Mécanique des milieux poreux en transformation finie: position des problèmes et méthodes de résolution', PhD Thesis, Ecole Nationale des Ponts et Chaussées.
58. Bourgeois E., Dormieux L. (1996), 'Consolidation of a non-linear poroelastic layer in finite deformations', *European Journal of Mechanical A/Solids*, **15**, (4), 575–598.
59. Bouteca M., Sarda J.-P. (1995), 'Experimental measurements of thermoporoelastic coefficients', in *Mechanics of Porous Media*, ed., Charlez P. Balkema, Rofferdam.
60. Bowen R.M. (1968), 'Thermochemistry of reacting materials', *Journal of Chemical Physics*, **49**, 1625–1637.
61. Bowen R.M. (1976), 'Theory of mixtures', in *Continuum Physics*, Vol. III, ed. Eringen A.C., Academic Press, New York, 1–127.
62. Campanella R.G., Mitchell J.K. (1968), 'Influence of temperature variations on soil behaviour', *Journal of the Soil Mechanics and Foundation Division*, ASCE, **94(SM3)**, 709–734.
63. Carmeliet J. (1999), 'Coupling of damage and fluid-solid interactions in quasi-brittle nonsaturated porous materials', *Proceedings of the IUTAM Symposium on Theoretical and Numerical Methods in Continuum Mechanics of Porous Materials*, ed. Ehlers W., Kluwer Academic, Dordrecht, 307–312.
64. Chateau X., Dormieux L. (1995), 'Homogénéisation d'un milieux poreux non saturé: lemme de Hill et applications', *Comptes Rendus de l'Academie des Sciences*, **320**, Série II b, 627–634.
65. Chateau X., Dormieux L. (2002). 'Micro mechanics of saturated and unsaturated porous media', *International Journal of Numerical and Analytical Methods in Geomechanics*, **26**, 831–844.
66. Chen J., Hopmans J.W., Grismer M.E. (1999), 'Parameter estimation of two-fluid capillary pressure-saturation and permeability functions', *Advances in Water Resources*, **22**, (5), 479–493.
67. Cheng A.H.-D., Detournay E. (1988), 'A direct boundary element method for plane strain poroelasticity', *International Journal of Numerical and Analytical Methods in Geomechanics*, **12**, 551–572.
68. Coussy O. (1989a), 'Thermodynamics of saturated porous solids in finite deformation', *European Journal of Mechanics*, A/Solids, **8**, 1–14.
69. Coussy O. (1989b), 'A general theory of thermoporoelastoplasticity for saturated materials', *Transport in Porous Media*, **4**, 281–293.

70. Coussy O., Dangla P. (2002), 'Approche énergétique du comportement des sols non saturés', *Mécanique des Sols Non Saturés*, ed. Coussy O. and Fleureau J.-M., Hermès, Paris.

71. Coussy O., Dangla P., Lassabatère T., Baroghel-Bouny V. (2003), 'The equivalent pore pressure and the swelling and shrinkage of cement based materials', *Concrete Science Engineering*, **36** (in press).

72. Coussy O., Detournay E., Dormieux L. (1998), 'From mixture theories to Biot's theory', *International Journal of Solids and Structures*, **35**, (34–35), 4619–4635.

73. Coussy O., Eymard R. (2003), 'Non-linear binding and the diffusion-migration test', *Transport in Porous Media*, **1770**, 1–24.

74. Coussy O., Eymard R., Lassabatère T. (1998), 'Constitutive modelling of unsaturated drying deformable materials', *Journal of Engineering Mechanics*, **124**, (6), 658–667.

75. Coussy O., Ulm F.-J. (1996), 'Creep and plasticity due to chemo-mechanical couplings', *Archive of Applied Mechanics*, **66**, 523–535.

76. Cowin S.C. (1998), 'Bone fluid poroelasticity', *Poromechanics, A tribute to M.A. Biot*, Proceedings of the Biot Conference on Poromechanics, ed. J.-F. Thymus *et al.*, Balkema, Rofferdam.

77. Dangla P., Coussy O. (1998), 'Non-linear poroelasticity for unsaturated porous materials: an energy approach', *Poromechanics, A tribute to M. A. Biot*, Proceedings of the Biot Conference on Poromechanics, ed. J.-F. Thymus *et al.*, Balkema, Rofferdam, 59–64.

78. Dangla P., Coussy O., Eymard R. (1998), 'A vanishing diffusion process in unsaturated soils', *International Journal of Non-linear Mechanics*, **33**, (6), 1027–1037.

79. Dangla P., Coussy O., Olchitzky E., Imbert C. (2000), 'Non-linear thermomechanical couplings in unsaturated clay barriers', in *Proceedings of IUTAM Symposium on Theoretical and Numerical Methods in Continuum Mechanics of Porous Materials*, ed. W. Ehlers, Kluwer Academic, Dordrecht.

80. Desrues J. (1984), 'La localisation de la déformation dans les milieux granulaires, Thèse de Doctorat d'Etat, Grenoble University.

81. Detournay E., Cheng A.H.-D. (1988), 'Poroelastic response of a borehole in non-hydrostatic stress field', *International Journal of Rock Mechanics Mining Science and Geomechanical Abstracts*, **25**, 171–182.

82. de Vries D.A., Kruger A.J. (1966), 'On the value of the diffusion coefficient of water vapour in air', *Phénomènes de transport avec changement de phase dans les milieux poreux ou colloïdaux*, ed. CNRS, 561–572.

83. Dopler T., Modaressi A., Michaud V. (2000), 'Simulation of metal composite isothermal infiltration processing', *Metallurgical and Materials Transactions B*, **31B**, 225–234.

84. Dormieux L., Barboux P., Coussy O., Dangla P. (1995), 'A macroscopic modelling for the swelling phenomenon of a saturated clay', *European Journal of Mechanics A/Solids*, **14**, (6), 981–1004.

85. Dormieux L., Lemarchand E., Coussy O. (2003), 'Macroscopic and micromechanical approaches to the modelling of the osmotic swelling in clays', *Transport in Porous Media.*, Special Issue, guest ed. Huyghe J., Vol. 50, 1–2.

86. Dormieux L., Molinari A., Kondo D. (2002), 'Micromechanical approach to the behavior of poroelastic materials', *Journal of the Mechanics and Physics of Solids*, **50**, 2203–2231.

87. Drucker D.C., Prager W. (1952), 'Soil mechanics and plastic analysis design', *Quarterly Journal of Applied Mathematics*, **10**, 157–165.

88. Fauchet B., Coussy O., Carrère A., Tardieu B. (1991), 'Poroplastic analysis of concrete dams and their foundations', *Dam Engineering*, **2**, 165–192.

89. Garagash D., Detournay E. (1997), 'An analysis of the influence of the pressurization rate on the borehole breakdown pressure', *Journal of Solids and Structures*, **34**, (24), 3099–3118.

90. Gassmann F. (1951), 'Uber die Elastizität Poroser Medien', *Vierteljarsschrift der Naturforschenden Gesellschaft in Zurich*, **96**, 1–23.

91. Hashin Z. (1983), 'Analysis of composite materials - A survey', *Journal of Applied Mechanics*, **50**, 481–504.

92. Heukamp F., Ulm F.-J. (2002), 'Chemomechanics of calcium leaching of cement-based materials at different scales: the role of CH-dissolution and C-S-H-degradation on strength and durability performance of materials and structures', MIT-CEE Report, **R002–03** (*PhD Thesis*, Massachusetts Institute of Technology).

93. Heukamp F.H., Ulm F.-J., Germaine J.T. (2002), 'Residual design strength of cement-based materials for nuclear waste storage systems', *Nuclear Engineering Design*, **211**, (1), 51–60.

94. Keister J.C., Kasting G.B. (1986), 'Ionic mass transport through a homogeneous membrane in the presence of a uniform electric field', *Journal of Membrane Science*, **29**, 155–167.

95. Kerbouche R., Shao J.-F., Skozylas F., Henry J.-P. (1995), 'On the poroplastic behavior of porous rocks', *European Journal of Mechanics, A/Solids*, **14**, (4), 3577–3587.

96. Klinkenberg L.J. (1941), 'The permeability of porous media to liquids and gases', *Drilling and production practices*, American Petroleum Institute, New York, 200–214.

97. Kümpel H.-J. (1991), 'Poroelasticity: parameters reviewed', *Geophysies Journal International*, **105**, 783–799.

98. Loret B., Harireche O. (1991), 'Acceleration waves, flutter instabilities and stationary discontinuities in inelastic porous media',*Journal of Mechanics and Physics of Solids*, **39**, 569–606.

99. Luckner L., van Genutchen M.Th., Nielsen D.R. (1989), 'A consistent set of parametric models for the two-phase flow of immiscible fluids in the subsurface', *Water Resources Research*, **25**, (10), 2187–2193.

100. McTigue D.F. (1986), 'Thermoelastic response of fluid-saturated porous rock', *Journal of Geophysical Research*, **91**, 9533–9542.

101. Mainguy M., Coussy O. (2000), 'Propagation fronts during calcium leaching and chloride penetration', *Journal of Engineering Mechanics*, ASCE, **126**, (3), 250–278.

102. Mainguy M., Coussy O., Baroghel-Bouny V. (2001), 'The role of air pressure in the drying of weakly permeable materials', *Journal of Engineering Mechanics*, ASCE, **127**, (6), 582–592.

103. Mainguy M., Ulm F.-J., Heukamp F.H. (2001), 'Similarity properties of demineralization and degradation of cracked porous media', *International Journal of Solids and Structures*, **38**, 7079–7100.

104. Mandel J. (1953), 'Consolidation des sols (étude mathématique)', *Géotechnique*, **3**, 287–299.

105. Melean Y., Broseta D., Hasmy A., Blossey R. (2003), 'Dispersion of imbibition fronts', *European Physics Letters*, **62**, (4), 505–511.

106. Millington R.J. (1959), 'Gas diffusion in porous media', *Science*, **130**, 100–102.

107. Nabarro F.N. (1948), 'Report of a Conference on Strength of Solids', *Physical Society*, London, 75.

108. Parker J.C., Lenhard R.J., Kuppusamy T. (1987), 'A parametric model for constitutive properties governing multiphase flow in porous media', *Water Resources Research*, **23**, (4), 618–624.

109. Plona T.J., Johnson D.L. (1980), 'Experimental study of two bulk compressional modes in water-saturated porous structures', *Ultrasonics Symposium*, IEEE, 868–872.

110. Rice J.R. (1975), 'On the stability of dilatant hardening for saturated rock masses', *Journal of Geophysical Research*, **80**, 1531–1536.

111. Rice J.R., Cleary M.P. (1976), 'Some basic stress diffusion solutions for fluid-saturated elastic porous media with compressible constituents', *Journal of Geophysical Research*, **14**, 227–241.

112. Roscoe K.H., Schofield A.N., Thurairajah A. (1963), 'Yielding of clays in states wetter than critical', *Geotechnique*, **8**, 22–53.

113. Roscoe K.H., Schofield A.N., Wroth C.P. (1958), 'On the yielding of soils', *Geotechnique*, **13**, 211–240.

114. Rossi P., Van Mier J.G.M., Toutlemonde F., Le Maou F., Boulay C. (1994), 'Effect of loading rate on the strength of concrete subjected to uniaxial tension', *Materials and Structures*, **27**, 260–264.

115. Rudnicki J.W., Rice J.R. (1975), 'Conditions of the localization of deformation in pressure-sensitive materials', *Journal of Mechanics and Physics of Solids*, **23**, 371–394.

116. Savins J.G. (1969), 'Non-Newtonian flow through porous media', *Industrial and Engineering Chemistry*, 18–47.

117. Skempton A.W. (1954), 'The pore pressure coefficients A and B', *Geotechnique*, **4**, 143–152.

118. Staverman A.J., Schwarzl F. (1952), 'Thermodynamics of viscoelastic behaviour', *Proceedings of the Koninklijke Nederlandse Akademie van Wettenshappen*, Series **B**, Vol. 55.

119. Sultan N., Delage P., Cui Y.J. (2002), 'Temperature effects on the volume change behaviour of Boom clay', *Engineering Geology*, **64**, (2–3), 135–145.

120. Terzaghi K. (1923), 'Die Berechnung der Durchlassigkeitsziffer des Tones aus dem Verlauf der hydrodynamischen Spannungserscheinungen', *Sitzung berichte. Akademie der Wissenschaften, Wien Mathematiesch-Naturwissenschaftliche Klasse, Abteilung IIa*, **132**, 105–124.

121. Truc O., Ollivier J.-P., Nilsson L.O. (2000), 'Numerical simulation of multi-species diffusion', *Materials and Structures*, **33**, 566–573.

122. Ulm F.-J., Coussy O. (1996), 'Strength growth as chemo-plastic hardening in early age concrete', *Journal of Engineering Mechanics*, **122**, (12), 1123–1131.

123. Ulm F.-J., Coussy O. (1998), 'Couplings in early-age concrete: from material modeling to structural design', *Journal of Solids and Structures*, **35**, (31–32), 4295–4311.

124. van Genuchten M.Th. (1980), 'A closed-form equation for predicting the hydraulic conductivity of unsaturated soils', *Soil Science Society of America Journal*, **44**, 892–898.

125. Vincké O., Boutéca M., Piau J.-M., Fourmaintraux D. (1998), 'Study of the effective stress at failure', *Poromechanics, A tribute to M.A. Biot*, Proceedings of the First Biot Conference, ed. J.-F. Thymus *et al.*, Balkema, Rofferdam.

126. Wittmann F.H. (1968), 'Surface tension, shrinkage and strength of hardened cement paste', *Material and Structures*, **6**, 547–552.

127. Xi Y., Bažant Z.P., Jennings H.M. (1994), 'Moisture diffusion in cementitious materials. Moisture capacity and diffusivity', *Advanced Cement Based Materials*, **1**, 258–266.

Index

Poromechanics O. Coussy

ISBN 0-470-84920-7

Printed in the USA/Agawam, MA
March 25, 2010
540351.010